Medicinal Chemistry: An Introduction

Medicinal Chemistry: An Introduction

Edited by
Oakely Webster

Larsen & Keller
www.larsen-keller.com

Medicinal Chemistry: An Introduction
Edited by Oakely Webster
ISBN: 978-1-63549-181-4 (Hardback)

Larsen & Keller

Published by Larsen and Keller Education,
5 Penn Plaza,
19th Floor,
New York, NY 10001, USA

Cataloging-in-Publication Data

Medicinal chemistry : an introduction / edited by Oakely Webster.
p. cm.
Includes bibliographical references and index.
ISBN 978-1-63549-181-4
1. Pharmaceutical chemistry. 2. Drugs. 3. Chemistry. 4. Medicine. I. Webster, Oakely.
RS403 .M43 2017
615.19--dc23

Printed and bound in China.

For more information regarding Larsen and Keller Education and its products, please visit the publisher's website www.larsen-keller.com

Table of Contents

Preface

This book elucidates the concepts and innovative models around prospective developments with respect to medicinal chemistry. It talks in detail about the various concepts and important aspects of this field. Medicinal chemistry refers to the amalgamation of synthetic organic chemistry with pharmacology in order to create and design better medicinal drugs. It combines branches like molecular biology, pharmacology, human medicine, toxicology, chemical biology, structural biology, etc. Most of the topics introduced in this text cover new techniques and the applications of medicinal biology. While understanding the long-term perspectives of the topics, the book makes an effort in highlighting their impact as a modern tool for the growth of the discipline. The text is an essential guide for both academicians and those who wish to pursue this discipline further.

To facilitate a deeper understanding of the contents of this book a short introduction of every chapter is written below:

Chapter 1- Medicinal chemistry is a discipline of chemistry that is the blend of synthetic organic chemistry and pharmacology. The compounds mostly used in medicinal chemistry are organic compounds. This chapter will provide an integrated understanding of medicinal chemistry.

Chapter 2- Specialization in the field of medicine has naturally caused changes in the manufacture of pharmaceutical medicine. Pharmacology is a branch of medicine that deals with drug actions and pharmacognosy studies those medicinal drugs which are derived from plants. The following content is a compilation of the various branches of medicine that forms an integral part of the broader subject matter.

Chapter 3- Generally the compounds used in medicinal chemistry are organic compounds. Some of the examples given of organic compounds are atorvastatin and ETFE. Inorganic compounds and lithium are also useful in medicinal chemistry. The succeeding section is a compilation of the various compound used in medicinal chemistry.

Chapter 4- A drug that can be used for medicinal purposes is known as a medicinal drug. Medicinal drugs are either used for a prescribed duration or on a regular basis if the patient is diagnosed with a chronic disorder. The text will provide an integrated understanding of medicinal drugs.

Chapter 5- This chapter provides a plethora of allied fields of medicinal chemistry. They allow for more complex developments in the field of medicinal chemistry and provide a better comprehension of medicinal chemistry. Chemistry, organic chemistry, biochemistry, chemical biology and organometallic chemistry are the associated fields that are listed in this chapter.

Chapter 6- Medicine has evolved over a period of thousands years. Prehistoric medicine integrated herbalism, animal organs and healing through faith. Modern medicine replaced ancient medicine through the discovery and promotion of scientific biomedical research. Newer trends in the field of medicine seek to eradicate and control disease through the development of pharmaceutical medicine. This chapter will provide an integrated understanding of the evolution of medicine.

I owe the completion of this book to the never-ending support of my family, who supported me throughout the project.

Editor

1

Introduction to Medicinal Chemistry

Medicinal chemistry is a discipline of chemistry that is the blend of synthetic organic chemistry and pharmacology. The compounds mostly used in medicinal chemistry are organic compounds. This chapter will provide an integrated understanding of medicinal chemistry.

Medicinal Chemistry

Medicinal chemistry and pharmaceutical chemistry are disciplines at the intersection of chemistry, especially synthetic organic chemistry, and pharmacology and various other biological specialties, where they are involved with design, chemical synthesis and development for market of pharmaceutical agents, or bio-active molecules (drugs).

Compounds used as medicines are most often organic compounds, which are often divided into the broad classes of small organic molecules (e.g., atorvastatin, fluticasone, clopidogrel) and "biologics" (infliximab, erythropoietin, insulin glargine), the latter of which are most often medicinal preparations of proteins (natural and recombinant antibodies, hormones, etc.). Inorganic and organometallic compounds are also useful as drugs (e.g., lithium and platinum-based agents such as lithium carbonate and cis-platin as well as gallium).

In particular, medicinal chemistry in its most common practice —focusing on small organic molecules—encompasses synthetic organic chemistry and aspects of natural products and computational chemistry in close combination with chemical biology, enzymology and structural biology, together aiming at the discovery and development of new therapeutic agents. Practically speaking, it involves chemical aspects of identification, and then systematic, thorough synthetic alteration of new chemical entities to make them suitable for therapeutic use. It includes synthetic and computational aspects of the study of existing drugs and agents in development in relation to their bioactivities (biological activities and properties), i.e., understanding their structure-activity relationships (SAR). Pharmaceutical chemistry is focused on quality aspects of medicines and aims to assure fitness for purpose of medicinal products.

At the biological interface, medicinal chemistry combines to form a set of highly interdisciplinary sciences, setting its organic, physical, and computational emphases alongside biological areas such as biochemistry, molecular biology, pharmacognosy and pharmacology, toxicology and veterinary and human medicine; these, with project management, statistics, and pharmaceutical business practices, systematically oversee altering identified chemical agents such that after pharmaceutical formulation, they are safe and efficacious, and therefore suitable for use in treatment of disease.

Medicinal Chemistry in the Path of Drug Discovery

Discovery

Discovery is the identification of novel active chemical compounds, often called "hits", which are typically found by assay of compounds for a desired biological activity. Initial hits can come from repurposing existing agents toward a new pathologic processes, and from observations of biologic effects of new or existing natural products from bacteria, fungi, plants, etc. In addition, hits also routinely originate from structural observations of small molecule "fragments" bound to therapeutic targets (enzymes, receptors, etc.), where the fragments serve as starting points to develop more chemically complex forms by synthesis. Finally, hits also regularly originate from *en-masse* testing of chemical compounds against biological targets, where the compounds may be from novel synthetic chemical libraries known to have particular properties (kinase inhibitory activity, diversity or drug-likeness, etc.), or from historic chemical compound collections or libraries created through combinatorial chemistry. While a number of approaches toward the identification and development of hits exist, the most successful techniques are based on chemical and biological intuition developed in team environments through years of rigorous practice aimed solely at discovering new therapeutic agents.

Hit to Lead and Lead Optimization

Further chemistry and analysis is necessary, first to identify the "triage" compounds that do not provide series displaying suitable SAR and chemical characteristics associated with long-term potential for development, then to improve remaining hit series with regard to the desired primary activity, as well as secondary activities and physiochemical properties such that the agent will be useful when administered in real patients. In this regard, chemical modifications can improve the recognition and binding geometries (pharmacophores) of the candidate compounds, and so their affinities for their targets, as well as improving the physicochemical properties of the molecule that underlie necessary pharmacokinetic/pharmacodynamic (PK/PD), and toxicologic profiles (stability toward metabolic degradation, lack of geno-, hepatic, and cardiac toxicities, etc.) such that the chemical compound or biologic is suitable for introduction into animal and human studies.

Process Chemistry and Development

The final synthetic chemistry stages involve the production of a lead compound in suitable quantity and quality to allow large scale animal testing, and then human clinical trials. This involves the optimization of the synthetic route for bulk industrial production, and discovery of the most suitable drug formulation. The former of these is still the bailiwick of medicinal chemistry, the latter brings in the specialization of formulation science (with its components of physical and polymer chemistry and materials science). The synthetic chemistry specialization in medicinal chemistry aimed at adaptation and optimization of the synthetic route for industrial scale syntheses of hundreds of kilograms or more is termed process synthesis, and involves thorough knowledge of acceptable synthetic practice in the context of large scale reactions (reaction thermodynamics, economics, safety, etc.). Critical at this stage is the transition to more stringent GMP requirements for material sourcing, handling, and chemistry.

Training in Medicinal Chemistry

Medicinal chemistry is by nature an interdisciplinary science, and practitioners have a strong background in organic chemistry, which must eventually be coupled with a broad understanding of biological concepts related to cellular drug targets. Scientists in medicinal chemistry work are principally industrial scientists, working as part of an interdisciplinary team that uses their chemistry abilities, especially, their synthetic abilities, to use chemical principles to design effective therapeutic agents. The length of training is intense with practitioners often required to attain a 4-year bachelor's followed by a 4-6 year Ph.D. in organic chemistry. Most train-ing regimens include a postdoctoral fellowship period of 2 or more years after receiving a Ph.D. in chemistry making the length of training ranging from *10-12* years of college education. However, employment opportunities at the Master's level also exist in the pharmaceutical industry, and at that and the Ph.D. level there are further opportunities for employment in academia and government. Many medicinal chemists, particularly in academia and research, also earn a Pharm.D (doctor of pharmacy). Some of these PharmD/PhD researchers are RPh's (Registered Pharmacists).

Graduate level programs in medicinal chemistry can be found in traditional medicinal chemistry or pharmaceutical sciences departments, both of which are traditionally associated with schools of pharmacy, and in some chemistry departments. However, the majority of working medicinal chemists have graduate degrees (MS, but especially Ph.D.) in organic chemistry, rather than medicinal chemistry, and the preponderance of positions are in discovery, where the net is necessarily cast widest, and most broad synthetic activity occurs.

In discovery of small molecule therapeutics, an emphasis on training that provides for breadth of synthetic experience and "pace" of bench operations is clearly present (e.g., for individuals with pure synthetic organic and natural products synthesis in Ph.D. and post-doctoral positions, ibid.). In the medicinal chemistry specialty areas associated with the design and synthesis of chemical libraries or the execution of process chemistry aimed at viable commercial syntheses (areas generally with fewer opportunities), training paths are often much more varied (e.g., including focused training in physical organic chemistry, library-related syntheses, etc.).

As such, most entry-level workers in medicinal chemistry, especially in the U.S., do not have formal training in medicinal chemistry but receive the necessary medicinal chemistry and pharmacologic background after employment—at entry into their work in a pharmaceutical company, where the company provides its particular understanding or model of "medichem" training through active involvement in practical synthesis on therapeutic projects. (The same is somewhat true of computational medicinal chemistry specialties, but not to the same degree as in synthetic areas.)

Pharmacy

Pharmacy is the science and technique of preparing and dispensing drugs. It is a health profession that links health sciences with chemical sciences and aims to ensure the safe and effective use of pharmaceutical drugs.

The scope of pharmacy practice includes more traditional roles such as compounding and dispensing medications, and it also includes more modern services related to health care, including clinical services, reviewing medications for safety and efficacy, and providing drug information.

Pharmacists, therefore, are the experts on drug therapy and are the primary health professionals who optimize use of medication for the benefit of the patients.

The Apothecary or The Chemist by Gabriël Metsu (c. 1651–67)

An establishment in which pharmacy (in the first sense) is practiced is called a pharmacy (this term is more common in the United States) or a chemist's (which is more common in Great Britain). In the United States and Canada, drugstores commonly sell drugs, as well as miscellaneous items such as confectionery, cosmetics, office supplies, and magazines and occasionally refreshments and groceries.

The word *pharmacy* is derived from its root word *pharma*, which had first been used sometime in the 15th–17th centuries. However, the original Greek roots from *pharmakos* imply sorcery or even poison. In addition to pharma responsibilities, the pharma offered general medical advice and a range of services that are now performed solely by other specialist practitioners, such as surgery and midwifery. The pharma (as it was referred to) often operated through a retail shop which, in addition to ingredients for medicines, sold tobacco and patent medicines. Often the place that did this was called an apothecary and several languages have this as the dominant term, though their practices are more akin to a modern pharmacy, in English the term apothecary would today be seen as outdated or only approproriate if herbal remedies were on offer to a large extent. The pharmas also used many other herbs not listed. The Greek word *Pharmakeia* derives from *pharmakon*, meaning "drug", "medicine" (or "poison").

In its investigation of herbal and chemical ingredients, the work of the pharma may be regarded as a precursor of the modern sciences of chemistry and pharmacology, prior to the formulation of the scientific method.

Disciplines

The field of pharmacy can generally be divided into three primary disciplines:

- Pharmaceutics
- Medicinal Chemistry and Pharmacognosy
- Pharmacy Practice

Pharmacy, tacuinum sanitatis casanatensis (14th century)

The boundaries between these disciplines and with other sciences, such as biochemistry, are not always clear-cut. Often, collaborative teams from various disciplines (pharmacists and other scientists) work together toward the introduction of new therapeutics and methods for patient care. However, pharmacy is not a basic or biomedical science in its typical form. Medicinal chemistry is also a distinct branch of synthetic chemistry combining pharmacology, organic chemistry, and chemical biology.

Pharmacology is sometimes considered as the 4th discipline of pharmacy. Although pharmacology is essential to the study of pharmacy, it is not specific to pharmacy. Both disciplines are distinct. Those who wish to practice both pharmacy (patient oriented) and pharmacology (a biomedical science requiring the scientific method) receive separate training and degrees unique to either discipline.

Pharmacoinformatics is considered another new discipline, for systematic drug discovery and development with efficiency and safety.

Professionals

The World Health Organization estimates that there are at least 2.6 million pharmacists and other pharmaceutical personnel worldwide.

Pharmacists

Convent pharmacy exhibited at the Museo nazionale della scienza e della tecnologia Leonardo da Vinci of Milan.

Pharmacists are healthcare professionals with specialised education and training who perform various roles to ensure optimal health outcomes for their patients through the quality use of medicines. Pharmacists may also be small-business proprietors, owning the pharmacy in which they practice. Since pharmacists know about the mode of action of a particular drug, and its metabolism and physiological effects on the human body in great detail, they play an important role in optimisation of a drug treatment for an individual.

Pharmacists are represented internationally by the International Pharmaceutical Federation (FIP). They are represented at the national level by professional organisations such as the Royal Pharmaceutical Society in the UK, the Pharmaceutical Society of Australia (PSA), the Canadian Pharmacists Association (CPhA),the Pakistan Pharmacists Association (PPA), and the American Pharmacists Association (APhA).

In some cases, the representative body is also the registering body, which is responsible for the regulation and ethics of the profession.

In the United States, specializations in pharmacy practice recognized by the Board of Pharmacy Specialties include: cardiovascular, infectious disease, oncology, pharmacotherapy, nuclear, nutrition, and psychiatry. The Commission for Certification in Geriatric Pharmacy certifies pharmacists in geriatric pharmacy practice. The American Board of Applied Toxicology certifies pharmacists and other medical professionals in applied toxicology.

Pharmacy Technicians

Pharmacy technicians support the work of pharmacists and other health professionals by performing a variety of pharmacy related functions, including dispensing prescription drugs and other medical devices to patients and instructing on their use. They may also perform administrative duties in pharmaceutical practice, such as reviewing prescription requests with medics's offices and insurance companies to ensure correct medications are provided and payment is received.

A Pharmacy Technician in the UK is considered a health care professional and often does not work under the direct supervision of a pharmacist (if employed in a hospital pharmacy) but instead is supervised and managed by other senior pharmacy technicians. In the UK the role of a PhT has grown and responsibility has been passed on to them to manage the pharmacy department and specialised areas in pharmacy practice allowing pharmacists the time to specialise in their expert field as medication consultants spending more time working with patients and in research. A pharmacy technician once qualified has to register as a professional on the General Pharmaceutical Council (GPhC) register. The GPhC is the governing body for pharmacy health care professionals and this is who regulates the practice of pharmacists and pharmacy technicians.

In the US, pharmacy technicians perform their duties under supervision of pharmacists. Although they may perform, under supervision, most dispensing, compounding and other tasks, they are not generally allowed to perform the role of counseling patients on the proper use of their medications.

History

The earliest known compilation of medicinal substances was the *Sushruta Samhita*, an Indian

Ayurvedic treatise attributed to Sushruta in the 6th century BC. However, the earliest text as preserved dates to the 3rd or 4th century AD.

Physician and Pharmacist, illustration from *Medicinarius* (1505) by Hieronymus Brunschwig.

Many Sumerian (late 6th millennium BC – early 2nd millennium BC) cuneiform clay tablets record prescriptions for medicine.

Ancient Egyptian pharmacological knowledge was recorded in various papyri such as the *Ebers Papyrus* of 1550 BC, and the *Edwin Smith Papyrus* of the 16th century BC.

Dioscorides, *De Materia Medica*, Byzantium, 15th century

In Ancient Greece, Diocles of Carystus (4th century BC) was one of several men studying the medicinal properties of plants. He wrote several treatises on the topic. The Greek physician Pedanius Dioscorides is famous for writing a five volume book in his native Greek in the 1st century AD. The Latin translation *De Materia Medica* (*Concerning medical substances*) was used a basis for many medieval texts, and was built upon by many middle eastern scientists during the Islamic Golden Age. The title coined the term *materia medica*.

The earliest known Chinese manual on materia medica is the *Shennong Bencao Jing* (*The Divine Farmer's Herb-Root Classic*), dating back to the 1st century AD. It was compiled during the Han dynasty and was attributed to the mythical Shennong. Earlier literature included lists of prescriptions for specific ailments, exemplified by a manuscript "Recipes for 52 Ailments", found in the

Mawangdui tomb, sealed in 168 BC. Further details on Chinese pharmacy can be found in the Pharmacy in China article.

In Japan, at the end of the Asuka period (538–710) and the early Nara period (710–794), the men who fulfilled roles similar to those of modern pharmacists were highly respected. The place of pharmacists in society was expressly defined in the Taihō Code (701) and re-stated in the Yōrō Code (718). Ranked positions in the pre-Heian Imperial court were established; and this organizational structure remained largely intact until the Meiji Restoration (1868). In this highly stable hierarchy, the pharmacists—and even pharmacist assistants—were assigned status superior to all others in health-related fields such as physicians and acupuncturists. In the Imperial household, the pharmacist was even ranked above the two personal physicians of the Emperor.

There is a stone sign for a pharmacy with a tripod, a mortar, and a pestle opposite one for a doctor in the Arcadian Way in Ephesus near Kusadasi in Turkey. The current Ephesus dates back to 400 BC and was the site of the Temple of Artemis, one of the seven wonders of the world.

In Baghdad the first pharmacies, or drug stores, were established in 754, under the Abbasid Caliphate during the Islamic Golden Age. By the 9th century, these pharmacies were state-regulated.

The advances made in the Middle East in botany and chemistry led medicine in medieval Islam substantially to develop pharmacology. Muhammad ibn Zakarīya Rāzi (Rhazes) (865–915), for instance, acted to promote the medical uses of chemical compounds. Abu al-Qasim al-Zahrawi (Abulcasis) (936–1013) pioneered the preparation of medicines by sublimation and distillation. His *Liber servitoris* is of particular interest, as it provides the reader with recipes and explains how to prepare the `simples' from which were compounded the complex drugs then generally used. Sabur Ibn Sahl (d 869), was, however, the first physician to initiate pharmacopoedia, describing a large variety of drugs and remedies for ailments. Al-Biruni (973–1050) wrote one of the most valuable Islamic works on pharmacology, entitled *Kitab al-Saydalah* (*The Book of Drugs*), in which he detailed the properties of drugs and outlined the role of pharmacy and the functions and duties of the pharmacist. Avicenna, too, described no less than 700 preparations, their properties, modes of action, and their indications. He devoted in fact a whole volume to simple drugs in *The Canon of Medicine*. Of great impact were also the works by al-Maridini of Baghdad and Cairo, and Ibn al-Wafid (1008–1074), both of which were printed in Latin more than fifty times, appearing as *De Medicinis universalibus et particularibus* by 'Mesue' the younger, and the *Medicamentis simplicibus* by 'Abenguefit'. Peter of Abano (1250–1316) translated and added a supplement to the work of al-Maridini under the title *De Veneris*. Al-Muwaffaq's contributions in the field are also pioneering. Living in the 10th century, he wrote *The foundations of the true properties of Remedies*, amongst others describing arsenious oxide, and being acquainted with silicic acid. He made clear distinction between sodium carbonate and potassium carbonate, and drew attention to the poisonous nature of copper compounds, especially copper vitriol, and also lead compounds. He also describes the distillation of sea-water for drinking.

In Europe pharmacy-like shops began to appear during the 12th century. In 1240 emperor Frederic II issued a decree by which the physician's and the apothecary's professions were separated. The first pharmacy in Europe (still working) was opened in 1241 in Trier, Germany.

Sign of the Town Hall Pharmacy in Tallinn, operating continuously from at least 1422

In Europe there are old pharmacies still operating in Dubrovnik, Croatia, located inside the Franciscan monastery, opened in 1317; and in the Town Hall Square of Tallinn, Estonia, dating from at least 1422. The oldest is claimed to have been set up in 1221 in the Church of Santa Maria Novella in Florence, Italy, which now houses a perfume museum. The medieval Esteve Pharmacy, located in Llívia, a Catalan enclave close to Puigcerdà, also now a museum, dates back to the 15th century, keeping albarellos from the 16th and 17th centuries, old prescription books and antique drugs.

The mortar and pestle, one of the internationally recognized symbols to represent the pharmacy profession

Typical American drug store with a soda fountain, about 1905

Drug store restoration ca. 1920 at Collingsworth County Museum and Art Center across from the courthouse in Wellington, Texas

Types of Pharmacy Practice Areas

Pharmacists practice in a variety of areas including community pharmacies, hospitals, clinics, extended care facilities, psychiatric hospitals, and regulatory agencies. Pharmacists can specialize in various areas of practice including but not limited to: hematology/oncology, infectious diseases, ambulatory care, nutrition support, drug information, critical care, pediatrics, etc.

Community Pharmacy

19th-century Italian pharmacy

Classic symbols at the wall of a former German pharmacy

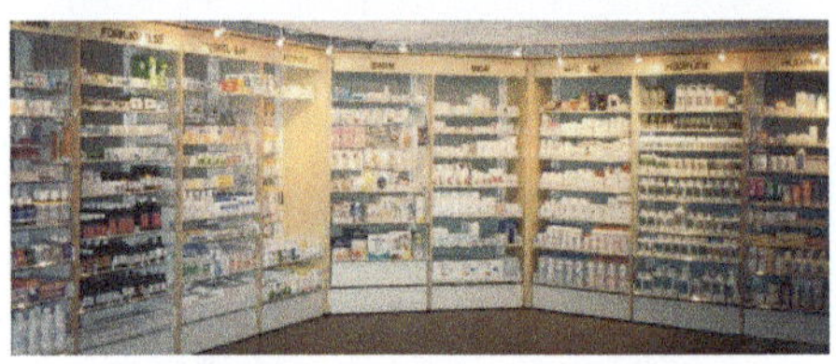

Modern pharmacy in Norway

A pharmacy (commonly the chemist in Australia, New Zealand and the UK; or drugstore in North America; retail pharmacy in industry terminology; or Apothecary, historically) is the place where most pharmacists practice the profession of pharmacy. It is the community pharmacy where the dichotomy of the profession exists—health professionals who are also retailers.

Community pharmacies usually consist of a retail storefront with a dispensary where medications are stored and dispensed. According to Sharif Kaf al-Ghazal, the opening of the first drugstores are recorded by Muslim pharmacists in Baghdad in 754.

In most countries, the dispensary is subject to pharmacy legislation; with requirements for storage conditions, compulsory texts, equipment, etc., specified in legislation. Where it was once the case that pharmacists stayed within the dispensary compounding/dispensing medications, there has been an increasing trend towards the use of trained pharmacy technicians while the pharmacist spends more time communicating with patients. Pharmacy technicians are now more dependent upon automation to assist them in their new role dealing with patients' prescriptions and patient safety issues.

Pharmacies are typically required to have a pharmacist on-duty at all times when open. It is also often a requirement that the owner of a pharmacy must be a registered pharmacist, although this is not the case in all jurisdictions, such that many retailers (including supermarkets and mass merchandisers) now include a pharmacy as a department of their store.

Likewise, many pharmacies are now rather grocery store-like in their design. In addition to medicines and prescriptions, many now sell a diverse arrangement of additional items such as cosmetics, shampoo, office supplies, confections, snack foods, durable medical equipment, greeting cards, and provide photo processing services.

Hospital Pharmacy

Pharmacies within hospitals differ considerably from community pharmacies. Some pharmacists in hospital pharmacies may have more complex clinical medication management issues whereas pharmacists in community pharmacies often have more complex business and customer relations issues.

Because of the complexity of medications including specific indications, effectiveness of treatment regimens, safety of medications (i.e., drug interactions) and patient compliance issues (in the hospital and at home) many pharmacists practicing in hospitals gain more education and training after pharmacy school through a pharmacy practice residency and sometimes followed by another residency in a specific area. Those pharmacists are often referred to as clinical pharmacists and they often specialize in various disciplines of pharmacy. For example, there are pharmacists who specialize in hematology/oncology, HIV/AIDS, infectious disease, critical care, emergency medicine, toxicology, nuclear pharmacy, pain management, psychiatry, anti-coagulation clinics, herbal medicine, neurology/epilepsy management, pediatrics, neonatal pharmacists and more.

Hospital pharmacies can often be found within the premises of the hospital. Hospital pharmacies usually stock a larger range of medications, including more specialized medications, than would be feasible in the community setting. Most hospital medications are unit-dose, or a single dose of medicine. Hospital pharmacists and trained pharmacy technicians compound sterile products for patients including total parenteral nutrition (TPN), and other medications given intravenously. This is a complex process that requires adequate training of personnel, quality assurance of products, and adequate facilities. Several hospital pharmacies have decided to outsource high risk preparations and some other compounding functions to companies who specialize in compounding. The high cost of medications and drug-related technology, combined with the potential impact of medications and pharmacy services on patient-care outcomes and patient safety, make it imperative that hospital pharmacies perform at the highest level possible.

Clinical Pharmacy

Pharmacists provide direct patient care services that optimizes the use of medication and promotes health, wellness, and disease prevention. Clinical pharmacists care for patients in all health care settings, but the clinical pharmacy movement initially began inside hospitals and clinics. Clinical pharmacists often collaborate with physicians and other healthcare professionals to improve pharmaceutical care. Clinical pharmacists are now an integral part of the interdisciplinary approach to patient care. They often participate in patient care rounds for drug product selection.

The clinical pharmacist's role involves creating a comprehensive drug therapy plan for patient-specific problems, identifying goals of therapy, and reviewing all prescribed medications prior to dispensing and administration to the patient. The review process often involves an evaluation of the appropriateness of the drug therapy (e.g., drug choice, dose, route, frequency, and duration of therapy) and its efficacy. The pharmacist must also monitor for potential drug interactions, adverse drug reactions, and assess patient drug allergies while designing and initiating a drug therapy plan.

Ambulatory Care Pharmacy

Since the emergence of modern clinical pharmacy, ambulatory care pharmacy practice has emerged as a unique pharmacy practice setting. Ambulatory care pharmacy is based primarily on pharmacotherapy services that a pharmacist provides in a clinic. Pharmacists in this setting often do not dispense drugs, but rather see patients in office visits to manage chronic disease states.

In the U.S. federal health care system (including the VA, the Indian Health Service, and NIH) ambulatory care pharmacists are given full independent prescribing authority. In some states such North Carolina and New Mexico these pharmacist clinicians are given collaborative prescriptive and diagnostic authority. In 2011 the board of Pharmaceutical Specialties approved ambulatory care pharmacy practice as a separate board certification. The official designation for pharmacists who pass the ambulatory care pharmacy specialty certification exam will be Board Certified Ambulatory Care Pharmacist and these pharmacists will carry the initials BCACP.

Compounding Pharmacy

Compounding is the practice of preparing drugs in new forms. For example, if a drug manufacturer only provides a drug as a tablet, a compounding pharmacist might make a medicated lollipop that contains the drug. Patients who have difficulty swallowing the tablet may prefer to suck the medicated lollipop instead.

Another form of compounding is by mixing different strengths (g,mg,mcg) of capsules or tablets to yield the desired amount of medication indicated by the physician, physician assistant, Nurse Practitioner, or clinical pharmacist practitioner. This form of compounding is found at community or hospital pharmacies or in-home administration therapy.

Compounding pharmacies specialize in compounding, although many also dispense the same non-compounded drugs that patients can obtain from community pharmacies.

Consultant Pharmacy

Consultant pharmacy practice focuses more on medication regimen review (i.e. "cognitive services") than on actual dispensing of drugs. Consultant pharmacists most typically work in nursing homes, but are increasingly branching into other institutions and non-institutional settings.

Traditionally consultant pharmacists were usually independent business owners, though in the United States many now work for several large pharmacy management companies (primarily Omnicare, Kindred Healthcare and PharMerica). This trend may be gradually reversing as consultant pharmacists begin to work directly with patients, primarily because many elderly people are now taking numerous medications but continue to live outside of institutional settings. Some community pharmacies employ consultant pharmacists and/or provide consulting services.

The main principle of consultant pharmacy is developed by Hepler and Strand in 1990.

Internet Pharmacy

Canisters of pills from a mail order pharmacy.

Since about the year 2000, a growing number of Internet pharmacies have been established worldwide. Many of these pharmacies are similar to community pharmacies, and in fact, many of them are actually operated by brick-and-mortar community pharmacies that serve consumers online and those that walk in their door. The primary difference is the method by which the medications are requested and received. Some customers consider this to be more convenient and private method rather than traveling to a community drugstore where another customer might overhear about the drugs that they take. Internet pharmacies (also known as online pharmacies) are also recommended to some patients by their physicians if they are homebound.

While most Internet pharmacies sell prescription drugs and require a valid prescription, some Internet pharmacies sell prescription drugs without requiring a prescription. Many customers order drugs from such pharmacies to avoid the "inconvenience" of visiting a doctor or to obtain medications which their doctors were unwilling to prescribe. However, this practice has been criticized as potentially dangerous, especially by those who feel that only doctors can reliably assess contraindications, risk/benefit ratios, and an individual's overall suitability for use of a medication. There also have been reports of such pharmacies dispensing substandard products.

Of particular concern with Internet pharmacies is the ease with which people, youth in particular, can obtain controlled substances (e.g., Vicodin, generically known as hydrocodone) via the Internet without a prescription issued by a doctor/practitioner who has an established doctor-patient relationship. There are many instances where a practitioner issues a prescription, brokered by an Internet server, for a controlled substance to a "patient" s/he has never met. In the United States, in order for a prescription for a controlled substance to be valid, it must be issued for a legitimate medical purpose by a licensed practitioner acting in the course of legitimate doctor-patient relationship. The filling pharmacy has a corresponding responsibility to ensure that the prescription is valid. Often, individual state laws outline what defines a valid patient-doctor relationship.

Canada is home to dozens of licensed Internet pharmacies, many of which sell their lower-cost prescription drugs to U.S. consumers, who pay one of the world's highest drug prices. In recent years, many consumers in the US and in other countries with high drug costs, have turned to licensed Internet pharmacies in India, Israel and the UK, which often have even lower prices than in Canada.

In the United States, there has been a push to legalize importation of medications from Canada and other countries, in order to reduce consumer costs. While in most cases importation of prescription medications violates Food and Drug Administration (FDA) regulations and federal laws, enforcement is generally targeted at international drug suppliers, rather than consumers. There is no known case of any U.S. citizens buying Canadian drugs for personal use with a prescription, who has ever been charged by authorities.

Veterinary Pharmacy

Veterinary pharmacies, sometimes called *animal pharmacies*, may fall in the category of hospital pharmacy, retail pharmacy or mail-order pharmacy. Veterinary pharmacies stock different varieties and different strengths of medications to fulfill the pharmaceutical needs of animals. Because the needs of animals, as well as the regulations on veterinary medicine, are often very different from those related to people, veterinary pharmacy is often kept separate from regular pharmacies.

Nuclear Pharmacy

Nuclear pharmacy focuses on preparing radioactive materials for diagnostic tests and for treating certain diseases. Nuclear pharmacists undergo additional training specific to handling radioactive materials, and unlike in community and hospital pharmacies, nuclear pharmacists typically do not interact directly with patients.

Military Pharmacy

Military pharmacy is an entirely different working environment due to the fact that technicians perform most duties that in a civilian sector would be illegal. State laws of Technician patient counseling and medication checking by a pharmacist do not apply.

Pharmacy Informatics

Pharmacy informatics is the combination of pharmacy practice science and applied information science. Pharmacy informaticists work in many practice areas of pharmacy, however, they may also work in information technology departments or for healthcare information technology vendor companies. As a practice area and specialist domain, pharmacy informatics is growing quickly to meet the needs of major national and international patient information projects and health system interoperability goals. Pharmacists in this area are trained to participate in medication management system development, deployment and optimization.

Specialty Pharmacy

Specialty pharmacies supply high cost injectable, oral, infused, or inhaled medications that are used for chronic and complex disease states such as cancer, hepatitis, and rheumatoid arthritis. Unlike a traditional community pharmacy where prescriptions for any common medication can be brought in and filled, specialty pharmacies carry novel medications that need to be properly stored, administered, carefully monitored, and clinically managed. In addition to supplying these drugs, specialty pharmacies also provide lab monitoring, adherence counseling, and assist patients with cost-containment strategies needed to obtain their expensive specialty drugs. It is currently the fastest growing sector of the pharmaceutical industry with 19 of 28 newly FDA approved medications in 2013 being specialty drugs.

Due to the demand for clinicians who can properly manage these specific patient populations, the Specialty Pharmacy Certification Board has developed a new certification exam to certify specialty pharmacists. Along with the 100 question computerized multiple-choice exam, pharmacists must also complete 3,000 hours of specialty pharmacy practice within the past three years as well as 30 hours of specialty pharmacist continuing education within the past two years.

Issues in Pharmacy

Separation of Prescribing from Dispensing

In most jurisdictions (such as the United States), pharmacists are regulated separately from physicians. These jurisdictions also usually specify that *only* pharmacists may supply scheduled pharmaceuticals to the public, and that pharmacists cannot form business partnerships with physicians or give them "kickback" payments. However, the American Medical Association (AMA) Code of Ethics provides that physicians may dispense drugs within their office practices as long as there is no patient exploitation and patients have the right to a written prescription that can be filled elsewhere. 7 to 10 percent of American physicians practices reportedly dispense drugs on their own.

In some rural areas in the United Kingdom, there are dispensing physicians who are allowed to both prescribe and dispense prescription-only medicines to their patients from within their practices. The law requires that the GP practice be located in a designated rural area and that there is also a specified, minimum distance (currently 1.6 kilometres) between a patient's home and the nearest retail pharmacy. This law also exists in Austria for general physicians if the nearest pharmacy is more than 4 kilometers away, or where none is registered in the city.

In other jurisdictions (particularly in Asian countries such as China, Malaysia, and Singapore), doctors are allowed to dispense drugs themselves and the practice of pharmacy is sometimes integrated with that of the physician, particularly in traditional Chinese medicine.

In Canada it is common for a medical clinic and a pharmacy to be located together and for the ownership in both enterprises to be common, but licensed separately.

One of a chain of pharmacies in Mexico City, Mexico, named "Doctor Discount," March 2010.

The reason for the majority rule is the high risk of a conflict of interest and/or the avoidance of absolute powers. Otherwise, the physician has a financial self-interest in "diagnosing" as many conditions as possible, and in exaggerating their seriousness, because he or she can then sell more medications to the patient. Such self-interest directly conflicts with the patient's interest in obtaining cost-effective medication and avoiding the unnecessary use of medication that may have side-effects. This system reflects much similarity to the checks and balances system of the U.S. and many other governments.

A campaign for separation has begun in many countries and has already been successful (as in Korea). As many of the remaining nations move towards separation, resistance and lobbying from dispensing doctors who have pecuniary interests may prove a major stumbling block (e.g. in Malaysia).

The Future of Pharmacy

In the coming decades, pharmacists are expected to become more integral within the health care system. Rather than simply dispensing medication, pharmacists are increasingly expected to be compensated for their patient care skills. In particular, Medication Therapy Management (MTM) includes the clinical services that pharmacists can provide for their patients. Such services include the thorough analysis of all medication (prescription, non-prescription, and herbals) currently being taken by an individual. The result is a reconciliation of medication and patient education resulting in increased patient health outcomes and decreased costs to the health care system.

This shift has already commenced in some countries; for instance, pharmacists in Australia receive remuneration from the Australian Government for conducting comprehensive Home Medicines Reviews. In Canada, pharmacists in certain provinces have limited prescribing rights (as in Alberta and British Columbia) or are remunerated by their provincial government for expanded services such as medications reviews (Medschecks in Ontario). In the United Kingdom, pharmacists who undertake additional training are obtaining prescribing rights and this is because of pharmacy education. They are also being paid for by the government for medicine use reviews. In Scotland the pharmacist can write prescriptions for Scottish registered patients of their regular medications, for the majority of drugs, except for controlled drugs, when the patient is unable to see their doctor, as could happen if they are away from home or the doctor is unavailable. In the United States, pharmaceutical care or clinical pharmacy has had an evolving influence on the practice of pharmacy. Moreover, the Doctor of Pharmacy (Pharm. D.) degree is now required before entering practice and some pharmacists now complete one or two years of residency or fellowship training following graduation. In addition, consultant pharmacists, who traditionally operated primarily in nursing homes are now expanding into direct consultation with patients, under the banner of "senior care pharmacy."

In addition to patient care, pharmacies will be a focal point for medical adherence initiatives. There is enough evidence to show that integrated pharmacy based initiatives significantly impact adherence for chronic patients. For example, a study published in NIH shows "pharmacy based interventions improved patients' medication adherence rates by 2.1 percent and increased physicians' initiation rates by 38 percent, compared to the control group".

Various Pharmacy Practice Areas

Community Pharmacy

A pharmacy - in Polish 'apteka' - near *Grunwaldzki Square* in Szczecin, Poland

A community pharmacy is a healthcare facility that emphasizes providing pharmaceutical services to a specific community. Typically, a registered pharmacist dispenses medicine from the community pharmacy.

Duties

The International Pharmaceutical Federation has declared their vision of a community-based pharmacist:

- An expert in pharmaceutical care, pharmacotherapy and health promotion
- A professional communicator with patients, other healthcare providers and decision makers
- Deliver high quality products, services and communication
- Document their work and communicate the outcome to professional colleagues.

Responsibilities

Community-based pharmacists' responsibilities include: checking and dispensing of prescription drugs, providing advice on drug selection and usage to doctors and other health professionals and counseling patients in health promotion, disease prevention and the proper use of medicines.

In most countries regulations govern how dispensaries may operate, with specific requirements for storage conditions, equipment and record keeping.

Clinical Roles

It is becoming more common for pharmacists to take on extended roles that provide more clinical care directly to patients as part of a primary care team. There are around 11,400 Community Pharmacies in England. Many are open for extended hours in the evenings and weekends and they are accessible without appointment.

United Kingdom

In the English NHS there are 438 million visits nationally to community pharmacies a year for health related reasons. More than 1 billion prescription items were dispensed in 2012. More than £14 billion is spent on prescribed drugs annually. Under the 2005 NHS Community Pharmacy contract all community pharmacists in England and Wales provide:

- Dispensing Service
- Repeat Dispensing Service: This allows the patient to collect regular repeat prescription medicines for an agreed period of time, without having to go back to their GP
- Disposal of Unwanted Medicines.
- Promotion of Healthy Lifestyles.
- Signposting to other Services.
- Support for Self-Care: advice on treating minor illnesses and long term conditions.

Widely available Advanced Services:

- Medicines Use Review & Prescription Intervention.
- New Medicine Service.For patients who have been given their first prescription for a medicine to treat Asthma and COPD, Type 2 diabetes, Hypertension or Anticoagulant Therapy.
- Appliance Use Review Service.
- Flu Vaccination.

Enhanced Services which are not available unless locally commissioned:

- Minimising Adverse Effects and Admissions Related to Medicines.

- Discharge and Transfer Planning.
- Managing Dental Pain.

Support Staff

To help pharmacists be able to take on extended roles, it is common for them to work as part of a team that can include pharmacy technicians, dispensing assistants and counter assistants.

Ownership

In parts of mainland Europe, the pharmacist is required to own the pharmacy of which she or he is the licensee. Under this arrangement, a pharmacist can be the operator of only a single outlet. In the UK, 60% of all community pharmacies are owned by companies that own multiple pharmacies.

In the United States, more than 25% of independent owners have ownership in two or more pharmacies.

Most of New Zealand's community pharmacies are owner-operated. In Australia, pharmacists recognise the need to integrate professional pharmacy services into the health system to meet the changing health care needs of the population.

Hospital Pharmacy

Hospital pharmacy in Thailand

Hospital pharmacies are pharmacies usually found within the premises of a hospital. Hospital pharmacies usually stock a larger range of medications, including more specialized and investigational medications (medicines that are being studied, but have not yet been approved), than would be feasible in the community setting. Hospital pharmacies typically provide medications for the hospitalized patients only, and are not retail establishments. They typically do not provide prescription service to the public. Some hospitals do have retail pharmacies within them (il-lustration), which sell over-the-counter as well as prescription medications to the public, but these are not the actual hospital pharmacy.

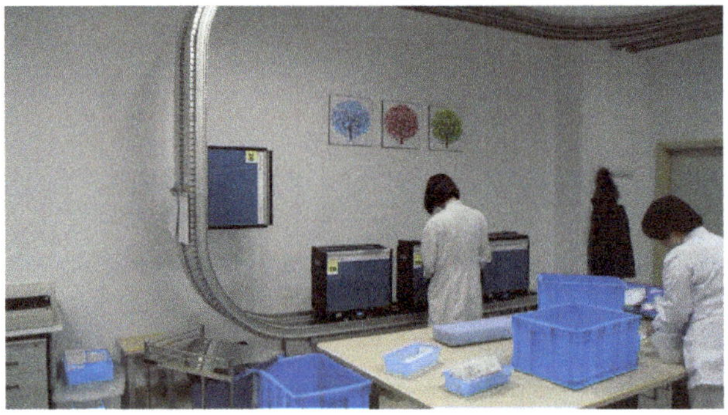

Electric track vehicle system for hospitals, type Telelift

which types of drugs in provide in hospital pharmacy

Pharmacy Logistic in Hospitals

Hospital pharmacies provide a huge quantity of medications per day which is allocated to the wards and to intensive care unit according to medication schedule. Larger hospitals use automated transport systems for dispatch of medications.

Sterile Production

Hospital pharmacist and trained pharmacy technicians compound sterile products for patients including total parenteral nutrition (TPN), and other medications given intravenously e.g. neonatal antibiotics and chemotherapy. This is a complex process that requires adequate training of personnel, quality assurance of products, and adequate facilities. Some hospital pharmacies have decided to outsource high-risk preparations and some other compounding functions to companies that specialize in compounding.

Clinical Pharmacy

Clinical pharmacy is the branch of pharmacy in which pharmacists provide patient care that optimizes the use of medication and promotes health, wellness, and disease prevention. Clinical pharmacists care for patients in all health care settings but the clinical pharmacy movement initially began inside hospitals and clinics. Clinical pharmacists often work in collaboration with physicians, nurse practitioners, and other healthcare professionals.

Education and Credentialing

Clinical pharmacists have extensive education in the biomedical, pharmaceutical, socio-behavioural and clinical sciences. Most clinical pharmacists have a Doctor of Pharmacy (Pharm.D.) degree and many have completed one or more years of post-graduate training (for example, a general and/or specialty pharmacy residency). In the United States, clinical pharmacists can choose to become Board-certified through the Board of Pharmacy Specialties (BPS), which was organized in 1976 as an independent certification agency of the American Pharmacists Association. The BPS certifies pharmacists in the following specialities: ambulatory care pharmacy, critical care pharmacy, nuclear pharmacy, nutrition support pharmacy, oncology pharmacy, paediatric pharmacy, pharmacotherapy, and psychiatric pharmacy.

Role in the Health Care System

Within the system of health care, clinical pharmacists are experts in the therapeutic use of medications. They routinely provide medication therapy evaluations and recommendations to patients and other health care professionals. Clinical pharmacists are a primary source of scientifically valid information and advice regarding the safe, appropriate, and cost-effective use of medications. Clinical pharmacists are also making themselves more readily available to the public. In the past, access to a clinical pharmacist was limited to hospitals, clinics, or educational institutions. However, clinical pharmacists are making themselves available through a medication information hotline, and reviewing medication lists, all in an effort to prevent medication errors in the foreseeable future. In the United Kingdom, clinical pharmacists are routinely involved in the direct care of patients within hospitals, and increasingly, in doctors surgeries. They also develop post registration professional education, professional curricula for workforce development, provide expertise on the use of medicines to national organizations such as NICE, the Department of Health, and the MHRA, and develop medicines guidelines for use in therapeutic areas.

Clinical pharmacists interact directly with patients in several different ways. They use their knowledge of medication (including dosage, drug interactions, side effects, expense, effectiveness, etc.) to determine if a medication plan is appropriate for their patient. If it is not, the pharmacist will consult the primary physician to ensure that the patient is on the proper medication plan. The pharmacist also works to educate their patients on the importance of taking and finishing their medications. Studies conducted into Pharmacist-led Chronic Disease Management show that it was associated with effects similar to usual care and might improve physiological goal attainment.

In some states, clinical pharmacists are given prescriptive authority under protocol with a medical provider, and their scope of practice is constantly evolving. In the United Kingdom clinical pharmacists are given independent prescriptive authority.

Basic components of clinical pharmacy practice include: Prescribing drugs, administering drugs, monitoring prescriptions, managing drug use, and counselling patients.

History of Pharmacy

The history of pharmacy as an independent science dates back to the first third of the 19th century. Before then, pharmacy evolved from antiquity as part of medicine.

Prehistoric Pharmacy

Paleopharmacological studies attest to the use of medicinal plants in pre-history.

Antiquity

Many Sumerian cuneiform tablets record prescriptions for medicine. Ancient Egyptian pharmacological knowledge was recorded in various papyri such as the *Ebers Papyrus* of 1550 BC, and the *Edwin Smith Papyrus* of the 16th century BC.

In Ancient Greece, according to Edward Kremers and Glenn Sonnedecker, "before, during and after the time of Hippocrates there was a group of experts in medicinal plants. Probably the most important representative of these *rhizotomoi* was Diocles of Carystus (4th century BC). He is considered to be the source for all Greek pharmacotherapeutic treatises between the time of Theophrastus and Dioscorides." The Greek physician Pedanius Dioscorides is famous for writing a five volume book in his native Greek ("Peri hules iatrikes") in the 1st century AD. The Latin translation *De Materia Medica* (*Concerning medical substances*) was used a basis for many medieval texts, and was built upon by many middle eastern scientists during the Islamic Golden Age. The title coined the term *materia medica*. There is a stone sign for a pharmacy with a tripod, a mortar, and a pestle opposite one for a doctor in the Arcadian Way in Ephesus near Kusadasi in Turkey.

The earliest known Chinese manual on materia medica is the *Shennong Bencao Jing* (*The Divine Farmer's Herb-Root Classic*), dating back to the 1st century AD. It was compiled during the Han dynasty and was attributed to the mythical Shennong. Earlier literature included lists of prescriptions for specific ailments, exemplified by a manuscript "Recipes for 52 Ailments", found in the Mawangdui tomb, sealed in 168 BC. Further details on Chinese pharmacy can be found in the Pharmacy in China article.

The earliest known compilation of medicinal substances in Indian traditional medicine dates to the 3rd or 4th century AD)(attributed to Sushruta, who is recorded as a physician of the 6th century BC).

In Japan, at the end of the Asuka period (538-710) and the early Nara period (710-794), the men who fulfilled roles similar to those of modern pharmacists were highly respected. The place of pharmacists in society was expressly defined in the Taihō Code (701) and re-stated in the Yōrō Code (718). Ranked positions in the pre-Heian Imperial court were established; and this organizational structure remained largely intact until the Meiji Restoration (1868). In this highly stable hierarchy, the pharmacists—and even pharmacist assistants—were assigned status superior to all others in health-related fields such as physicians and acupuncturists. In the Imperial household, the pharmacist was even ranked above the two personal physicians of the Emperor.

Middle Ages

In Baghdad the first pharmacies, or drug stores, were established in 754, under the Abbasid Caliphate during the Islamic Golden Age. By the 9th century, these pharmacies were state-regulated.

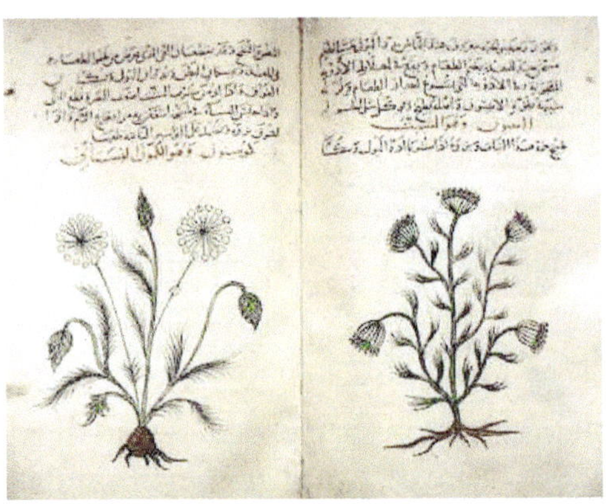

Arabic herbal medicine guidebook *De Materia Medica* of Dioscorides. Cumin & dill. c. 1334.

The advances made in the Middle East in botany and chemistry led medicine in medieval Islam substantially to develop pharmacology. Muhammad ibn Zakarīya Rāzi (Rhazes) (865-915), for instance, acted to promote the medical uses of chemical compounds. Abu al-Qasim al-Zahrawi (Abulcasis) (936-1013) pioneered the preparation of medicines by sublimation and distillation. His *Liber servitoris* is of particular interest, as it provides the reader with recipes and explains how to prepare the `simples' from which were compounded the complex drugs then generally used. Sabur Ibn Sahl (d 869), was, however, the first physician to initiate pharmacopoedia, describing a large variety of drugs and remedies for ailments. Al-Biruni (973-1050) wrote one of the most valuable Islamic works on pharmacology entitled *Kitab al-Saydalah* (*The Book of Drugs*), where he gave detailed knowledge of the properties of drugs and outlined the role of pharmacy and the functions and duties of the pharmacist. Ibn Sina (Avicenna), too, described no less than 700 preparations, their properties, mode of action and their indications. He devoted in fact a whole volume to simple drugs in *The Canon of Medicine*. Of great impact were also the works by al-Maridini of Baghdad and Cairo, and Ibn al-Wafid (1008–1074), both of which were printed in Latin more than fifty times, appearing as *De Medicinis universalibus et particularibus* by `Mesue' the younger, and the *Medicamentis simplicibus* by `Abenguefit'. Peter of Abano (1250–1316) translated and added a supplement to the work of al-Maridini under the title *De Veneris*. Al-Muwaffaq's contributions in the field are also pioneering. Living in the 10th century, he wrote *The foundations of the true properties of Remedies*, amongst others describing arsenious oxide, and being acquainted with silicic acid. He made clear distinction between sodium carbonate and potassium carbonate, and drew attention to the poisonous nature of copper compounds, especially copper vitriol, and also lead compounds. He also describes the distillation of sea-water for drinking.

In *Europe* pharmacy-like shops began to appear during the 12th century. In 1240 emperor Frederic II issued a decree by which the physician's and the apothecary's professions were separated.

In Europe there are old pharmacies still operating in Dubrovnik, Croatia located inside the Franciscan monastery, opened in 1317 ; and one in the Town Hall Square of Tallinn, Estonia dating from at least 1422.

The oldest pharmacy is claimed to be set up in 1221 in the Church of Santa Maria Novella in Florence, Italy, which now houses a perfume museum. The medieval Esteve Pharmacy, located in Llívia, a Catalan enclave close to Puigcerdà, is also now a museum dating back to the 15th century, keeping albarellos from the 16th and 17th centuries, old prescription books and antique drugs.

The Republic of Venice was the first State with health modern policies which requires that the nature of the drug is public. Actually thirteen secrets survive which were offered to sale to the Venetian Republic.

References

- Edward Kremers, Glenn Sonnedecker (1986). "Kremers and Urdang's History of pharmacy". Amer. Inst. History of Pharmacy. p.17. ISBN 0931292174.
- Ellis, Linda (2000). Archaeological Method and Theory: An Encyclopedia. Taylor & Francis. pp. 443–448. ISBN 978-0-8153-1305-2.
- "An Overview of the Clinical Pharmacist Practitioner in NC". North Carolina Association of Pharmacists. 23 May 2005. Archived from the original on May 23, 2005. Retrieved 28 April 2016.

- "How the health system is skewed towards harm for pill-popping pensioners". EconoTimes. 27 November 2015. Retrieved 30 November 2015.
- Gauld, Natalie (8 December 2010). "Sun, surf, snow and pharmacy- pharmacy practice in New Zealand". PJ online. Retrieved 8 July 2014.
- Cragg, Gordon M.; Newman, David J. (2013-06-01). "Natural products: A continuing source of novel drug leads". Biochimica et Biophysica Acta (BBA) - General Subjects. 1830 (6): 3670–3695. doi:10.1016/j.bbagen.2013.02.008. PMC 3672862. PMID 23428572.
- Wild, D. Specialty Pharmacy Continuum Carving a Specialty Niche Within the ACO Model." Volume 1 (Summer Issue). August, 2012.
- Hughes, Jp; Rees, S; Kalindjian, Sb; Philpott, Kl (2011-03-01). "Principles of early drug discovery". British Journal of Pharmacology. 162 (6): 1239–1249. doi:10.1111/j.1476-5381.2010.01127.x. ISSN 1476-5381. PMC 3058157. PMID 21091654.

2

Branches of Medicinal Chemistry

Specialization in the field of medicine has naturally caused changes in the manufacture of pharmaceutical medicine. Pharmacology is a branch of medicine that deals with drug actions and pharmacognosy studies those medicinal drugs which are derived from plants. The following content is a compilation of the various branches of medicine that forms an integral part of the broader subject matter.

Pharmacology

Pharmacology is the branch of medicine and biology concerned with the study of drug action, where a drug can be broadly defined as any man-made, natural, or endogenous (from within body) molecule which exerts a biochemical and/or physiological effect on the cell, tissue, organ, or organism (sometimes the word pharmacon is used as a term to encompass these endogenous and exogenous bioactive species). More specifically, it is the study of the interactions that occur between a living organism and chemicals that affect normal or abnormal biochemical function. If substances have medicinal properties, they are considered pharmaceuticals.

A variety of topics involved with pharmacology, including neuropharmacology, renal pharmacology, human metabolism, intracellular metabolism, and intracellular regulation

The field encompasses drug composition and properties, synthesis and drug design, molecular and cellular mechanisms, organ/systems mechanisms, signal transduction/cellular communication, molecular diagnostics, interactions, toxicology, chemical biology, therapy, and medical applications and antipathogenic capabilities. The two main areas of pharmacology are pharmacodynamics and pharmacokinetics. The former studies the effects of the drug on biological systems, and the latter the effects of biological systems on the drug. In broad terms, pharmacodynamics discusses the chemicals with biological receptors, and pharmacokinetics discusses the absorption, distribution, metabolism, and excretion (ADME) of chemicals from the biological systems. Pharmacology is not synonymous with pharmacy and the two terms are frequently confused. Pharmacology, a biomedical science, deals with the research, discovery, and characterization of chemicals which show biological effects and the elucidation of cellular and organismal function in relation to these chemicals. In contrast, pharmacy, a health services profession, is concerned with application of the principles learned from pharmacology in its clinical settings; whether it be in a dispensing or clinical care role. In either field, the primary contrast between the two are their distinctions between direct-patient care, for pharmacy practice, and the science-oriented research field, driven by pharmacology.

The origins of clinical pharmacology date back to the Middle Ages in Avicenna's *The Canon of Medicine*, Peter of Spain's *Commentary on Isaac*, and John of St Amand's *Commentary on the Antedotary of Nicholas*. Clinical pharmacology owes much of its foundation to the work of William Withering. Pharmacology as a scientific discipline did not further advance until the mid-19th century amid the great biomedical resurgence of that period. Before the second half of the nineteenth century, the remarkable potency and specificity of the actions of drugs such as morphine, quinine and digitalis were explained vaguely and with reference to extraordinary chemical powers and affinities to certain organs or tissues. The first pharmacology department was set up by Rudolf Buchheim in 1847, in recognition of the need to understand how therapeutic drugs and poisons produced their effects.

Early pharmacologists focused on natural substances, mainly plant extracts. Pharmacology developed in the 19th century as a biomedical science that applied the principles of scientific experimentation to therapeutic contexts. Today pharmacologists use genetics, molecular biology, chemistry, and other advanced tools to transform information about molecular mechanisms and targets into therapies directed against disease, defects or pathogens, and create methods for preventative care, diagnostics, and ultimately personalized medicine.

Divisions

The discipline of pharmacology can be divided into many sub disciplines each with a specific focus.

Clinical Pharmacology

Clinical pharmacology is the basic science of pharmacology with an added focus on the application of pharmacological principles and methods in the medical clinic and towards patient care and outcomes.

Neuropharmacology

Neuropharmacology is the study of the effects of medication on central and peripheral nervous system functioning.

Psychopharmacology

Psychopharmacology, also known as behavioral pharmacology, is the study of the effects of medication on the psyche (psychology), observing changed behaviors of the body and mind, and how molecular events are manifest in a measurable behavioral form. This is similar to the closely related ethnopharmacology. Psychopharmacology is an interdisciplinary field which studies behavioral effects of psychoactive drugs. It incorporates approaches and techniques from neuropharmacology, animal behavior and behavioral neuroscience, and is interested in the behavioral and neurobiological mechanisms of action of psychoactive drugs. Another goal of behavioral pharmacology is to develop animal behavioral models to screen chemical compounds with therapeutic potentials. People in this field (called behavioral pharmacologists) typically use small animals (e.g. rodents) to study psychotherapeutic drugs such as antipsychotics, antidepressants and anxiolytics, and drugs of abuse such as nicotine, cocaine, methamphetamine, etc. study of drugs which affect behavior. *Ethopharmacology* is a term which has been in use since the 1960s and derives from the Greek word "*ethos*" meaning character and "pharmacology" the study of drug actions and mechanism.

Cardiovascular Pharmacology

Cardiovascular pharmacology is the study of the effects of drugs on the entire cardiovascular system, including the heart and blood vessels.

Pharmacogenetics

Pharmacogenetics is clinical testing of genetic variation that gives rise to differing response to drugs.

Pharmacogenomics

Pharmacogenomics is the application of genomic technologies to drug discovery and further characterization of older drugs.

Pharmacoepidemiology

Pharmacoepidemiology is the study of the effects of drugs in large numbers of people.

Systems Pharmacology

Systems pharmacology is the application of systems biology principles to the field of pharmacology.

Toxicology

Toxicology is the study of the adverse effects, molecular targets, and characterization of drugs or any chemical substance in excess (including those beneficial in lower doses).

Theoretical Pharmacology

Theoretical pharmacology is a relatively new and rapidly expanding field of research activity in which many of the techniques of computational chemistry, in particular computational quantum chemistry and the method of molecular mechanics, are proving to be of great value. Theoretical pharmacologists aim at rationalizing the relation between the activity of a particular drug, as observed experimentally, and its structural features as derived from computer experiments. They aim to find structure—activity relations. Furthermore, on the basis of the structure of a given organic molecule, the theoretical pharmacologist aims at predicting the biological activity of new drugs that are of the same general type as existing drugs. More ambitiously, it aims to predict entirely new classes of drugs, tailor-made for specific purposes.

Posology

Posology is the study of how medicines are dosed. It also depends upon various factors including age, climate, weight, sex, and time of administration. It is derived from the Greek words posos meaning how much and logos meaning science.

Pharmacognosy

Pharmacognosy is a branch of pharmacology dealing especially with the composition, use, and development of medicinal substances of biological origin and especially medicinal substances obtained from plants.

Environmental Pharmacology

Environmental pharmacology is a new discipline. Focus is being given to understand gene–environment interaction, drug-environment interaction and toxin-environment interaction. There is a close collaboration between environmental science and medicine in addressing these issues, as healthcare itself can be a cause of environmental damage or remediation. Human health and ecology are intimately related. Demand for more pharmaceutical products may place the public at risk through the destruction of species. The entry of chemicals and drugs into the aquatic ecosystem is a more serious concern today. In addition, the production of some illegal drugs pollutes drinking water supply by releasing carcinogens. This field is intimately linked with Public Health fields.

Dental Pharmacology

Dental pharmacology relates to the study of drugs commonly used in the treatment of dental disease.

Scientific Background

The study of chemicals requires intimate knowledge of the biological system affected. With the knowledge of cell biology and biochemistry increasing, the field of pharmacology has also changed substantially. It has become possible, through molecular analysis of receptors, to design chemicals that act on specific cellular signaling or metabolic pathways by affecting sites directly on cell-surface receptors (which modulate and mediate cellular signaling pathways controlling cellular function).

A chemical has, from the pharmacological point-of-view, various properties. Pharmacokinetics describes the effect of the body on the chemical (e.g. half-life and volume of distribution), and pharmacodynamics describes the chemical's effect on the body (desired or toxic).

When describing the pharmacokinetic properties of the chemical that is the active ingredient or active pharmaceutical ingredient (API), pharmacologists are often interested in *L-ADME*:

- Liberation – How is the API disintegrated (for solid oral forms (breaking down into smaller particles)), dispersed, or dissolved from the medication?
- Absorption – How is the API absorbed (through the skin, the intestine, the oral mucosa)?
- Distribution – How does the API spread through the organism?

- Metabolism – Is the API converted chemically inside the body, and into which substances. Are these active (as well)? Could they be toxic?
- Excretion – How is the API excreted (through the bile, urine, breath, skin)?

Medication is said to have a narrow or wide *therapeutic index* or *therapeutic window*. This describes the ratio of desired effect to toxic effect. A compound with a narrow therapeutic index (close to one) exerts its desired effect at a dose close to its toxic dose. A compound with a wide therapeutic index (greater than five) exerts its desired effect at a dose substantially below its toxic dose. Those with a narrow margin are more difficult to dose and administer, and may require therapeutic drug monitoring (examples are warfarin, some antiepileptics, aminoglycoside antibiotics). Most anti-cancer drugs have a narrow therapeutic margin: toxic side-effects are almost always encountered at doses used to kill tumors.

Medicine Development and Safety Testing

Development of medication is a vital concern to medicine, but also has strong economical and political implications. To protect the consumer and prevent abuse, many governments regulate the manufacture, sale, and administration of medication. In the United States, the main body that regulates pharmaceuticals is the Food and Drug Administration and they enforce standards set by the United States Pharmacopoeia. In the European Union, the main body that regulates pharmaceuticals is the EMA and they enforce standards set by the European Pharmacopoeia.

The metabolic stability and the reactivity of a library of candidate drug compounds have to be assessed for drug metabolism and toxicological studies. Many methods have been proposed for quantitative predictions in drug metabolism; one example of a recent computational method is SPORCalc. If the chemical structure of a medicinal compound is altered slightly, this could slightly or dramatically alter the medicinal properties of the compound depending on the level of alteration as it relates to the structural composition of the substrate or receptor site on which it exerts its medicinal effect, a concept referred to as the structural activity relationship (SAR). This means that when a useful activity has been identified, chemists will make many similar compounds called analogues, in an attempt to maximize the desired medicinal effect(s) of the compound. This development phase can take anywhere from a few years to a decade or more and is very expensive.

These new analogues need to be developed. It needs to be determined how safe the medicine is for human consumption, its stability in the human body and the best form for delivery to the desired organ system, like tablet or aerosol. After extensive testing, which can take up to 6 years, the new medicine is ready for marketing and selling.

As a result of the long time required to develop analogues and test a new medicine and the fact that of every 5000 potential new medicines typically only one will ever reach the open market, this is an expensive way of doing things, often costing over 1 billion dollars. To recoup this outlay pharmaceutical companies may do a number of things:

- Carefully research the demand for their potential new product before spending an outlay of company funds.

- Obtain a patent on the new medicine preventing other companies from producing that medicine for a certain allocation of time.

Drug Legislation and Safety

In the United States, the Food and Drug Administration (FDA) is responsible for creating guidelines for the approval and use of drugs. The FDA requires that all approved drugs fulfill two requirements:

1. The drug must be found to be effective against the disease for which it is seeking approval (where 'effective' means only that the drug performed better than placebo or competitors in at least two trials).
2. The drug must meet safety criteria by being subject to animal and controlled human testing.

Gaining FDA approval usually takes several years to attain. Testing done on animals must be extensive and must include several species to help in the evaluation of both the effectiveness and toxicity of the drug. The dosage of any drug approved for use is intended to fall within a range in which the drug produces a therapeutic effect or desired outcome.

The safety and effectiveness of prescription drugs in the U.S. is regulated by the federal Prescription Drug Marketing Act of 1987.

The Medicines and Healthcare products Regulatory Agency (MHRA) has a similar role in the UK.

Education

Students of pharmacology are trained as biomedical scientists, studying the effects of drugs on living organisms. This can lead to new drug discoveries, as well as a better understanding of the way in which the human body works.

Students of pharmacology must have detailed working knowledge of aspects in physiology, pathology and chemistry. During a typical degree they will cover areas such as (but not limited to) biochemistry, cell biology, basic physiology, genetics & the Central Dogma, medical microbiology, neuroscience, and depending on the department's interests, bio-organic chemistry and/or chemical biology.

Modern Pharmacology is highly interdisciplinary. Graduate programs accept students from most biological and chemical backgrounds. With the increasing drive towards biophysical and computational research to describe systems, pharmacologists may even consider themselves mainly physical scientists. In many instances, Analytical Chemistry is closely related to the studies and needs of pharmacological research. Therefore, many institutions will include pharmacology under a Chemistry or Biochemistry Department, especially if a separate Pharmacology Dept. does not exist. What makes an institutional department independent of another, or exist in the first place, is usually an artifact of historical times.

Whereas a pharmacy student will eventually work in a pharmacy dispensing medications, a pharmacologist will typically work within a laboratory setting. Careers for a pharmacologist include academic positions (medical and non-medical), governmental positions, private industrial positions, science writing, scientific patents and law, consultation, biotech and pharmaceutical employment, the alcohol industry, food industry, forensics/law enforcement, public health, and environmental/ecological sciences.

Pharmacognosy

Pharmacognosy is the study of medicinal drugs derived from plants or other natural sources. The American Society of Pharmacognosy defines pharmacognosy as "the study of the physical, chemical, biochemical and biological properties of drugs, drug substances or potential drugs or drug substances of natural origin as well as the search for new drugs from natural sources." It is also defined as the study of crude drugs.

Introduction

The ancient Egyptian polymath Imhotep is believed by James Henry Breasted to have extracted medicine from plants. The word "pharmacognosy" itself is derived, however, from two Greek words pharmakon (drug), and knowledge). The term "pharmacognosy" was used for the first time by the Austrian physician Schmidt in 1811 and 1815 by Crr. Anotheus Seydler in a work titled *Analecta Pharmacognostica.*

Originally—during the 19th century and the beginning of the 20th century—"pharmacognosy" was used to define the branch of medicine or commodity sciences (*Warenkunde* in German) which deals with drugs in their crude, or unprepared, form. Crude drugs are the dried, unprepared material of plant, animal or mineral origin, used for medicine. The study of these materials under the name *pharmakognosie* was first developed in German-speaking areas of Europe, while other language areas often used the older term *materia medica* taken from the works of Galen and Dioscorides. In German the term *drogenkunde* ("science of crude drugs") is also used synonymously.

As late as the beginning of the 20th century, the subject had developed mainly on the botanical side, being particularly concerned with the description and identification of drugs both in their whole state and in powder form. Such branches of pharmacognosy are still of fundamental importance, particularly for pharmacopoeial identification and quality control purposes, but rapid development in other areas has enormously expanded the subject. The advent of the 21st century brought a renaissance of pharmacognosy and its conventional botanical approach has been broadened up to molecular and metabolomic level.

Although most pharmacognostic studies focus on plants and medicines derived from plants, other types of organisms are also regarded as pharmacognostically interesting, in particular, various types of microbes (bacteria, fungi, etc.), and, recently, various marine organisms.

In addition to the previously mentioned definition, the American Society of Pharmacognosy also defines pharmacognosy as "the study of natural product molecules (typically secondary metabolites) that are useful for their medicinal, ecological, gustatory, or other functional properties." Other definitions are more encompassing, drawing on a broad spectrum of biological subjects, including botany, ethnobotany, marine biology, microbiology, herbal medicine, chemistry, biotechnology, phytochemistry, pharmacology, pharmaceutics, clinical pharmacy and pharmacy practice.

- medical ethnobotany: the study of the traditional use of plants for medicinal purposes;
- ethnopharmacology: the study of the pharmacological qualities of traditional medicinal substances;
- the study of phytotherapy (the medicinal use of plant extracts); and
- phytochemistry, the study of chemicals derived from plants (including the identification of new drug candidates derived from plant sources).
- zoopharmacognosy, the process by which animals self-medicate, by selecting and using plants, soils, and insects to treat and prevent disease.
- marine pharmacognosy, the study of chemicals derived from marine organisms.

At the 9th congress of Italian society of pharmacognosy it was stated that current return of phyto-therapy was clearly reflected by the increased market of such products. In 1998 the latest figures available for Europe, the total OTC market for herbal medicinal products reached a figure of $6 billion, with consumption for Germany of $2.5 billion, France $1.6 billion and Italy $600 million. In the US, where the use of herbal products has never been as prevalent as in continental Europe, the market for all herb sales reached a peak in 1998 of $700 billion. This welcomed the scientific investigation of a rigorous nature.

The plant kingdom still holds many species of plants containing substances of medicinal value which have yet to be discovered. Large numbers of plants are constantly being screened for their possible pharmacological value.

Biological Background

The carotenoids in primrose produce bright red, yellow and orange shades. On average, people consuming diets rich in carotenoids from natural foods, such as fruits and vegetables, are healthier and have lower mortality from a number of chronic illnesses

All plants produce chemical compounds as part of their normal metabolic activities. These phyto-chemicals are divided into (1) primary metabolites such as sugars and fats, which are found in all plants; and (2) secondary metabolites—compounds which are found in a smaller range of plants, serving a more specific function. For example, some secondary metabolites are toxins used to deter predation and others are pheromones used to attract insects for pollination. It is these secondary metabolites and pigments that can have therapeutic actions in humans and which can be refined to produce drugs—examples are inulin from the roots of dahlias, quinine from the cinchona, morphine and codeine from the poppy, and digoxin from the foxglove.

Plants synthesize a bewildering variety of phytochemicals but most are derivatives of a few biochemical motifs:

- Alkaloids are a class of chemical compounds containing a nitrogen ring. Alkaloids are produced by a large variety of organisms, including bacteria, fungi, plants, and animals, and are part of the group of natural products (also called secondary metabolites). Many alkaloids can be purified from crude extracts by acid-base extraction. Many alkaloids are toxic to other organisms. They often have pharmacological effects and are used as medications, as recreational drugs, or in entheogenic rituals. Examples are the local anesthetic and stimulant cocaine; the psychedelic psilocin; the stimulant caffeine; nicotine; the analgesic morphine; the antibacterial berberine; the anticancer compound vincristine; the antihypertension agent reserpine; the cholinomimeric galatamine; the spasmolysis agent atropine; the vasodilator vincamine; the anti-arhythmia compound quinidine; the anti-asthma therapeutic ephedrine; and the antimalarial drug quinine. Although alkaloids act on a diversity of metabolic systems in humans and other animals, they almost uniformly invoke a bitter taste.

- Polyphenols (also known as phenolics) are compounds that contain phenol rings. The anthocyanins that give grapes their purple color, the isoflavones, the phytoestrogens from soy and the tannins that give tea its astringency are phenolics.

- Glycosides are molecules in which a sugar is bound to a non-carbohydrate moiety, usually a small organic molecule. Glycosides play numerous important roles in living organisms. Many plants store chemicals in the form of inactive glycosides. These can be activated by enzyme hydrolysis, which causes the sugar part to be broken off, making the chemical available for use. Many such plant glycosides are used as medications. In animals and humans, poisons are often bound to sugar molecules as part of their elimination from the body. An example is the cyanoglycosides in cherry pits that release toxins only when bitten by a herbivore.

- Terpenes are a large and diverse class of organic compounds, produced by a variety of plants, particularly conifers, which are often strong smelling and thus may have had a protective function. They are the major components of resin, and of turpentine produced from resin. (The name "terpene" is derived from the word "turpentine"). Terpenes are major biosynthetic building blocks within nearly every living creature. Steroids, for example, are derivatives of the triterpene squalene. When terpenes are modified chemically, such as by oxidation or rearrangement of the carbon skeleton, the resulting compounds are generally referred to as *terpenoids*. Terpenes and terpenoids are the primary constituents of the essential oils of many types of plants and flowers. Essential oils are used widely as natural flavor additives for food, as fragrances in perfumery, and in traditional and alternative medicines such as aromatherapy. Synthetic variations and derivatives of natural terpenes and terpenoids also greatly expand the variety of aromas used in perfumery and flavors used in food additives. Vitamin A is an example of a terpene. The fragrance of rose and lavender is due to monoterpenes. The carotenoids produce the reds, yellows and oranges of pumpkin, corn and tomatoes.

A consortium of plant molecular researchers at Washington State University, the Donald Danforth Plant Science Center, the National Center for Genome Resources, and the University of Illinois at Chicago began an NIH-sponsored study of over thirty medicinal plant species late 2009. The initial work, to develop a sequence reference for the transcriptome of each, has led to the development of the Medicinal Plant Transcriptomics Database.

Issues in Phytotherapy

The part of pharmacognosy focusing on the use of crude extracts or semi-pure mixtures originating from nature, namely phytotherapy, is probably the best known and also the most debated area in pharmacognosy. Although phytotherapy is sometimes considered as alternative medicine, when critically conducted, it can be considered the scientific study on the effects and clinical use of herbal medicines. Consequently, herbal products might also become officially approved for clinical application as botanical drugs (e.g., Veregen (sinecatechins), a green tea leaves extract, approved for use by FDA).

Constituents and Drug Synergysm

One characteristic of crude drug material is that constituents may have an opposite, moderating or enhancing effect. Hence, the final effect of any crude drug material will be a product of the interactions between the constituents and the effect of each constituent on its own. To effectively study the existence and effect of such interactions, scientific studies must examine the effect that multiple constituents, given concurrently, have on the system. Herbalists assert that as phytopharmaceuticals rely upon synergy for their activities, plants with high levels of active constituents like ginsenosides or hypericin may not correlate with the strength of the herbs. In phytopharmaceuticals or herbal medicine, the therapeutic effects of herbs cannot be determined unless its active ingredient or cofactors are identified or the herb is administered as a whole. One way to indicate strength is standardization to one or several marker compound that are believed to be mainly responsible for the biological effects. However, many herbalists believe that the active ingredient in a plant is the plant itself.

Herb and Drug Interactions

A study of herb drug interactions indicated that the vast majority of drug interactions occurred in four classes of drugs, the chief class being blood thinners, but also including protease inhibitors, cardiac glycosides and the immuno-suppressant ciclosporin.

Natural Products Chemistry

Most bioactive compounds of natural origin are secondary metabolites, i.e., species-specific chemical agents that can be grouped into various categories. A typical protocol to isolate a pure chemical agent from natural origin is bioassay-guided fractionation, meaning step-by-step separation of extracted components based on differences in their physicochemical properties, and assessing the biological activity, followed by next round of separation and assaying. Typically, such work is initiated after a given crude drug formulation (typically prepared by solvent extraction of the natural material) is deemed "active" in a particular *in vitro* assay. If the end-goal of the work at hand is to identify which one(s) of the scores or hundreds of compounds are responsible for the observed *in vitro* activity, the path to that end is fairly straightforward: 1. fractionate the crude extract, e.g. by solvent partitioning or chromatography. 2. test the fractions thereby generated with *in vitro* assay. 3. repeat steps 1) and 2) until pure, active compounds are obtained. 4. determine structure(s) of active compound(s), typically by using spectroscopic methods. *In vitro* activity does not necessarily translate to activity in humans or other living systems.

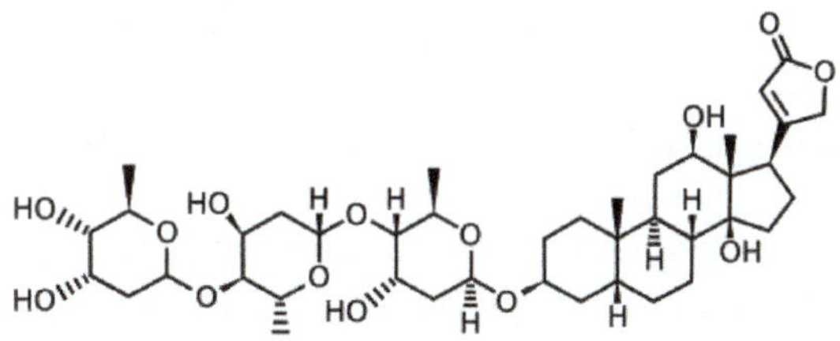

Digoxin is a purified cardiac glycoside that is extracted from the foxglove plant, *Digitalis lanata*. Digoxin is widely used in the treatment of various heart conditions, namely atrial fibrillation, atrial flutter and sometimes heart failure that cannot be controlled by other medication.

The most common means for fractionation are solvent-solvent partitioning and chromatographic techniques such as high-performance liquid chromatography (HPLC), medium-pressure liquid chromatography, "flash" chromatography, open-column chromatography, vacuum-liquid chromatography (VLC), thin-layer chromatography (TLC), with each technique being most appropriate for a given amount of starting material. Countercurrent chromatography (CCC) is particularly well-suited for bioassay-guided fractionation because, as an all-liquid separation technique, concern about irreversible loss or denaturation of active sample components is minimized. After isolation of a pure substance, the task of elucidating its chemical structure can be addressed. For this purpose, the most powerful methodologies available are nuclear magnetic resonance spectroscopy (NMR) and mass spectrometry (MS) . In the case of drug discovery efforts, structure elucidation of all components that are active *in vitro* is typically the end goal. In the case of phytotherapy research, the investigator may use *in vitro* BAGF as a tool to identify pharmacologically interesting or important components of the crude drug. The work does not stop after structural identification of *in vitro* actives, however. The task of "dissecting and reassembling" the crude drug one active component at a time, in order to achieve a mechanistic understanding of how it works in phytotherapy, is quite daunting. This is because it is simply too difficult, from cost, time, regulatory, and even scientific perspectives, to study experimental fractions of the crude drug in humans. *In vitro* assays are therefore used to identify chemical components of the crude drug that may rationally be expected to have a given pharmacological effect in humans, and to provide a rational basis for standardization of a crude drug formulation to be tested in [and sold/marketed to] humans.

Loss of Biodiversity

Farnsworth for example, has found that 25% of all prescriptions dispensed from community pharmacies in the United States from 1959 to 1980 contained active ingredients extracted from higher plants. In some countries in Asia and Africa 80% of the population relies on traditional medicine (including herbal medicine) for primary health care. Native American cultures have also relied on traditional medicine such as ceremonial smoking of tobacco, potlatch ceremonies, and herbalism, to name a few, prior to European colonization. Constituents of substances used by traditional healers, have rarely been incorporated into modern medicine. Quinine, physostigmine, d-tubocurarine, pilocarpine and ephedrine, have been demonstrated to have active effects Knowledge of traditional medicinal practices is disappearing, particularly in the Amazon, as native healers die out and are replaced by more modern medical practitioners. Botanists and pharmacologists are racing to learn these ancient practices, which, like the forest plants they employ, are also endangered

An explanation for some species loss is habitat lost due to invasive species introduction. Herbalist David Winston has suggested that a high proportion of nonnative species seen as invasive (kudzu, Japanese knotweed, mimosa, lonicera, St. Johnswort and purple loosestrife) may be harvested for the domestic herbal medicine market.

Species extinction is not only due to habitat loss. Overharvesting of medicinal species of plants and animals also contributes to species loss. This is particularly notable in the matter of Traditional Chinese Medicine where crude drugs of plant and animal origin are used with increasing demand. People with a stake in TCM often seek chemical and biological alternatives to endangered species because they realize that plants and animals lost from the wild are also lost to medicine forever but different cultural attitudes bedevil conservation efforts . Still conservation is not a new idea: Chinese advice against overexploitation of natural medicinal species dates from at least Mencius, a philosopher living in the 4th century BC.

Cooperation between Western conservationists and practitioners have been beset by cultural difficulties. Westerners may emphasise urgency in matters of conservation, while Chinese may wish for the products used in TCM to remain publicly available. One repeated fallacy is that rhinoceros horn is used as an aphrodisiac in TCM. It is, in fact, prescribed for fevers and convulsions by TCM practitioners. There are no peer-reviewed studies showing that this treatment is effective. In 1995 representatives of the oriental medicine communities in Asia met with conservationists at a symposium in Hong Kong, organized by TRAFFIC. The two groups established a clear willingness to cooperate through dialogue and mutual understanding. This has led to several meetings, including the 1997 First International Symposium on Endangered Species Used in Traditional East Asian Medicine, where China was among 136 nations to sign a formal resolution recognizing that the uncontrolled use of wild species in traditional medicine threatens their survival and the continuation of these medical practices. The resolution, drawn up by the UN Convention on International Trade in Endangered Species (CITES), aims to initiate new partnerships in conservation.

Sustainable Sources of Plant and Animal Drugs

As species face loss of habitat or overharvesting, there have been new issues to deal with in sourcing crude drugs. These include changes to the herb from farming practices, substitution of species or other plants altogether, adulteration and cross-pollination issues. For instance, ginseng which is field farmed may have significant problems with fungus, making contamination with fungicides an issue. This may be remedied with woods grown programs, but they are insufficient to produce enough ginseng to meet demand. The wildcrafted echinacea, black cohosh and American ginseng often rely upon old growth root, often in excess of 50 years of age and it is not clear that younger stock will have the same pharmaceutical effect. Black cohosh may be adulterated with the related

Chinese actea species, which is not the same. Ginseng may be replaced by ginseniodes from Jiaogulan which has been stated to have a different effect than the full panax root.

The problem may be exacerbated by the growth of pills and capsules as the preferred method of ingesting medication as they are cheaper and more available than traditional, individually tailored prescriptions of raw medicinals but the contents are harder to track. Seahorses are a case in point: Seahorses once had to be of a certain size and quality before they were accepted by practitioners and consumers. But declining availability of the preferred large, pale and smooth seahorses has been offset by the shift towards prepackaged medicines, which make it possible for TCM merchants to sell previously unused juvenile, spiny and dark-coloured animals. Today almost a third of the seahorses sold in China are prepackaged.

The farming of plant or animal species for medicinal purposes has caused difficulties. Rob Parry Jones and Amanda Vincent write:

- One solution is to farm medicinal animals and plants. Chinese officials have promoted this as a way of guaranteeing supplies as well as protecting endangered species. And there have been some successes—notably with plant species, such as American ginseng—which is used as a general tonic and for chronic coughs. Red deer, too, have for centuries been farmed for their antlers, which are used to treat impotence and general fatigue. But growing your own is not a universal panacea. Some plants grow so slowly that cultivation in not economically viable. Animals such as musk deer may be difficult to farm, and so generate little profit. Seahorses are difficult to feed and plagued by disease in captivity. Other species cannot be cultivated at all. Even when it works, farming usually fails to match the scale of demand. Overall, cultivated TCM plants in China supply less than 20 per cent of the required 1.6 million tonnes per annum. Similarly, China's demand for animal products such as musk and pangolin scales far exceeds supply from captive-bred sources.

- Farming alone can never resolve conservation concerns, as government authorities and those who use Chinese medicine realise. For a start, consumers often prefer ingredients taken from the wild, believing them to be more potent. This is reflected in the price, with wild oriental ginseng fetching up to 32 times as much as cultivated plants. Then there are welfare concerns. Bear farming in China is particularly controversial. Around 7600 captive bears have their bile "milked" through tubes inserted into their gall bladders. World Animal Protection states that bear bile farming "causes intense, unjustified suffering to bears". Chinese officials state that 10,000 wild bears would need to be killed each year to produce as much bile, making bear farming the more desirable option. World Animal Protection, however, states that "it is commonly believed in China that the bile from a wild bear is the most potent, and so farming bears for their bile cannot replace the demand for the product extracted from wild animals".

- One alternative to farming involves replacing medical ingredients from threatened species with manufactured chemical compounds. In general, this sort of substitution is difficult to achieve because the active ingredient is often not known. In addition, most TCM users believe that TCM compounds may act synergistically so several ingredients may interact to give the required effect. Thus TCM users often prefer the wild source. Tauro ursodeoxycholic acid, the active ingredient of bear bile, can be synthesised and is used by some Western doctors to treat gallstones, but many TCM consumers reject it as being inferior to the natural substance from wild animals.

Pharmacogenetics

Pharmacogenetics is the study of inherited genetic differences in drug metabolic pathways which can affect individual responses to drugs, both in terms of therapeutic effect as well as adverse effects. The term *pharmacogenetics* is often used interchangeably with the term *pharmacogenomics* which also investigates the role of acquired and inherited genetic differences in relation to drug response and drug behavior through a systematic examination of genes, gene products, and inter- and intra-individual variation in gene expression and function.

In oncology, *pharmacogenetics* historically is the study of germline mutations (e.g., single-nucleotide polymorphisms affecting genes coding for liver enzymes responsible for drug deposition and pharmacokinetics), whereas *pharmacogenomics* refers to somatic mutations in tumoral DNA leading to alteration in drug response (e.g., KRAS mutations in patients treated with anti-Her1 biologics).

Predicting Drug-drug Interactions

Much of current clinical interest is at the level of pharmacogenetics, involving variation in genes involved in drug metabolism with a particular emphasis on improving drug safety. The wider use of pharmacogenetic testing is viewed by many as an outstanding opportunity to improve prescribing safety and efficacy. Driving this trend are the 106,000 deaths and 2.2 Million serious events caused by adverse drug reactions in the US each year. As such ADRs are responsible for 5-7% of hospital admissions in the US and Europe, lead to the withdrawal of 4% of new medicines, and cost society an amount equal to the costs of drug treatment.

Comparisons of the list of drugs most commonly implicated in adverse drug reactions with the list of metabolizing enzymes with known polymorphisms found that drugs commonly involved in adverse drug reactions were also those that were metabolized by enzymes with known polymor-phisms.

Scientists and doctors are using this new technology for a variety of things, one being improving the efficacy of drugs. In psychology, we can predict quite accurately which anti-depressant a patient will best respond to by simply looking into their genetic code. This is a huge step from the previous practice of adjusting and experimenting with different medications to get the best response. Antidepressants also have a large percentage of unresponsive patients and poor prediction rate of ADRs (adverse drug reactions). In depressed patients, 30% are not helped by antidepressants. In psychopharmacological therapy, a patient must be on a drug for 2 weeks before the effects can be fully examined and evaluated. For a patient in that 30%, this could mean months of trying medications to find an antidote to their pain. Any assistance in predicting a patient's drug reaction to psychopharmacological therapy should be taken advantage of. Pharmacogenetics is a very useful and important tool in predicting which drugs will be effective in various patients. The drug Plavix blocks platelet reception and is the second best selling prescription drug in the world, however, it is known to warrant different responses among patients. GWAS studies have linked the gene CYP2C19 to those who cannot normally metabolize Plavix. Plavix is given to patients after receiving a stent in the coronary artery to prevent clotting.

Stent clots almost always result in heart attack or sudden death, fortunately it only occurs in 1 or 2% of the population. That 1 or 2% are those with the CYP2C19 SNP. This finding has been applied in at least two hospitals, Scripps and Vanderbilt University, where patients who are candidates for heart stents are screened for the CYP2C19 variants.

Another newfound use of pharmacogenetics involves the use of Vitamin E. The Technion Israel Institute of Technology observed that vitamin E can be used to in certain genotypes to lower the risk of cardiovascular disease in patients with diabetes, but in the same patients with another genotype, vitamin E can raise the risk of cardiovascular disease. A study was carried out, showing vitamin E is able to increase the function of HDL in those with the genotype haptoglobin 2-2 who suffer from diabetes. HDL is a lipoprotein that removes cholesterol from the blood and is associated with a reduced risk of atherosclerosis and heart disease. However, if you have the misfortune to possess the genotype haptoglobin 2-1, the study shows that this same treatment can drastically decrease your HDL function and cause cardiovascular disease.

Pharmacogenetics is a rising concern in clinical oncology, because the therapeutic window of most anticancer drugs is narrow and patients with impaired ability to detoxify drugs will undergo life-threatening toxicities. In particular, genetic deregulations affecting genes coding for DPD, UGT1A1, TPMT, CDA and CYP2D6 are now considered as critical issues for patients treated with 5-FU/capecitabine, irinotecan, mercaptopurine/azathioprine, gemcitabine/capecitabine/AraC and tamoxifen, respectively. The decision to use pharmacogenetic techniques is influenced by the relative costs of genotyping technologies and the cost of providing a treatment to a patient with an incompatible genotype. When available, phenotype-based approaches proved their usefulness while being cost-effective.

In the search for informative correlates of psychotropic drug response, pharmacogenetics has several advantages:

- The genotype of an individual is essentially invariable and remains unaffected by the treatment itself.
- Molecular biology techniques provide an accurate assessment of the genotype of an individual.
- There has been a dramatic increase in the amount of genomic information that is available. This information provides the necessary data for comprehensive studies of individual genes and broad investigation of genome-wide variation.
- The ease of accessibility to genotype information through peripheral blood or saliva sampling and advances in molecular techniques has increased the feasibility of DNA collection and genotyping in large-scale clinical trials.

History

The first observations of genetic variation in drug response date from the 1950s, involving the muscle relaxant suxamethonium chloride, and drugs metabolized by N-acetyltransferase. One in

3500 Caucasians has less efficient variant of the enzyme (butyrylcholinesterase) that metabolizes suxamethonium chloride. As a consequence, the drug's effect is prolonged, with slower recovery from surgical paralysis. Variation in the N-acetyltransferase gene divides people into "slow acetylators" and "fast acetylators", with very different half-lives and blood concentrations of such important drugs as isoniazid (antituberculosis) and procainamide (antiarrhythmic). As part of the inborn system for clearing the body of xenobiotics, the cytochrome P450 oxidases (CYPs) are heavily involved in drug metabolism, and genetic variations in CYPs affect large populations. One member of the CYP superfamily, CYP2D6, now has over 75 known allelic variations, some of which lead to no activity, and some to enhanced activity. An estimated 29% of people in parts of East Africa may have multiple copies of the gene, and will therefore not be adequately treated with standard doses of drugs such as the painkiller codeine (which is activated by the enzyme). The first study using Genome-wide association studies (GWAS) linked age-related macular degeneration (AMD) with a SNP located on chromosome 1 that increased one's risk of AMD. AMD is the most common cause of blindness, affecting more than seven million Americans. Until this study in 2005, we only knew about the inflammation of the retinal tissue causing AMD, not the genes responsible.

Thiopurines and TPMT (Thiopurine Methyl Transferase)

One of the earliest tests for a genetic variation resulting in a clinically important consequence was on the enzyme thiopurine methyltransferase (TPMT). TPMT metabolizes 6-mercaptopurine and azathioprine, two thiopurine drugs used in a range of indications, from childhood leukemia to autoimmune diseases. In people with a deficiency in TPMT activity, thiopurine metabolism must proceed by other pathways, one of which leads to the active thiopurine metabolite that is toxic to the bone marrow at high concentrations. Deficiency of TPMT affects a small proportion of people, though seriously. One in 300 people have two variant alleles and lack TPMT activity; these people need only 6-10% of the standard dose of the drug, and, if treated with the full dose, are at risk of severe bone marrow suppression. For them, genotype predicts clinical outcome, a prerequisite for an effective pharmacogenetic test. In 85-90% of affected people, this deficiency results from one of three common variant alleles. Around 10% of people are heterozygous - they carry one variant allele - and produce a reduced quantity of functional enzyme. Overall, they are at greater risk of adverse effects, although as individuals their genotype is not necessarily predictive of their clinical outcome, which makes the interpretation of a clinical test difficult. Recent research suggests that patients who are heterozygous may have a better response to treatment, which raises whether people who have two wild-type alleles could tolerate a higher therapeutic dose. The US Food and Drug Administration (FDA) have recently deliberated the inclusion of a recommendation for testing for TPMT deficiency to the prescribing information for 6-mercaptopurine and azathioprine. The information previously carried the warning that inherited deficiency of the enzyme could increase the risk of severe bone marrow suppression. It now caries the recommendation that people who develop bone marrow suppression while receiving 6-mercaptopurine or azathioprine be tested for TPMT deficiency.

Hepatitis C

A polymorphism near a human interferon gene is predictive of the effectiveness of an artificial interferon treatment for Hepatitis C. For genotype 1 hepatitis C treated with Pegylated interferon-alpha-2a or Pegylated interferon-alpha-2b (brand names Pegasys or PEG-Intron) combined with ribavirin, it has been shown that genetic polymorphisms near the human IL28B gene, encoding interferon lambda 3, are associated with significant differences in response to the treatment. Genotype 1 hepatitis C patients carrying certain genetic variant alleles near the IL28B gene are more probable to achieve sustained virological response after the treatment than others, and demonstrated that the same genetic variants are also associated with the natural clearance of the genotype 1 hepatitis C virus.

Integrating Pharmacogenetics into the Health Care System

Despite the many successes, most drugs are not tested using GWAS. However, it is estimated that over 25% of common medication have some type of genetic information that could be used in the medical field. If the use of personalized medicine is widely adopted and used, it will make medical trials more efficient. This will lower the costs that come about due to adverse drug side effects and prescription of drugs that have been proven ineffective in certain genotypes. It is very costly when a clinical trial is put to a stop by licensing authorities because of the small population who experiences adverse drug reactions. With the new push for pharmacogenetics, it is possible to develop and license a drug specifically intended for those who are the small population genetically at risk for adverse side effects.

The ability to test and analyze an individual's DNA to determine if the body can break down certain drugs through the biochemical pathways has application in all fields of medicine. Pharmacogenetics gives those in the health care industry a potential solution to help prevent the significant amount of deaths that occur each year due to drug reactions and side effects. The companies or laboratories that perform this testing can do so acrossed all categories or drugs whether it be for high blood pressure, gastrointestinal, urological, psychotropic or anti-anxiety drugs. Results can be presented showing which drugs the body is capable of breaking down normally versus the drugs the body cannot break down normally. This test only needs to be done once and can provide valuable information such as a summary of an individual's genetic polymorphisms, which could help in a situation such as being a patient in the emergency room.

Technological Advances

As the cost per genetic test decreases, the development of personalized drug therapies will increase. Technology now allows for genetic analysis of hundreds of target genes involved in medication metabolism and response in less than 24 hours for under $1,000. This a huge step towards bringing pharmacogenetic technology into everyday medical decisions. Likewise, companies like deCODE genetics, Navigenics and 23andMe offer genome scans. The companies use the same genotyping chips that are used in GWAS studies and provide customers with a write-up of individual risk for various traits and diseases and testing for 500,000 known SNPs. Costs range from $995 to $2500 and include updates with new data from studies as they become available. The more expensive packages even included a telephone session with a genetics counselor to discuss the results.

Ethics

Pharmacogenetics has become a controversial issue in the area of bioethics. It's a new topic to the medical field, as well as the public. This new technique will have a huge impact on society, influencing the treatment of both common and rare diseases. As a new topic in the medical field the ethics behind it are still not clear. However, ethical issues and their possible solutions are already being addressed.

There are three main ethical issues that have risen from pharmacogenetics. First, would there be a type equity at both drug development and the accessibility to tests. The concern of accessibility to the test is whether it is going to be available directly to patients via the internet, or over the counter. The second concern regards the confidentiality of storage and usage of genetic information. Thirdly, would patients have the control over being tested.

One concern that has risen is the ethical decision health providers must take with respect to educating the patient of the risks and benefits of medicine developed by this new technology. Pharmacogenetics is a new process that may increase the benefits of medicine while decreasing the risk. However clinicians have been unsuccessful in educating patients regarding the concept of benefits over risk. The Nuffield Council reported that patients and health professionals have adequate information about pharmacogenetics tests and medicine. Health care providers will also encounter an ethical decision in deciding to tell their patients that only certain individuals will benefit from the new medicine due to their genetic make-up. Another ethical concern is that patients who have not taken the test be able to have access to this type of medicine. If access is given by the doctor the medicine could negatively impact the patient's health. The ethical issues behind pharmacogenetics tests, as well as medicine, are still a concern and policies will need to be implemented in the future.

Pharmacogenetic Testing Companies

- Pathway Genomics
- 23andMe
- Genelex
- Kashi Clinical Laboratories
- MD Labs

Pharmacokinetics

Pharmacokinetics (from Ancient Greek *pharmakon* "drug" and *kinetikos* "moving, putting in mo-tion"; chemical kinetics), sometimes abbreviated as PK, is a branch of pharmacology dedicated to determining the fate of substances administered to a living organism. The substances of interest include pharmaceutical drugs, hormones, nutrients, and toxins. It attempts to analyze drug me-tabolism and to discover the fate of a drug from the moment that it is administered up to the point at which it is completely eliminated from the body. Pharmacokinetics describes how the body af-fects a specific drug after administration through the mechanisms of absorption and distribution, as well as the chemical changes of the substance in the body (e.g. by metabolic enzymes such as cytochrome P450 or glucuronosyltransferase enzymes), and the effects and routes of excretion of the metabolites of the drug. Pharmacokinetic properties of drugs may be affected by elements such as the site of administration and the dose of administered drug. These may affect the absorption rate.

Pharmacokinetics is often studied in conjunction with pharmacodynamics, the study of a drug's pharmacological effect on the body.

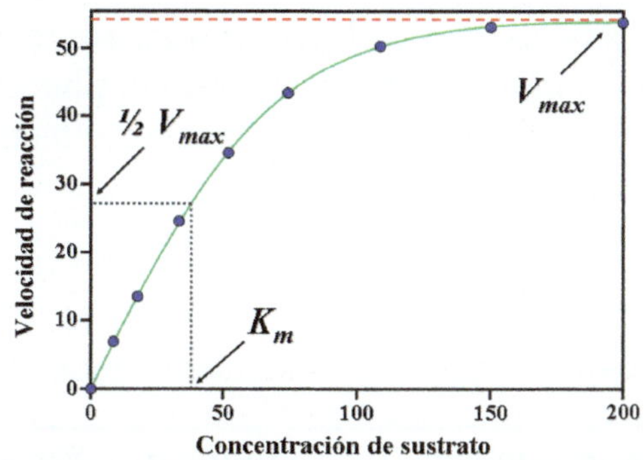

Graph that demonstrates the Michaelis-Menten kinetics model for the relationship between an enzyme and a substrate: one of the parameters studies in pharmacokinetics, where the substrate is a pharmaceutical drug.

A number of different models have been developed in order to simplify conceptualization of the many processes that take place in the interaction between an organism and a drug. One of these models, the multi-compartment model, gives the best approximation to reality; however, the complexity involved in using this type of model means that *monocompartmental models* and above all *two compartmental models* are the most-frequently used. The various compartments that the model is divided into are commonly referred to as the ADME scheme (also referred to as LADME if liberation is included as a separate step from absorption):

- Liberation - the process of release of a drug from the pharmaceutical formulation.
- Absorption - the process of a substance entering the blood circulation.
- Distribution - the dispersion or dissemination of substances throughout the fluids and tissues of the body.
- Metabolism (or biotransformation, or inactivation) – the recognition by the organism that a foreign substance is present and the irreversible transformation of parent compounds into daughter metabolites.
- Excretion - the removal of the substances from the body. In rare cases, some drugs irreversibly accumulate in body tissue.

The two phases of metabolism and excretion can also be grouped together under the title elimination. The study of these distinct phases involves the use and manipulation of basic concepts in order to understand the process dynamics. For this reason in order to fully comprehend the *kinetics* of a drug it is necessary to have detailed knowledge of a number of factors such as: the properties of the substances that act as excipients, the characteristics of the appropriate biological membranes and the way that substances can cross them, or the characteristics of the enzyme reactions that inactivate the drug.

All these concepts can be represented through mathematical formulas that have a corresponding graphical representation. The use of these models allows an understanding of the characteristics of a molecule, as well as how a particular drug will behave given information regarding some of its basic characteristics such as its acid dissociation constant (pKa), bioavailability and solubility, absorption capacity and distribution in the organism.

The model outputs for a drug can be used in industry (for example, in calculating bioequivalence when designing generic drugs) or in the clinical application of pharmacokinetic concepts. Clinical pharmacokinetics provides many performance guidelines for effective and efficient use of drugs for human-health professionals and in veterinary medicine.

Metrics

The following are the most commonly measured pharmacokinetic metrics:

Characteristic	Description	Example value	Symbol	Formula
Dose	Amount of drug administered.	500 mg	D	Design parameter
Dosing interval	Time between drug dose administrations.	24 h	τ	Design parameter
C_{max}	The peak plasma concentration of a drug after administration.	60.9 mg/L	C_{max}	Direct measurement
t_{max}	Time to reach C_{max}.	3.9 h	t_{max}	Direct measurement
C_{min}	The lowest (trough) concentration that a drug reaches before the next dose is administered.	27.7 mg/L	$C_{\text{min,ss}}$	Direct measurement
Volume of distribution	The apparent volume in which a drug is distributed (i.e., the parameter relating drug concentration to drug amount in the body).	6.0 L	V_{d}	$=\frac{D}{C_0}$
Concentration	Amount of drug in a given volume of plasma.	83.3 mg/L	C_0, C_{ss}	$=\frac{D}{V_{\text{d}}}$
Elimination half-life	The time required for the concentration of the drug to reach half of its original value.	12 h	$t_{\frac{1}{2}}$	$=\frac{\ln(2)}{k_{\text{e}}}$
Elimination rate constant	The rate at which a drug is removed from the body.	0.0578 h^{-1}	k_{e}	$=\frac{\ln(2)}{t_{\frac{1}{2}}}=\frac{CL}{V_{\text{d}}}$
Infusion rate	Rate of infusion required to balance elimination.	50 mg/h	k_{in}	$=C_{\text{ss}}\cdot CL$
Area under the curve	The integral of the concentration-time curve (after a single dose or in steady state).	1,320 mg/L·h	$AUC_{0-\infty}$	$=\int_0^{\infty} C\,dt$
			$AUC_{\tau,\text{ss}}$	$=\int_t^{t+\tau} C\,dt$

Clearance	The volume of plasma cleared of the drug per unit time.	0.38 L/h	CL	$= V_d \cdot k_e = \dfrac{D}{AUC}$
Bioavailability	The systemically available fraction of a drug.	0.8	f	$= \dfrac{AUC_{po} \cdot D_{iv}}{AUC_{iv} \cdot D_{po}}$
Fluctuation	Peak trough fluctuation within one dosing interval at steady state	41.8 %	$\%PTF$	$= \dfrac{C_{max,ss} - C_{min,ss}}{C_{av,ss}} \cdot 100$ where $C_{av,ss} = \dfrac{1}{\tau} AUC_{\tau,ss}$

In pharmacokinetics, *steady state* refers to the situation where the overall intake of a drug is fairly in dynamic equilibrium with its elimination. In practice, it is generally considered that steady state is reached when a time of 4 to 5 times the half-life for a drug after regular dosing is started.

The following graph depicts a typical time course of drug plasma concentration and illustrates main pharmacokinetic metrics:

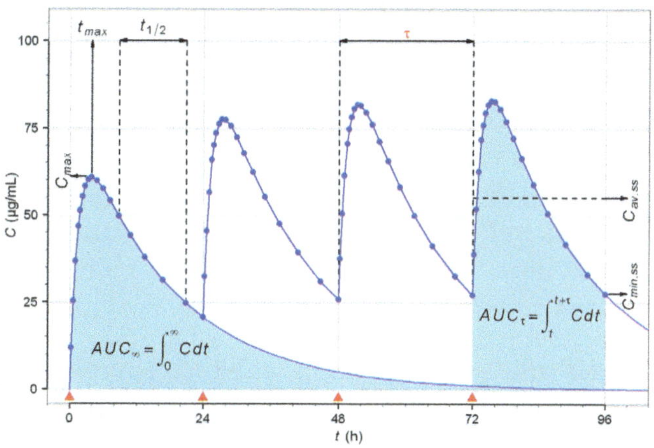

The time course of drug plasma concentrations over 96 hours following oral administrations every 24 hours. Note that the AUC in steady state equals AUC_∞ after the first dose.

Pharmacokinetic Models

Pharmacokinetic modelling is performed by noncompartmental or compartmental methods. Non-compartmental methods estimate the exposure to a drug by estimating the area under the curve of a concentration-time graph. Compartmental methods estimate the concentration-time graph using kinetic models. Noncompartmental methods are often more versatile in that they do not assume any specific compartmental model and produce accurate results also acceptable for bio-equivalence studies. The final outcome of the transformations that a drug undergoes in an organism and the rules that determine this fate depend on a number of interrelated factors. A number of functional models have been developed in order to simplify the study of pharmacokinetics. These models are based on a consideration of an organism as a number of related compartments. The simplest idea is to think of an organism as only one homogenous compartment. This *monocompartimental model* presupposes that blood plasma concentrations of the drug are a true reflection of the drug's concentration in other fluids or tissues and that the elimination of the drug is directly proportional to the drug's concentration in the organism (first order kinetics).

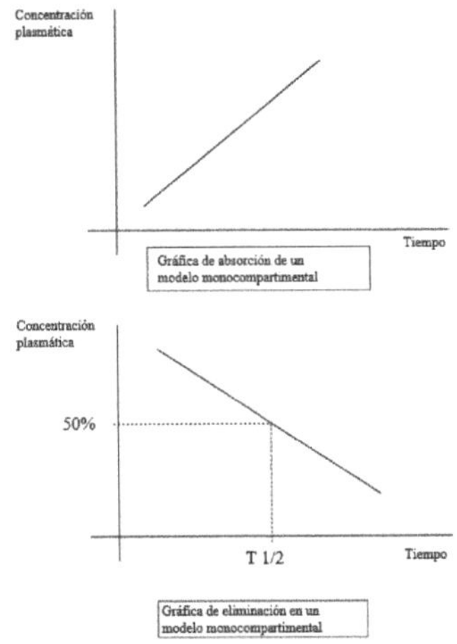

Graph representing the monocompartmental action model.

However, these models do not always truly reflect the real situation within an organism. For example, not all body tissues have the same blood supply, so the distribution of the drug will be slower in these tissues than in others with a better blood supply. In addition, there are some tissues (such as the brain tissue) that present a real barrier to the distribution of drugs, that can be breached with greater or lesser ease depending on the drug's characteristics. If these relative conditions for the different tissue types are considered along with the rate of elimination, the organism can be considered to be acting like two compartments: one that we can call the *central compartment* that has a more rapid distribution, comprising organs and systems with a well-developed blood supply; and a *peripheral compartment* made up of organs with a lower blood flow. Other tissues, such as the brain, can occupy a variable position depending on a drug's ability to cross the barrier that separates the organ from the blood supply.

This *two compartment model* will vary depending on which compartment elimination occurs in. The most common situation is that elimination occurs in the central compartment as the liver and kidneys are organs with a good blood supply. However, in some situations it may be that elimination occurs in the peripheral compartment or even in both. This can mean that there are three possible variations in the two compartment model, which still do not cover all possibilities.

This model may not be applicable in situations where some of the enzymes responsible for metabolizing the drug become saturated, or where an active elimination mechanism is present that is independent of the drug's plasma concentration. In the real world each tissue will have its own distribution characteristics and none of them will be strictly linear. If we label the drug's volume of distribution within the organism Vd_F and its volume of distribution in a tissue Vd_T the former will be described by an equation that takes into account all the tissues that act in different ways, that is:

$$Vd_F = Vd_{T1} + Vd_{T2} + Vd_{T3} + \ldots + Vd_{Tn}$$

This represents the *multi-compartment model* with a number of curves that express complicated equations in order to obtain an overall curve. A number of computer programs have been developed to plot these equations. However complicated and precise this model may be, it still does not truly represent reality despite the effort involved in obtaining various distribution values for a drug. This is because the concept of distribution volume is a relative concept that is not a true reflection of reality. The choice of model therefore comes down to deciding which one offers the lowest margin of error for the drug involved.

Noncompartmental Analysis

Noncompartmental PK analysis is highly dependent on estimation of total drug exposure. Total drug exposure is most often estimated by area under the curve (AUC) methods, with the trapezoidal rule (numerical integration) the most common method. Due to the dependence on the length of 'x' in the trapezoidal rule, the area estimation is highly dependent on the blood/plasma sampling schedule. That is, the closer time points are, the closer the trapezoids reflect the actual shape of the concentration-time curve.

Compartmental Analysis

Compartmental PK analysis uses kinetic models to describe and predict the concentration-time curve. PK compartmental models are often similar to kinetic models used in other scientific disciplines such as chemical kinetics and thermodynamics. The advantage of compartmental over some noncompartmental analyses is the ability to predict the concentration at any time. The disadvantage is the difficulty in developing and validating the proper model. Compartment-free modelling based on curve stripping does not suffer this limitation. The simplest PK compartmental model is the one-compartmental PK model with IV bolus administration and first-order elimination. The most complex PK models (called PBPK models) rely on the use of physiological information to ease development and validation.

Single-compartment Model

Linear pharmacokinetics is so-called because the graph of the relationship between the various factors involved (dose, blood plasma concentrations, elimination, etc.) gives a straight line or an approximation to one. For drugs to be effective they need to be able to move rapidly from blood plasma to other body fluids and tissues.

The change in concentration over time can be expressed as $C = C_{initial} * e^{-kelt}$

Multi-compartmental Models

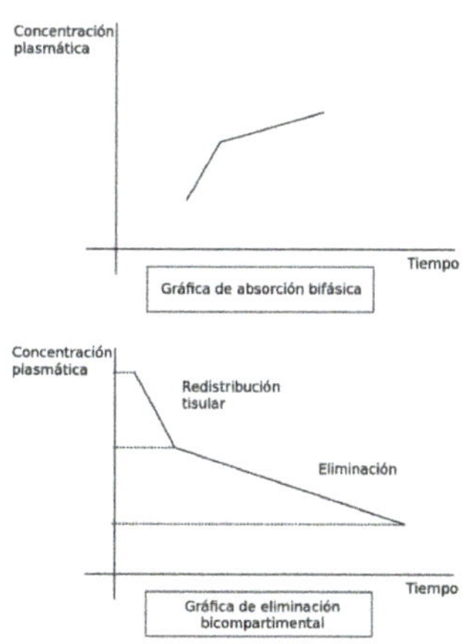

Graphs for absorption and elimination for a non-linear pharmacokinetic model.

The graph for the non-linear relationship between the various factors is represented by a curve, the relationships between the factors can then be found by calculating the dimensions of different areas under the curve. The models used in '*non linear pharmacokinetics* are largely based on Michaelis-Menten kinetics. A reaction's factors of non linearity include the following:

- Multiphasic absorption : Drugs injected intravenously are removed from the plasma through two primary mechanisms: (1) Distribution to body tissues and (2) metabolism + excretion of the drugs. The resulting decrease of the drug's plasma concentration follows a biphasic pattern.

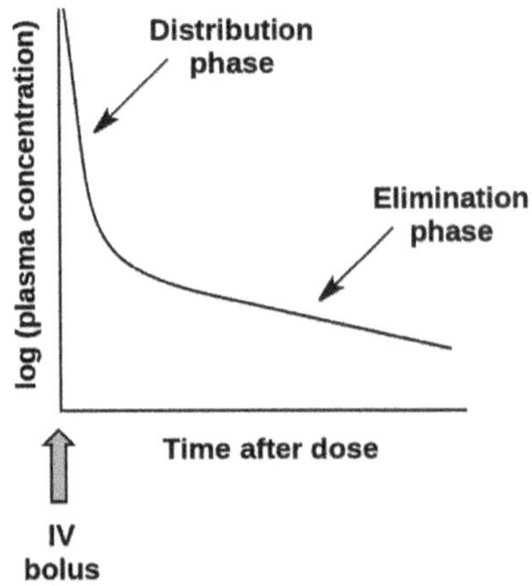

Plasma drug concentration vs time after an IV dose

  - Alpha phase: An initial phase of rapid decrease in plasma concentration. The decrease is primarily attributed to drug distribution from the central compartment (circulation) into the peripheral compartments (body tissues). This phase ends when a pseudo-equilibrium of drug concentration is established between the central and peripheral compartments.
  - Beta phase: A phase of gradual decrease in plasma concentration after the alpha phase. The decrease is primarily attributed to drug metabolism and excretion.
  - Additional phases (gamma, delta, etc.) are sometimes seen.

- A drug's characteristics make a clear distinction between tissues with high and low blood flow.
- Enzymatic saturation: When the dose of a drug whose elimination depends on biotransformation is increased above a certain threshold the enzymes responsible for its metabolism become saturated. The drug's plasma concentration will then increase disproportionately and its elimination will no longer be constant.
- Induction or enzymatic inhibition: Some drugs have the capacity to inhibit or stimulate their own metabolism, in negative or positive feedback reactions. As occurs with fluvoxamine, fluoxetine and phenytoin. As larger doses of these pharmaceuticals are administered the plasma concentrations of the unmetabolized drug increases and the elimination half-life increases. It is therefore necessary to adjust the dose or other treatment parameters when a high dosage is required.

- The kidneys can also establish active elimination mechanisms for some drugs, independent of plasma concentrations.

It can therefore be seen that non-linearity can occur because of reasons that affect the entire pharmacokinetic sequence: absorption, distribution, metabolism and elimination.

Bioavailability

At a practical level, a drug's bioavailability can be defined as the proportion of the drug that reaches its site of action. From this perspective the intravenous administration of a drug provides the greatest possible bioavailability, and this method is considered to yield a bioavailability of 1 (or 100%). Bioavailability of other delivery methods is compared with that of intravenous injection («absolute bioavailability») or to a standard value related to other delivery methods in a particular study («relative bioavailability»).

$$B_A = \frac{[ABC]_P . D_{IV}}{[ABC]_{IV} . D_P}$$

$$B_R = \frac{[ABC]_A . dose_B}{[ABC]_B . dose_A}$$

Once a drug's bioavailability has been established it is possible to calculate the changes that need to be made to its dosage in order to reach the required blood plasma levels. Bioavailability is therefore a mathematical factor for each individual drug that influences the administered dose. It is possible to calculate the amount of a drug in the blood plasma that has a real potential to bring about its effect using the formula: $De = B.Da$; where *De* is the effective dose, *B* bioavailability and *Da* the administered dose.

Therefore, if a drug has a bioavailability of 0.8 (or 80%) and it is administered in a dose of 100 mg, the equation will demonstrate the following:

$De = 0.8 \times 100\ mg = 80\ mg$

That is the 100 mg administered represents a blood plasma concentration of 80 mg that has the capacity to have a pharmaceutical effect.

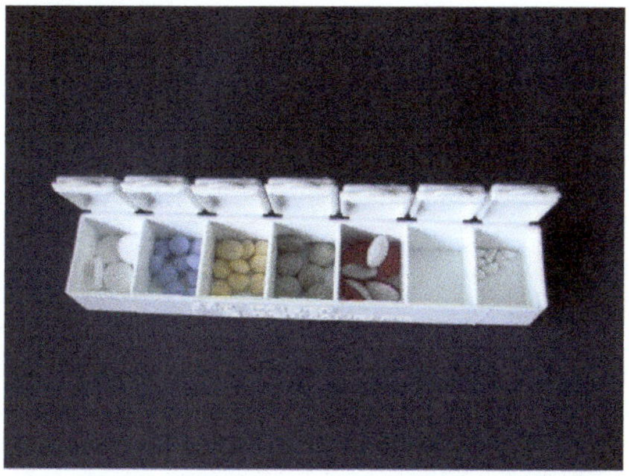

Different forms of tablets, which will have different pharmacokinetic behaviours after their administration.

This concept depends on a series of factors inherent to each drug, such as:

- Pharmaceutical form
- Chemical form
- Route of administration
- Stability
- Metabolism

These concepts, which are discussed in detail in their respective titled articles, can be mathematically quantified and integrated to obtain an overall mathematical equation:

$$De = Q.Da.B$$

where Q is the drug's purity.

$$Va = \frac{Da.B.Q}{\tau}$$

where Va is the drug's rate of administration and τ is the rate at which the absorbed drug reaches the circulatory system.

Finally, using the Henderson-Hasselbalch equation, and knowing the drug's pKa (pH at which there is an equilibrium between its ionized and non ionized molecules), it is possible to calculate the non ionized concentration of the drug and therefore the concentration that will be subject to absorption:

$$pH = pKa + log\frac{B}{A}$$

When two drugs have the same bioavailability, they are said to be biological equivalents or bioequivalents. This concept of bioequivalence is important because it is currently used as a yardstick in the authorization of generic drugs in many countries .

LADME

A number of phases occur once the drug enters into contact with the organism, these are described using the acronym LADME:

- Liberation of the active substance from the delivery system,
- Absorption of the active substance by the organism,
- Distribution through the blood plasma and different body tissues,
- Metabolism that is, inactivation of the xenobiotic substance, and finally
- Excretion or elimination of the substance or the products of its metabolism.

Some textbooks combine the first two phases as the drug is often administered in an active form, which means that there is no liberation phase. Others include a phase that combines distribution, metabolism and excretion into a «disposition phase». Other authors include the drug's toxicological aspect in what is known as *ADME-Tox* or *ADMET*.

Each of the phases is subject to physico-chemical interactions between a drug and an organism, which can be expressed mathematically. Pharmacokinetics is therefore based on mathematical equations that allow the prediction of a drug's behavior and which place great emphasis on the relationships between drug plasma concentrations and the time elapsed since the drug's administration.

Analysis

Bioanalytical Methods

Bioanalytical methods are necessary to construct a concentration-time profile. Chemical techniques are employed to measure the concentration of drugs in biological matrix, most often plasma. Proper bioanalytical methods should be selective and sensitive. For example, microscale thermophoresis can be used to quantify how the biological matrix/liquid affects the affinity of a drug to its target.

Mass Spectrometry

Pharmacokinetics is often studied using mass spectrometry because of the complex nature of the matrix (often plasma or urine) and the need for high sensitivity to observe concentrations after a low dose and a long time period. The most common instrumentation used in this application is LC-MS with a triple quadrupole mass spectrometer. Tandem mass spectrometry is usually employed for added specificity. Standard curves and internal standards are used for quantitation of usually a single pharmaceutical in the samples. The samples represent different time points as a pharmaceutical is administered and then metabolized or cleared from the body. Blank samples taken before administration are important in determining background and ensuring data integrity with such complex sample matrices. Much attention is paid to the linearity of the standard curve; however it is common to use curve fitting with more complex functions such as quadratics since the response of most mass spectrometers is less than linear across large concentration ranges.

There is currently considerable interest in the use of very high sensitivity mass spectrometry for microdosing studies, which are seen as a promising alternative to animal experimentation.

Population Pharmacokinetics

Population pharmacokinetics is the study of the sources and correlates of variability in drug concentrations among individuals who are the target patient population receiving clinically relevant doses of a drug of interest. Certain patient demographic, pathophysiological, and therapeutical features, such as body weight, excretory and metabolic functions, and the presence of other therapies, can regularly alter dose-concentration relationships. For example, steady-state concentrations of drugs eliminated mostly by the kidney are usually greater in patients suffering from renal failure than they are in patients with normal renal function receiving the same drug dosage. Population pharmacokinetics seeks to identify the measurable pathophysiologic factors that cause changes in the dose-concentration relationship and the extent of these changes so that, if such changes are associated with clinically significant shifts in the therapeutic index, dosage can be appropriately modified. An advantage of population pharmacokinetic modelling is its ability to analyse sparse data sets (sometimes only one concentration measurement per patient is available).

Software packages used in population pharmacokinetics modelling include NONMEM, which was developed at the UCSF and newer packages which incorporate GUIs like Monolix as well as graphical model building tools; Phoenix NLME.

Clinical Pharmacokinetics

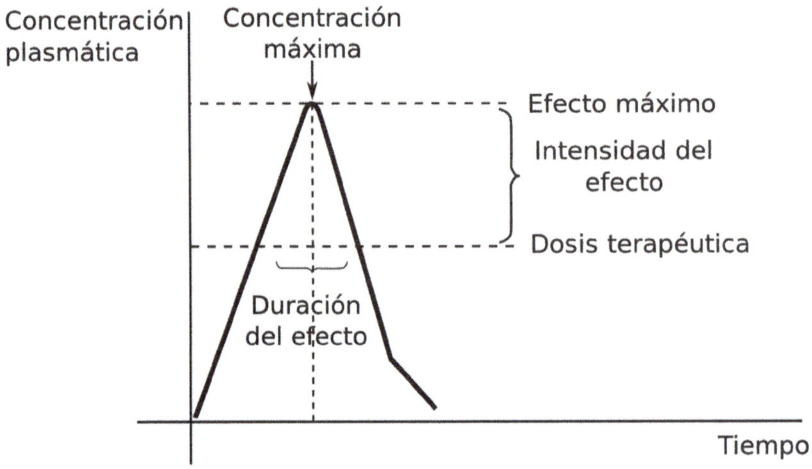

Basic graph for evaluating the therapeutic implications of pharmacokinetics.

Clinical pharmacokinetics (arising from the clinical use of population pharmacokinetics) is the direct application to a therapeutic situation of knowledge regarding a drug's pharmacokinetics and the characteristics of a population that a patient belongs to (or can be ascribed to).

An example is the relaunch of the use of ciclosporin as an immunosuppressor to facilitate organ transplant. The drug's therapeutic properties were initially demonstrated, but it was almost never used after it was found to cause nephrotoxicity in a number of patients. However, it was then realized that it was possible to individualize a patient's dose of ciclosporin by analysing the patients plasmatic concentrations (pharmacokinetic monitoring). This practice has allowed this drug to be used again and has facilitated a great number of organ transplants.

Clinical monitoring is usually carried out by determination of plasma concentrations as this data is usually the easiest to obtain and the most reliable. The main reasons for determining a drug's plasma concentration include:

- Narrow therapeutic range (difference between toxic and therapeutic concentrations)
- High toxicity
- High risk to life.

Drugs where pharmacokinetic monitoring is recommended include:

Medication for which monitoring is recommended.			
• Antiepileptic medication + Phenytoin + Carbamazepine + Valproic acid + Lamotrigine + Ethosuximide + Phenobarbital + Primidone	• Cardioactive medication + Digoxin + Lidocaine	• Immunosuppressor medication + Ciclosporin + Tacrolimus + Sirolimus + Everolimus + Mycophenolate	• Antibiotic medication + Gentamicin + Tobramycin + Amikacin + Vancomycin
• Bronchodilator medication + Theophylline	• Cytostatic medication + Methotrexate + 5-Fluorouracil + Irinotecan	* Antiviral (HIV) medication + Efavirenz + Tenofovir + Ritonavir	• Coagulation factors + Factor VIII, + Factor IX, + Factor VIIa, + Factor XI

Ecotoxicology

Ecotoxicology is the study of toxic effects on a wide range of organisms and includes considerations of toxicokinetics and toxicodynamics.

Various Aspects of Pharmacokinetic

Multi-compartment Model

A multi-compartment model is a type of mathematical model used for describing the way materials or energies are transmitted among the *compartments* of a system. Each compartment is assumed to be a homogeneous entity within which the entities being modelled are equivalent. For instance, in a pharmacokinetic model, the compartments may represent different sections of a body within which the concentration of a drug is assumed to be uniformly equal.

Hence a multi-compartment model is a lumped parameters model.

Multi-compartment models are used in many fields including pharmacokinetics, epidemiology, biomedicine, systems theory, complexity theory, engineering, physics, information science and social science. The circuits systems can be viewed as a multi-compartment model as well.

In systems theory, it involves the description of a network whose components are compartments that represent a population of elements that are equivalent with respect to the manner in which they process input signals to the compartment.

- Instant homogeneous distribution of materials or energies within a "compartment."
- The exchange rate of materials or energies among the compartments is related to the densities of these compartments.
- Usually, it is desirable that the materials do not undergo chemical reactions while transmitting among the compartments.
- When concentration of the cell is of interest, typically the volume is assumed to be constant over time, though this may not be totally true in reality.

Most commonly, the mathematics of multi-compartment models is simplified to provide only a single parameter—such as concentration—within a compartment.

Single-compartment Model

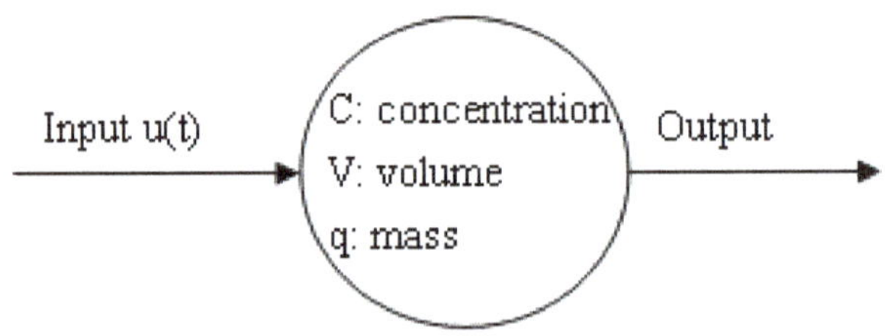

Possibly the simplest application of multi-compartment model is in the single-cell concentration monitoring. If the volume of a cell is *V*, the mass of solute is *q*, the input is *u*(*t*) and the secretion of the solution is proportional to the density of it within the cell, then the concentration of the solution *C''' within the cell over time is given by*

$$\frac{dq}{dt} = u(t) - kq$$

$$C = \frac{q}{V}$$

where *k* is the proportionality.

As the number of compartments increases, the model can be very complex and the solutions usually beyond ordinary calculation. Below shows a three-cell model with interlinks among each other.

The formulae for n-cell multi-compartment models become:

$$\dot{q}_1 = q_1k_{11} + q_2k_{12} + \cdots + q_nk_{1n} + u_1(t)$$

$$\dot{q}_2 = q_1k_{21} + q_2k_{22} + \cdots + q_nk_{2n} + u_2(t)$$

$$\dot{q}_n = q_1k_{n1} + q_2k_{n2} + \cdots + q_nk_{nn} + u_n(t)$$

Where

$0 = \sum_{i=1}^{n} k_{ij}$ for $j = 1, 2, \ldots, n$ (as the total 'contents' of all compartments is constant in a closed system)

Or in matrix forms:

$$\dot{\mathbf{q}} = \mathbf{K}\mathbf{q} + \mathbf{u}$$

Where $\mathbf{K} = \begin{bmatrix} k_{11} & k_{12} & \cdots & k_{1n} \\ k_{21} & k_{22} & \cdots & k_{2n} \\ \vdots & \vdots & \ddots & \vdots \\ k_{n1} & k_{n2} & \cdots & k_{nn} \end{bmatrix} \mathbf{q} = \begin{bmatrix} q_1 \\ q_2 \\ \vdots \\ q_n \end{bmatrix} \mathbf{u} = \begin{bmatrix} u_1(t) \\ u_2(t) \\ \vdots \\ u_n(t) \end{bmatrix}$ and $\begin{bmatrix} 1 & 1 & \cdots & 1 \end{bmatrix}\mathbf{K} = \begin{bmatrix} 0 & 0 & \cdots & 0 \end{bmatrix}$

(as the total 'contents' of all compartments is constant in a closed system)

In the special case of a closed system i.e. where $\mathbf{u} = 0$ then there is a general solution.

$$\mathbf{q} = c_1 e^{\lambda_1 t}\mathbf{v_1} + c_2 e^{\lambda_2 t}\mathbf{v_2} + \cdots + c_n e^{\lambda_n t}\mathbf{v_n}$$

Where λ_1, λ_2, ... and λ_n are the eigenvalues of $\mathbf{K}$; $\mathbf{v_1}$, $\mathbf{v_2}$, ... and $\mathbf{v_n}$ are the respective eigenvectors of $\mathbf{K}$; and c_1, c_2, ... and c_n are constants.

However it can be shown that given the above requirement to ensure the 'contents' of a closed system are constant, then for every pair of eigenvalue and eigenvector then either $\lambda = 0$ or $\begin{bmatrix} 1 & 1 & \cdots & 1 \end{bmatrix}\mathbf{v} = 0$ and also that one eigenvalue is 0, say

So

$$\mathbf{q} = c_1\mathbf{v_1} + c_2 e^{\lambda_2 t}\mathbf{v_2} + \cdots + c_n e^{\lambda_n t}\mathbf{v_n}$$

Where $\begin{bmatrix} 1 & 1 & \cdots & 1 \end{bmatrix}\mathbf{v_i} = 0$ for $\mathbf{i} = 2, 3, \ldots n$

This solution can be rearranged:

$$\mathbf{q} = \left[\mathbf{v_1}\begin{bmatrix} c_1 & 0 & \cdots & 0 \end{bmatrix} + \mathbf{v_2}\begin{bmatrix} 0 & c_2 & \cdots & 0 \end{bmatrix} + \ldots + \mathbf{v_n}\begin{bmatrix} 0 & 0 & \cdots & c_n \end{bmatrix}\right]\begin{bmatrix} 1 \\ e^{\lambda_2 t} \\ \vdots \\ e^{\lambda_n t} \end{bmatrix}$$

This somewhat inelegant equation demonstrates that all solutions of an *n-cell* multi-compartment model with constant or no inputs are of the form:

$$\mathbf{q} = \mathbf{A}\begin{bmatrix} 1 \\ e^{\lambda_2 t} \\ \vdots \\ e^{\lambda_n t} \end{bmatrix}$$

Where $\mathbf{A}$ is a *nxn* matrix and λ_2 , λ_3, , ... and λ_n are constants. Where

Model Topologies

Generally speaking, as the number of compartments increase, it is challenging both to find the algebraic and numerical solutions of the model. However, there are special cases of models, which rarely exist in nature, when the topologies exhibit certain regularities that the solutions become easier to find. The model can be classified according to the interconnection of cells and input/output characteristics:

1. Closed model: No sinks or source, lit. all $k_{oi} = 0$ and $u_i = 0$;
2. Open model: There are sinks or/and sources among cells.
3. Catenary model: All compartments are arranged in a chain, with each pool connecting only to its neighbors. This model has two or more cells.
4. Cyclic model: It's a special case of the catenary model, with three or more cells, in which the first and last cell are connected, i.e. $k_{1n} \neq 0$ or/and $k_{n1} \neq 0$.
5. Mammillary model: Consists of a central compartment with peripheral compartments connecting to it. There are no interconnections among other compartments.
6. Reducible model: It's a set of unconnected models. It bears great resemblance to the computer concept of forest as against trees.

Michaelis–Menten Kinetics

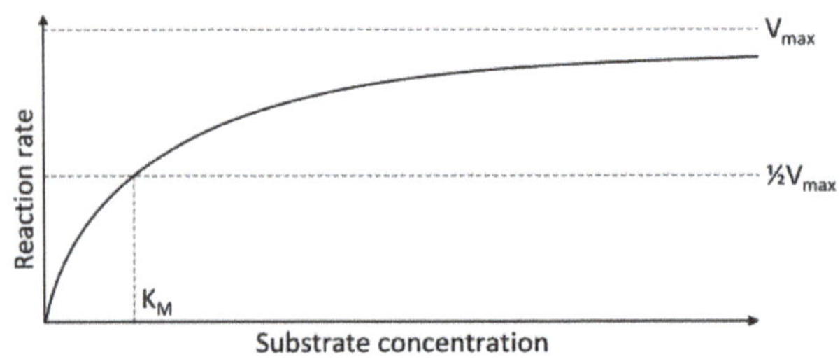

Michaelis–Menten saturation curve for an enzyme reaction showing the relation between the substrate concentration and reaction rate.

In biochemistry, Michaelis–Menten kinetics is one of the best-known models of enzyme kinetics. It is named after German biochemist Leonor Michaelis and Canadian physician Maud Menten. The model takes the form of an equation describing the rate of enzymatic reactions, by relating reaction rate v to $[S]$,, the concentration of a substrate S. Its formula is given by

$$v = \frac{d[P]}{dt} = \frac{V_{\max}[S]}{K_{\mathrm{M}} + [S]}.$$

This equation is called Michaelis–Menten equation. Here, $V_{\max}$ represents the maximum rate achieved by the system, at maximum saturation of the substrate concentration. The Michaelis constant K_{M} is the substrate concentration at which the reaction rate is half of $V_{\max}$. Biochemical reactions involving a single substrate are often assumed to follow Michaelis–Menten kinetics, without regard to the model's underlying assumptions.

Model

In 1903, French physical chemist Victor Henri found that enzyme reactions were initiated by a bond (more generally, a binding interaction) between the enzyme and the substrate. His work was taken up by German biochemist Leonor Michaelis and Canadian physician Maud Menten, who investigated the kinetics of an enzymatic reaction mechanism, invertase, that catalyzes the hydrolysis of sucrose into glucose and fructose. In 1913, they proposed a mathematical model of the reaction. It involves an enzyme E binding to a substrate S to form a ES complex, which in turn is converted into a product P and the enzyme. This may be represented schematically as

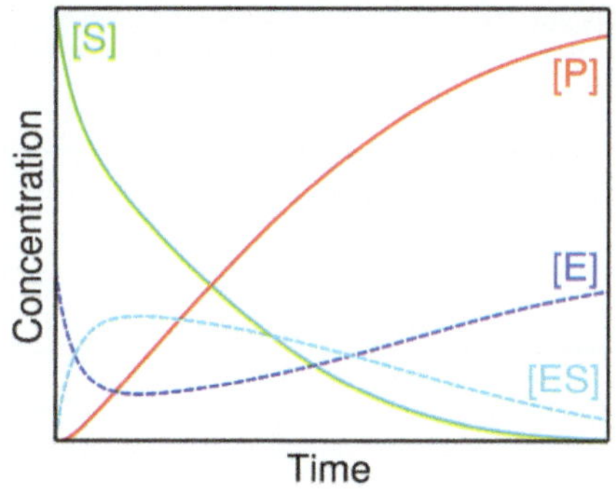

Change in concentrations over time for enzyme E, substrate S, complex ES and product P

$$E + S \underset{k_r}{\overset{k_f}{\rightleftarrows}} ES \xrightarrow{k_{cat}}]E + P$$

where k_f, k_r, , and k_{cat} denote the rate constants, and the double arrows between S and ES represent the fact that enzyme-substrate binding is a reversible process.

Under certain assumptions – such as the enzyme concentration being much less than the substrate concentration – the rate of product formation is given by

$$v = \frac{d[P]}{dt} = V_{\max} \frac{[S]}{K_M + [S]} = k_{cat}[E]_0 \frac{[S]}{K_M + [S]}.$$

The reaction order depends on the relative size of the two terms in the denominator. At low substrate concentration so that $[S] \ll K_M$ $v = k_{cat}[E]_0 \frac{[S]}{K_M}$. Under these conditions the reaction rate varies

linearly with substrate concentration $[S]$ (first-order kinetics). However at higher $[S]$ with $[S] \gg K_M$, the reaction becomes independent of $[S]$ (zero-order kinetics) and asymptotically approaches its maximum rate $V_{max} = k_{cat}[E]_0$, where $[E]_0$ is the initial enzyme concentration. This rate is attained when all enzyme is bound to substrate. k_{cat}, , the turnover number, is the maximum number of substrate molecules converted to product per enzyme molecule per second. Further addition of substrate does not increase the rate which is said to be saturated.

The Michaelis constant K_M is the $[S]$ at which the reaction rate is at half-maximum, and is an inverse measure of the substrate's affinity for the enzyme—as a small K_M indicates high affinity, meaning that the rate will approach V_{max} with lower $[S]$ than those reactions with a larger K_M. The value of K_M. is dependent on both the enzyme and the substrate, as well as conditions such as temperature and pH.

The model is used in a variety of biochemical situations other than enzyme-substrate interaction, including antigen-antibody binding, DNA-DNA hybridization, and protein-protein interaction. It can be used to characterise a generic biochemical reaction, in the same way that the Langmuir equation can be used to model generic adsorption of biomolecular species. When an empirical equation of this form is applied to microbial growth, it is sometimes called a Monod equation.

Applications

Parameter values vary widely between enzymes:

Enzyme	(M)	(1/s)	(1/M*s)
Chymotrypsin	1.5×10^{-2}	0.14	9.3
Pepsin	3.0×10^{-4}	0.50	1.7×10^{3}
Tyrosyl-tRNA synthetase	9.0×10^{-4}	7.6	8.4×10^{3}
Ribonuclease	7.9×10^{-3}	7.9×10^{2}	1.0×10^{5}
Carbonic anhydrase	2.6×10^{-2}	4.0×10^{5}	1.5×10^{7}
Fumarase	5.0×10^{-6}	8.0×10^{2}	1.6×10^{8}

The constant k_{cat} / K_M (catalytic efficiency) is a measure of how an enzyme efficiently converts a substrate into product. Diffusion limited enzymes, such as fumarase, work at the theoretical upper limit of $10^8 - 10^{10}$ /M*s, limited by diffusion of substrate into the active site.

Michaelis–Menten kinetics have also been applied to a variety of spheres outside of biochemical reactions, including alveolar clearance of dusts, the richness of species pools, clearance of blood alcohol, the photosynthesis-irradiance relationship, and bacterial phage infection.

Derivation

Applying the law of mass action, which states that the rate of a reaction is proportional to the product of the concentrations of the reactants (i.e.[E][S]), gives a system of four non-linear ordinary differential equations that define the rate of change of reactants with time $t^{[}$

$$\frac{d[E]}{dt} = -k_f[E][S] + k_r[ES] + k_{cat}[ES]$$

$$\frac{d[S]}{dt} = -k_f[E][S] + k_r[ES]$$

$$\frac{d[ES]}{dt} = k_f[E][S] - k_r[ES] - k_{cat}[ES]$$

$$\frac{d[P]}{dt} = k_{cat}[ES].$$

In this mechanism, the enzyme E is a catalyst, which only facilitates the reaction, so that its total concentration, free plus combined, $[E]+[ES]=[E]_0$ is a constant. This conservation law can also be observed by adding the first and third equations above.

Equilibrium Approximation

In their original analysis, Michaelis and Menten assumed that the substrate is in instantaneous chemical equilibrium with the complex, which implies

$$k_f[E][S] = k_r[ES].$$

From the enzyme conservation law, we obtain

$$[E]=[E]_0-[ES].$$

Combining the two expressions above, gives us

$$k_f([E]_0-[ES])[S]=k_r[ES].$$

Upon simplification, we get

$$[ES]=\frac{[E]_0[S]}{K_d+[S]}$$

where $K_d=k_r/k_f$ is the dissociation constant for the enzyme-substrate complex. Hence the velocity v of the reaction – the rate at which P is formed – is

$$v=\frac{d[P]}{dt}=\frac{V_{max}[S]}{K_d+[S]}$$

where $V_{max}=k_{cat}[E]_0$ is the maximum reaction velocity.

Quasi-steady-state Approximation

An alternative analysis of the system was undertaken by British botanist G. E. Briggs and British geneticist J. B. S. Haldane in 1925. They assumed that the concentration of the intermediate complex does not change on the time-scale of product formation – known as the quasi-steady-state assumption or pseudo-steady-state-hypothesis. Mathematically, this assumption means $k_f[E][S]=k_r[ES]+k_{cat}[ES]$. Combining this relationship with the enzyme conservation law, the concentration of complex is

$$[ES]=\frac{[E]_0[S]}{K_M+[S]}$$

where

$$K_M=\frac{k_r+k_{cat}}{k_f}$$

is known as the Michaelis constant, where k_r, k_{cat}, and k_f are, respectively, the constants for substrate unbinding, conversion to product, and binding to the enzyme. Hence the velocity v of the reaction is

$$v=\frac{d[P]}{dt}=\frac{V_{max}[S]}{K_M+[S]}.$$

Assumptions and Limitations

The first step in the derivation applies the law of mass action, which is reliant on free diffusion. However, in the environment of a living cell where there is a high concentration of proteins, the cytoplasm often behaves more like a gel than a liquid, limiting molecular movements and altering reaction rates. Whilst the law of mass action can be valid in heterogeneous environments, it is more appropriate to model the cytoplasm as a fractal, in order to capture its limited-mobility kinetics.

The resulting reaction rates predicted by the two approaches are similar, with the only difference being that the equilibrium approximation defines the constant as K_d, whilst the quasi-steady-state approximation uses K_M. However, each approach is founded upon a different assumption. The

Michaelis–Menten equilibrium analysis is valid if the substrate reaches equilibrium on a much faster time-scale than the product is formed or, more precisely, that

$$\epsilon_d = \frac{k_{cat}}{k_r} \ll 1.$$

By contrast, the Briggs–Haldane quasi-steady-state analysis is valid if

$$\epsilon_m = \frac{[E]_0}{[S]_0 + K_M} \ll 1.$$

Thus it holds if the enzyme concentration is much less than the substrate concentration. Even if this is not satisfied, the approximation is valid if K_M is large.

In both the Michaelis–Menten and Briggs–Haldane analyses, the quality of the approximation improves as ȯ decreases. However, in model building, Michaelis–Menten kinetics are often invoked without regard to the underlying assumptions.

It is also important to remember that, while irreversibility is a necessary simplification in order to yield a tractable analytic solution, in the general case product formation is not in fact irreversible. The enzyme reaction is more correctly described as

In general, the assumption of irreversibility is a good one in situations where one of the below is true:

1. The concentration of substrate(s) is very much larger than the concentration of products:

$[S] \gg [P]$.

This is true under standard *in vitro* assay conditions, and is true for many *in vivo* biological reactions, particularly where the product is continually removed by a subsequent reaction.

2. The energy released in the reaction is very large, that is

$\Delta G \ll 0$.

In situations where neither of these two conditions hold (that is, the reaction is low energy and a substantial pool of product(s) exists), the Michaelis–Menten equation breaks down, and more complex modelling approaches explicitly taking the forward and reverse reactions into account must be taken to understand the enzyme biology.

Determination of Constants

The typical method for determining the constants V_{max} and K_M involves running a series of enzyme assays at varying substrate concentrations $[S]$, and measuring the initial reaction rate v_0. 'Initial' here is taken to mean that the reaction rate is measured after a relatively short time period, during which it is assumed that the enzyme-substrate complex has formed, but that the substrate concentration held approximately constant, and so the equilibrium or quasi-steady-state approximation remain valid. By plotting reaction rate against concentration, and using nonlinear regression of the Michaelis–Menten equation, the parameters may be obtained.

Before computing facilities to perform nonlinear regression became available, graphical methods involving linearisation of the equation were used. A number of these were proposed, including the Eadie–Hofstee diagram, Hanes–Woolf plot and Lineweaver–Burk plot; of these, the Hanes–Woolf plot is the most accurate. However, while useful for visualization, all three methods distort the error structure of the data and are inferior to nonlinear regression. Nonetheless, their use can still be found in modern literature.

In 1997 Santiago Schnell and Claudio Mendoza suggested a closed form solution for the time course kinetics analysis of the Michaelis–Menten kinetics based on the solution of the Lambert W function. Namely:

$$\frac{[S]}{K_M} = W(F(t))$$

where W is the Lambert W function and

$$F(t) = \frac{[S]_0}{K_M} \exp\left(\frac{[S]_0}{K_M} - \frac{V_{max}}{K_M} t\right).$$

The above equation has been used to estimate V_{max} and K_M from time course data.

Role of Substrate Unbinding

The Michaelis-Menten equation has been used to predict the rate of product formation in enzymatic reactions for more than a century. Specifically, it states that the rate of an enzymatic reaction will increase as substrate concentration increases, and that increased unbinding of enzyme-substrate complexes will decrease the reaction rate. While the first prediction is well established, the second has never been tested experimentally. To determine whether an increased rate of unbinding does in fact decrease the reaction rate, Shlomi Reuveni *et al.* mathematically analyzed the effect of enzyme-substrate unbinding on enzymatic reactions at the single-molecule level. According to the study, unbinding of an enzyme from a substrate can reduce the rate of product formation under some conditions, but may also have the opposite effect. As substrate concentrations increase, a tipping point can be reached where an increase in the unbinding rate results in an increase, rather than a decrease, of the reaction rate. The results indicate that enzymatic reactions can behave in ways that violate the classical Michaelis-Menten equation, and that the role of unbinding in enzymatic catalysis still remains to be determined experimentally.

ADME

ADME is an abbreviation in pharmacokinetics and pharmacology for "absorption, distribution, metabolism, and excretion," and describes the disposition of a pharmaceutical compound within an organism. The four criteria all influence the drug levels and kinetics of drug exposure to the tissues and hence influence the performance and pharmacological activity of the compound as a drug.

Components

Absorption/Administration

For a compound to reach a tissue, it usually must be taken into the bloodstream - often via mucous surfaces like the digestive tract (intestinal absorption) - before being taken up by the target cells. Factors such as poor compound solubility, gastric emptying time, intestinal transit time, chemical instability in the stomach, and inability to permeate the intestinal wall can all reduce the extent to which a drug is absorbed after oral administration. Absorption critically determines the compound's bioavailability. Drugs that absorb poorly when taken orally must be administered in some less desirable way, like intravenously or by inhalation (e.g. zanamivir). Routes of administration is an important consideration

Distribution

The compound needs to be carried to its effector site, most often via the bloodstream. From there, the compound may distribute into muscle and organs, usually to differing extents. After entry into the systemic circulation, either by intravascular injection or by absorption from any of the various extracellular sites, the drug is subjected to numerous distribution processes that tend to lower its plasma concentration.

Distribution is defined as the reversible transfer of a drug between one compartment to another. Some factors affecting drug distribution include regional blood flow rates, molecular size, polarity and binding to serum proteins, forming a complex. Distribution can be a serious problem at some natural barriers like the blood–brain barrier.

Metabolism

Compounds begin to break down as soon as they enter the body. The majority of small-molecule drug metabolism is carried out in the liver by redox enzymes, termed cytochrome P450 enzymes. As metabolism occurs, the initial (parent) compound is converted to new compounds called metabolites. When metabolites are pharmacologically inert, metabolism deactivates the administered dose of parent drug and this usually reduces the effects on the body. Metabolites may also be phar-macologically active, sometimes more so than the parent drug .

Excretion

Compounds and their metabolites need to be removed from the body via excretion, usually through the kidneys (urine) or in the feces. Unless excretion is complete, accumulation of foreign substances can adversely affect normal metabolism.

There are three main sites where drug excretion occurs. The kidney is the most important site and it is where products are excreted through urine. Biliary excretion or fecal excretion is the process that initiates in the liver and passes through to the gut until the products are finally excreted along with waste products or feces. The last main method of excretion is through the lungs (e.g. anesthetic gases).

Excretion of drugs by the kidney involves 3 main mechanisms:

- Glomerular filtration of unbound drug.
- Active secretion of (free & protein-bound) drug by transporters (e.g. anions such as urate, penicillin, glucuronide, sulfate conjugates) or cations such as choline, histamine.
- Filtrate 100-fold concentrated in tubules for a favorable concentration gradient so that it may be secreted by passive diffusion and passed out through the urine.

Toxicity

Sometimes, the potential or real toxicity of the compound is taken into account (ADME-Tox or ADMET).

Computational chemists try to predict the ADME-Tox qualities of compounds through methods like QSPR or QSAR.

The route of administration critically influences ADME.

Pharmacodynamics

Pharmacodynamics is the study of the biochemical and physiologic effects of drugs (especially pharmaceutical drugs). The effects can include those manifested within animals (including humans), microorganisms, or combinations of organisms (for example, infection). Pharmacodynamics is the study of how a drug affects an organism, whereas pharmacokinetics is the study of how the organism affects the drug. Both together influence dosing, benefit, and adverse effects. Pharmacodynamics is sometimes abbreviated as PD and pharmacokinetics as PK, especially in combined reference (for example, when speaking of PK/PD models).

Pharmacodynamics places particular emphasis on dose–response relationships, that is, the relationships between drug concentration and effect. One dominant example is drug-receptor interactions as modeled by

$$L + R \rightleftharpoons L{\cdot}R$$

where *L*, *R*, and *LR* represent ligand (drug), receptor, and ligand-receptor complex concentrations, respectively. This equation represents a simplified model of reaction dynamics that can be studied mathematically through tools such as free energy maps.

Effects on the Body

The majority of drugs either(a) mimic or inhibit normal physiological/biochemical processes or inhibit pathological processes in animals or(b) inhibit vital processes of endo- or ectoparasites and microbial organisms. There are 7 main drug actions:

- stimulating action through direct receptor agonism and downstream effects
- depressing action through direct receptor agonism and downstream effects (ex.: inverse agonist)
- blocking/antagonizing action (as with silent antagonists), the drug binds the receptor but does not activate it
- stabilizing action, the drug seems to act neither as a stimulant or as a depressant (ex.: some drugs possess receptor activity that allows them to stabilize general receptor activation, like buprenorphine in opioid dependent individuals or aripiprazole in schizophrenia, all depending on the dose and the recipient)

- exchanging/replacing substances or accumulating them to form a reserve (ex.: glycogen storage)
- direct beneficial chemical reaction as in free radical scavenging
- direct harmful chemical reaction which might result in damage or destruction of the cells, through induced toxic or lethal damage (cytotoxicity or irritation)

Desired Activity

The desired activity of a drug is mainly due to successful targeting of one of the following:

- Cellular membrane disruption
- Chemical reaction with downstream effects
- Interaction with enzyme proteins
- Interaction with structural proteins
- Interaction with carrier proteins
- Interaction with ion channels
- Ligand binding to receptors:
    - Hormone receptors
    - Neuromodulator receptors
    - Neurotransmitter receptors

General anesthetics were once thought to work by disordering the neural membranes, thereby altering the Na^+ influx. Antacids and chelating agents combine chemically in the body. Enzyme-substrate binding is a way to alter the production or metabolism of key endogenous chemicals, for example aspirin irreversibly inhibits the enzyme prostaglandin synthetase (cyclooxygenase) thereby preventing inflammatory response. Colchicine, a drug for gout, interferes with the function of the structural protein tubulin, while Digitalis, a drug still used in heart failure, inhibits the activity of the carrier molecule, Na-K-ATPase pump. The widest class of drugs act as ligands which bind to receptors which determine cellular effects. Upon drug binding, receptors can elicit their normal action (agonist), blocked action (antagonist), or even action opposite to normal (inverse agonist).

In principle, a pharmacologist would aim for a target plasma concentration of the drug for a desired level of response. In reality, there are many factors affecting this goal. Pharmacokinetic factors determine peak concentrations, and concentrations cannot be maintained with absolute consistency because of metabolic breakdown and excretory clearance. Genetic factors may exist which would alter metabolism or drug action itself, and a patient's immediate status may also affect indicated dosage.

Undesirable Effects

Undesirable effects of a drug include:

- Increased probability of cell mutation (carcinogenic activity)

- A multitude of simultaneous assorted actions which may be deleterious
- Interaction (additive, multiplicative, or metabolic)
- Induced physiological damage, or abnormal chronic conditions

Therapeutic Window

The therapeutic window is the amount of a medication between the amount that gives an effect (effective dose) and the amount that gives more adverse effects than desired effects. For instance, medication with a small pharmaceutical window must be administered with care and control, e.g. by frequently measuring blood concentration of the drug, since it easily loses effects or gives adverse effects.

Duration of Action

The *duration of action* of a drug is the length of time that particular drug is effective. Duration of action is a function of several parameters including plasma half-life, the time to equilibrate between plasma and target compartments, and the off rate of the drug from its biological target.

Receptor Binding and Effect

The binding of ligands (drug) to receptors is governed by the *law of mass action* which relates the large-scale status to the rate of numerous molecular processes. The rates of formation and un-formation can be used to determine the equilibrium concentration of bound receptors. The *equilibrium dissociation constant* is defined by:

$$L + R \leftrightarrow L{\cdot}R \qquad K_d = \frac{[L][R]}{[L{\cdot}R]}$$

where *L*=ligand, *R*=receptor, square brackets [] denote concentration. The fraction of bound receptors is

$$Fraction\ Bound = \frac{[L{\cdot}R]}{[R]+[L{\cdot}R]} = \frac{1}{1+\frac{K_d}{[L]}}$$

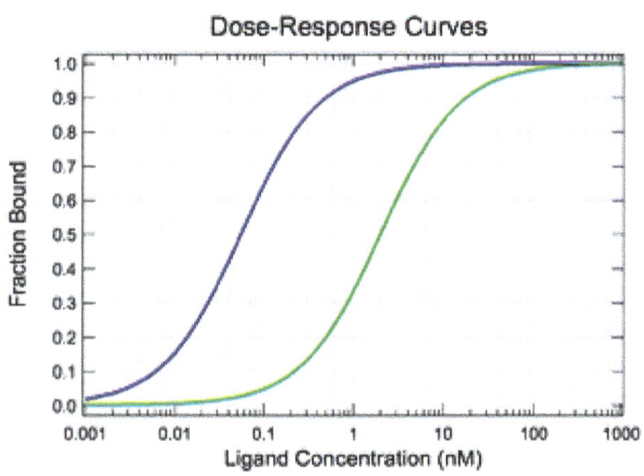

Semi-log plots of two agonists with different K_d. The blue curve represents the ligand with greater potency.

This expression is one way to consider the effect of a drug, in which the response is related to the fraction of bound receptors. The fraction of bound receptors is known as occu-pancy. The relationship between occupancy and pharmacological response is usually nonlinear. This explains the so-called *receptor reserve* phenomenon i.e. the concentration producing 50% occupancy is typically higher than the concentration producing 50% of maximum response. More precisely, receptor reserve refers to a phenomenon whereby stimulation of only a fraction of the whole receptor population apparently elicits the maximal effect achievable in a particular tissue.

The simplest interpretation of receptor reserve is that it is a model that states there are excess receptors on the cell surface than what is necessary for full effect. Taking a more sophisticated approach, receptor reserve is an integrative measure of the response-inducing capacity of an agonist (in some receptor models it is termed intrinsic efficacy or intrinsic activity) and of the signal amplification capacity of the corresponding receptor (and its downstream signaling pathways). Thus, the existence (and magnitude) of receptor reserve depends on the agonist (efficacy), tissue (signal amplification ability) and measured effect (pathways activated to cause signal amplification). As receptor reserve is very sensitive to agonist's intrinsic efficacy, it is usually defined only for full (high-efficacy) agonists.

Often the response is determined as a function of *log[L]* to consider many orders of magnitude of concentration. However, there is no biological or physical theory which relates effects to the log of concentration. It is just convenient for graphing purposes. It is useful to note that 50% of the receptors are bound when $[L]=K_d$.

The graph shown represents the conc-response for two hypothetical receptor agonists, plotted in a semi-log fashion. The curve toward the left represents a higher potency (potency arrow does not indicate direction of increase) since lower concentrations are needed for a given response. The effect increases as a function of concentration.

Multicellular Pharmacodynamics

The concept of pharmacodynamics has been expanded to include Multicellular Pharmacodynamics (MCPD). MCPD is the study of the static and dynamic properties and relationships between a set of drugs and a dynamic and diverse multicellular four-dimensional organization. It is the study of the workings of a drug on a minimal multicellular system (mMCS), both *in vivo* and *in silico*. Networked Multicellular Pharmacodynamics (Net-MCPD) further extends the concept of MCPD to model regulatory genomic networks together with signal transduction pathways, as part of a complex of interacting components in the cell.

Toxicodynamics

Pharmacokinetics and pharmacodynamics are termed toxicokinetics and toxicodynamics in the field of ecotoxicology. Here, the focus is on toxic effects on a wide range of organisms. The corresponding models are called toxicokinetic-toxicodynamic models (more details can be found at Ecotoxmodels.org).

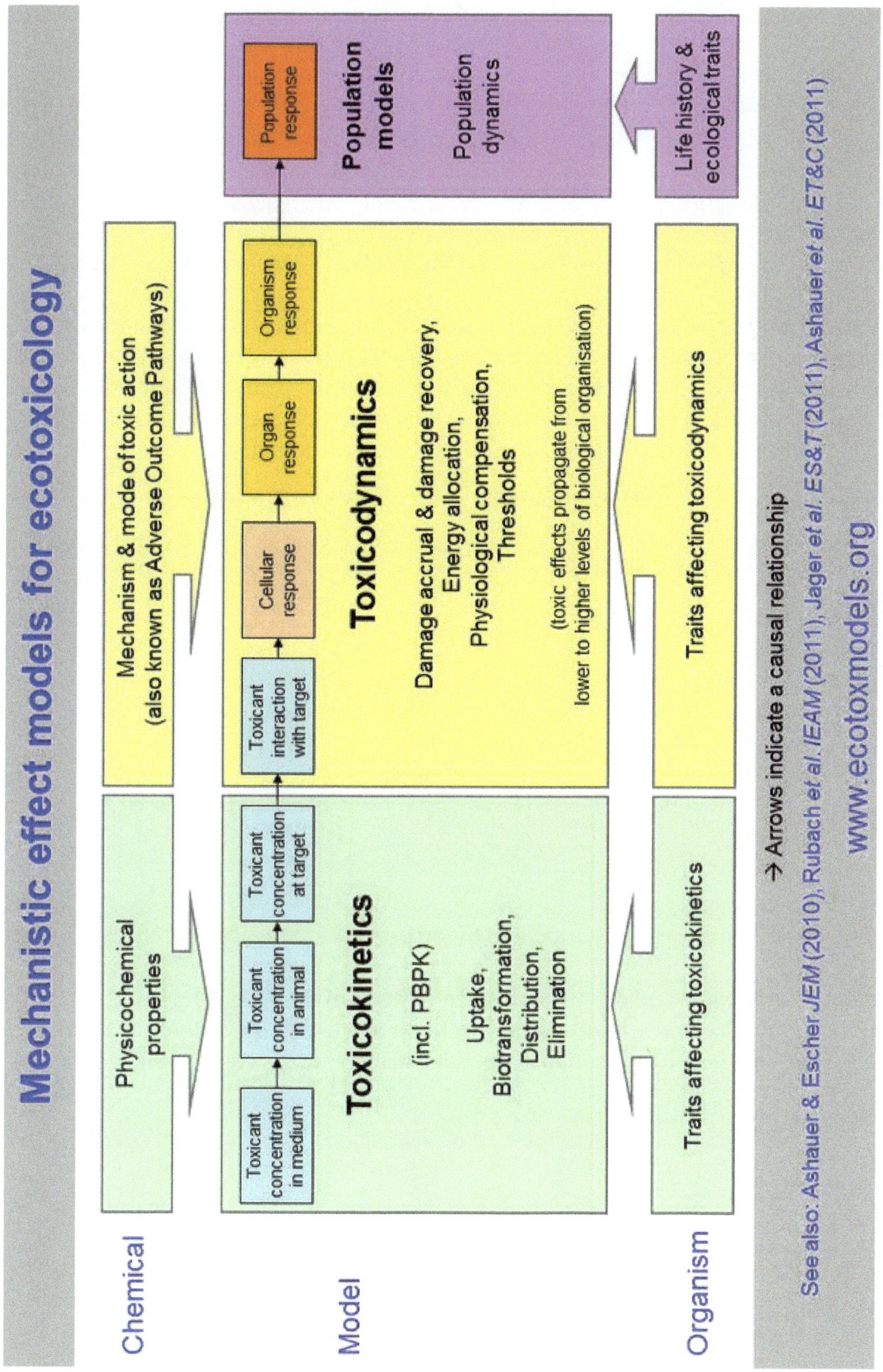
Mechanistic effect models for ecotoxicology
Chemical
Model
Organism
Physicochemical properties
Mechanism & mode of toxic action (also known as Adverse Outcome Pathways)
Toxicant concentration in medium
Toxicant concentration in animal
Toxicant concentration at target
Toxicant interaction with target
Cellular response
Organ response
Organism response
Population response
Toxicokinetics
(incl. PBPK)
Uptake, Biotransformation, Distribution, Elimination
Toxicodynamics
Damage accrual & damage recovery, Energy allocation, Physiological compensation, Thresholds
(toxic effects propagate from lower to higher levels of biological organisation)
Population models
Population dynamics
Traits affecting toxicokinetics
Traits affecting toxicodynamics
Life history & ecological traits
→ Arrows indicate a causal relationship
See also: Ashauer & Escher JEM (2010), Rubach et al. IEAM (2011), Jager et al. ES&T (2011), Ashauer et al. ET&C (2011)
www.ecotoxmodels.org

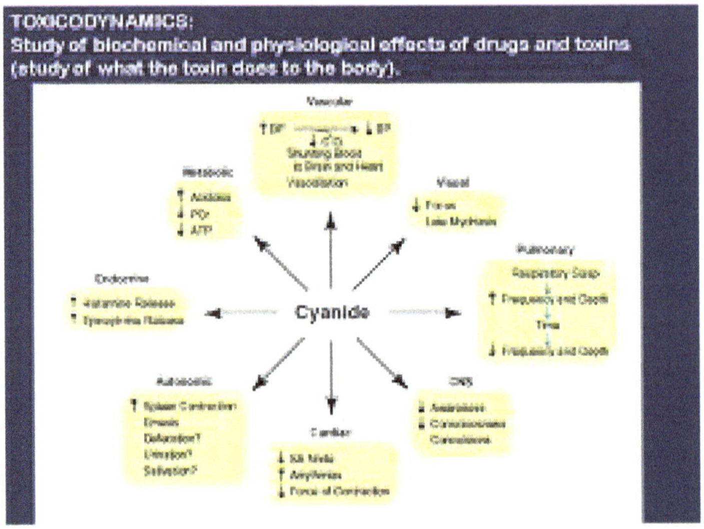
TOXICODYNAMICS:
Study of biochemical and physiological effects of drugs and toxins (study of what the toxin does to the body).
Vascular
Metabolic
Visual
Pulmonary
Endocrine
Cyanide
Autonomic
CNS
Cardiac

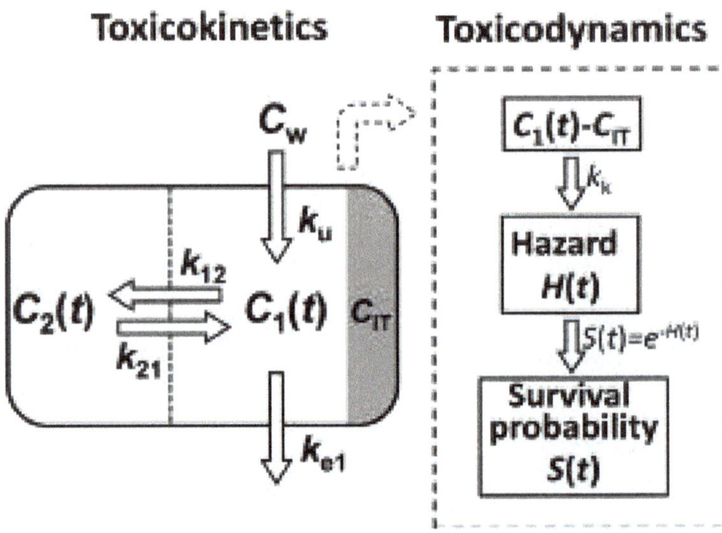
Toxicokinetics
Toxicodynamics
C_w
k_u
k_{12}
$C_2(t)$
$C_1(t)$
C_{IT}
k_{21}
k_{e1}
$C_1(t)$-C_{IT}
k_k
Hazard
$H(t)$
$S(t)=e^{-H(t)}$
Survival probability
$S(t)$

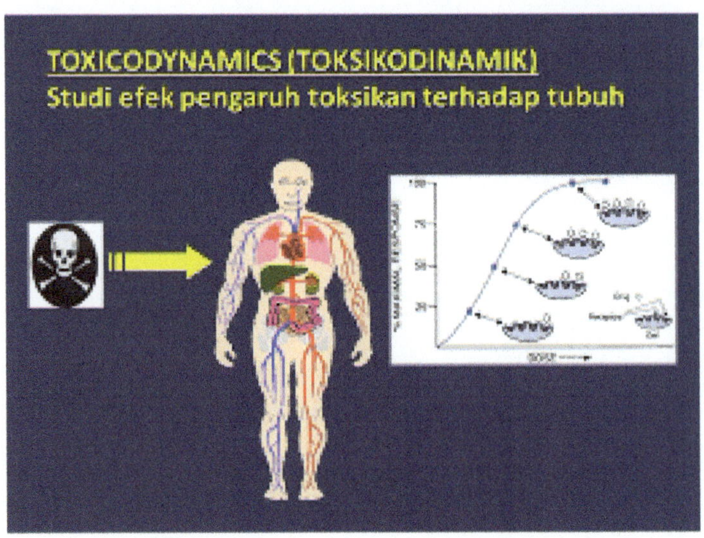
TOXICODYNAMICS (TOKSIKODINAMIK)
Studi efek pengaruh toksikan terhadap tubuh

References

- Newton, David; Alasdair Thorpe; Chris Otter (2004). Revise A2 Chemistry. Heinemann Educational Publishers. p. 1. ISBN 0-435-58347-6.
- Nagle, Hinter; Barbara Nagle (2005). Pharmacology: An Introduction. Boston: McGraw Hill. ISBN 0-07-312275-0
- The Creative Destruction of Medicine: How the Digital Revolution Will Create Better Health Care. New York: Basic Books. 2012. ISBN 0-465-02550-1.
- Kathleen Knights; Bronwen Bryant (2002). Pharmacology for Health Professionals. Amsterdam: Elsevier. ISBN 0-7295-3664-5.
- Koch HP, Ritschel WA (1986). "Liberation". Synopsis der Biopharmazie und Pharmakokinetik (in German). Landsberg, München: Ecomed. pp. 99–131. ISBN 3-609-64970-4.
- Weiner, Daniel; Johan Gabrielsson (2000). "PK24 - Non-linear kinetics - flow II". Pharmacokinetic/pharmacodynamic data analysis: concepts and applications. Apotekarsocieteten. pp. 527–36. ISBN 91-86274-92-9.
- Michael E. Winter, Mary Anne Koda-Kimple, Lloyd Y. Young, Emilio Pol Yanguas Farmacocinética clínica básica Ediciones Díaz de Santos, 1994 pgs. 8-14 ISBN 84-7978-147-5, 9788479781477 (in Spanish)
- Joaquín Herrera Carranza Manual de farmacia clínica y Atención Farmacéutica (in Spanish). Published by Elsevier España, 2003; page 159. ISBN 84-8174-658-4
- Lehninger, A.L.; Nelson, D.L.; Cox, M.M. (2005). Lehninger principles of biochemistry. New York: W.H. Freeman. ISBN 978-0-7167-4339-2.
- Evans, W. C., Linear Systems, Compartmental Modeling, and Estimability Issues in IAQ Studies, in Tichenor, B., Characterizing Sources of Indoor Air Pollution and Related Sink Effects, ASTM STP 1287, pp. 239–262, 1996 (ISBN 0-8031-2030-3).
- Director, Dr Soram Khalsa Medical; Institute, East-West Medical Research (2015-06-28). "Pharmacogenetics: What It Is And Why You Need to Know". The Huffington Post. Retrieved 2016-10-05.
- Reuveni, Shlomi; Urbakh, Michael; Klafter, Joseph (2014). "Role of Substrate Unbinding in Michaelis-Menten Enzymatic Reactions". Proceedings of the National Academy of Sciences. 111 (12): 4391–4396. Bibcode:2014PNAS..111.4391R. doi:10.1073/pnas.1318122111.

3

Compounds used in Medicinal Chemistry

Generally the compounds used in medicinal chemistry are organic compounds. Some of the examples given of organic compounds are atorvastatin and ETFE. Inorganic compounds and lithium are also useful in medicinal chemistry. The succeeding section is a compilation of the various compound used in medicinal chemistry.

Organic Compound

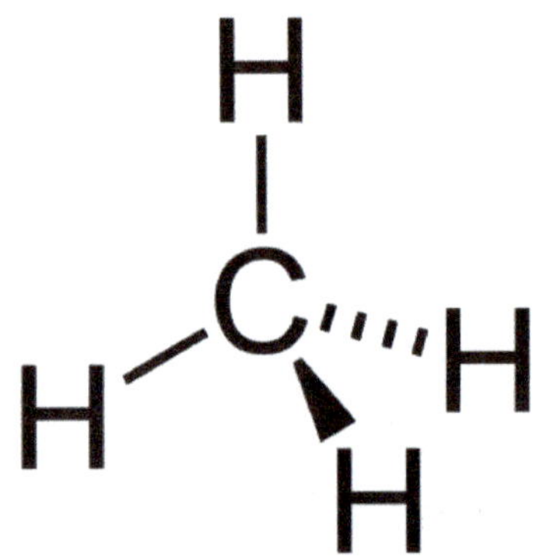

Methane is one of the simplest organic compounds

An organic compound is any member of a large class of gaseous, liquid, or solid chemical compounds whose molecules contain carbon. For historical reasons discussed below, a few types of carbon-containing compounds, such as carbides, carbonates, simple oxides of carbon (such as CO and CO_2), and cyanides are considered inorganic. The distinction between *organic* and *inorganic* carbon compounds, while "useful in organizing the vast subject of chemistry... is somewhat arbitrary".

Organic chemistry is the science concerned with all aspects of organic compounds. Organic synthesis is the methodology of their preparation.

History

Vitalism

The word *organic* is historical, dating to the 1st century. For many centuries, Western alchemists believed in vitalism. This is the theory that certain compounds could be synthesized only from their classical elements—earth, water, air, and fire—by the action of a "life-force" (*vis vitalis*) that only organisms possessed. Vitalism taught that these "organic" compounds were fundamentally different from the "inorganic" compounds that could be obtained from the elements by chemical manipulation.

Vitalism survived for a while even after the rise of modern atomic theory and the replacement of the Aristotelian elements by those we know today. It first came under question in 1824, when Friedrich Wöhler synthesized oxalic acid, a compound known to occur only in living organisms, from cyanogen. A more decisive experiment was Wöhler's 1828 synthesis of urea from the inorganic salts potassium cyanate and ammonium sulfate. Urea had long been considered an "organic" compound, as it was known to occur only in the urine of living organisms. Wöhler's experiments were followed by many others, where increasingly complex "organic" substances were produced from "inorganic" ones without the involvement of any living organism.

Modern Classification

Even though vitalism has been discredited, scientific nomenclature retains the distinction between *organic* and *inorganic* compounds. The modern meaning of *organic compound* is any compound that contains a significant amount of carbon—even though many of the organic compounds known today have no connection to any substance found in living organisms.

There is no single "official" definition of an organic compound. Some textbooks define an organic compound as one that contains one or more C-H bonds. Others include C-C bonds in the definition. Others state that if a molecule contains carbon—it is organic.

Even the broader definition of "carbon-containing molecules" requires the exclusion of carbon-containing alloys (including steel), a relatively small number of carbon-containing compounds, such as metal carbonates and carbonyls, simple oxides of carbon and cyanides, as well as the allotropes of carbon and simple carbon halides and sulfides, which are usually considered inorganic.

The "C-H" definition excludes compounds that are historically and practically considered organic. Neither urea nor oxalic acid is organic by this definition, yet they were two key compounds in the vitalism debate. The IUPAC Blue Book on organic nomenclature specifically mentions urea and oxalic acid. Other compounds lacking C-H bonds that are also traditionally considered organic include benzenehexol, mesoxalic acid, and carbon tetrachloride. Mellitic acid, which contains no C-H bonds, is considered a possible organic substance in Martian soil. C-C bonds are found in most organic compounds, except some small molecules like methane and methanol, which have only one carbon atom in their structure.

The "C-H bond-only" rule also leads to somewhat arbitrary divisions in sets of carbon-fluorine compounds, as, for example, Teflon is considered by this rule to be "inorganic", whereas Tefzel is considered to be organic. Likewise, many Halons are considered inorganic, whereas the rest are considered organic. For these and other reasons, most sources believe that C-H compounds are only a subset of "organic" compounds.

In summary, most carbon-containing compounds are organic, and almost all organic compounds contain at least a C-H bond or a C-C bond. A compound does not need to contain C-H bonds to be considered organic (e.g., urea), but most organic compounds do.

Classification

Organic compounds may be classified in a variety of ways. One major distinction is between natural and synthetic compounds. Organic compounds can also be classified or subdivided by the presence of heteroatoms, e.g., organometallic compounds, which feature bonds between carbon and a metal, and organophosphorus compounds, which feature bonds between carbon and a phosphorus.

Another distinction, based on the size of organic compounds, distinguishes between small molecules and polymers.

Natural Compounds

Natural compounds refer to those that are produced by plants or animals. Many of these are still extracted from natural sources because they would be more expensive to produce artificially. Examples include most sugars, some alkaloids and terpenoids, certain nutrients such as vitamin B_{12}, and, in general, those natural products with large or stereoisometrically complicated molecules present in reasonable concentrations in living organisms.

Further compounds of prime importance in biochemistry are antigens, carbohydrates, enzymes, hormones, lipids and fatty acids, neurotransmitters, nucleic acids, proteins, peptides and amino acids, lectins, vitamins, and fats and oils.

Synthetic Compounds

Compounds that are prepared by reaction of other compounds are known as "synthetic". They may be either compounds that already are found in plants or animals or those that do not occur naturally.

Most polymers (a category that includes all plastics and rubbers), are organic synthetic or semi-synthetic compounds.

Biotechnology

Several compounds are industrially manufactured utilizing the biochemistry of organisms such as bacteria and yeast. Two examples are ethanol and insulin. Regularly, the DNA of the organism is altered to express desired compounds that are often not ordinarily produced by that organism. Sometimes the biotechnologically engineered compounds were never present in nature in the first place.

Nomenclature

The IUPAC nomenclature of organic compounds slightly differs from the CAS nomenclature.

Databases

- The *CAS* database is the most comprehensive repository for data on organic compounds. The search tool *SciFinder* is offered.
- The *Beilstein database* contains information on 9.8 million substances, covers the scientific literature from 1771 to the present, and is today accessible via Reaxys. Structures and a large diversity of physical and chemical properties is available for each substance, with reference to original literature.

- *PubChem* contains 18.4 million entries on compounds and especially covers the field of medicinal chemistry.

There is a great number of more specialized databases for diverse branches of organic chemistry.

Structure Determination

Today, the main tools are proton and carbon-13 NMR spectroscopy, IR Spectroscopy, Mass spectrometry, UV/Vis Spectroscopy and X-ray crystallography.

Example of Organic Compounds

Atorvastatin

Atorvastatin, marketed under the trade name Lipitor among others, is a member of the drug class known as statins, which are used primarily as a lipid-lowering agent and for prevention of events associated with cardiovascular disease. Like all statins, atorvastatin works by inhibiting HMG-CoA reductase, an enzyme found in liver tissue that plays a key role in production of cholesterol in the body.

The synthesis of atorvastatin took place in August 1985 at Warner-Lambert's Parke-Davis Pharmaceutical Research facility in Ann Arbor, Michigan. by Bruce Roth, leading a team of 18 scientists researching atherosclerosis. From 1996 to 2012 under the trade name Lipitor, atorvastatin became the world's best-selling drug of all time, with more than US$125 billion in sales over approximately 14.5 years.

Medical Uses

The primary uses of atorvastatin is for the treatment of dyslipidemia and the prevention of cardiovascular disease:

Dyslipidemia

- Hypercholesterolemia (heterozygous familial and nonfamilial) and mixed dyslipidemia (Fredrickson types IIa and IIb) to reduce total cholesterol, LDL-C, apo-B, triglycerides levels, and CRP as well as increase HDL levels.
- Heterozygous familial hypercholesterolemia in pediatric patients
- Homozygous familial hypercholesterolemia
- Hypertriglyceridemia (Fredrickson Type IV)
- Primary dysbetalipoproteinemia (Fredrickson Type III)
- Combined hyperlipidemia

Cardiovascular Disease

- Primary prevention of heart attack, stroke, and need for revascularization procedures in patients who have risk factors such as age, smoking, high blood pressure, low HDL-C, and a family history of early heart disease, but have not yet developed clinically evident coronary heart disease.

- Secondary prevention of myocardial infarction, stroke, unstable angina, and revascularization in people with established coronary heart disease.
- Myocardial infarction and stroke prophylaxis in patients with type II diabetes

Other Uses

There have been recent studies suggesting that high-dose statin therapy plays a plaque-stabilizing role in patients suffering from acute coronary syndrome and thrombotic stroke.

Administration

Atorvastatin may be used in combination with bile acid sequestrants and ezetimibe to increase the reduction in cholesterol levels. However, It is not recommended to combine statin drug treatment with certain other cholesterol-lowering drugs, particularly fibrates, because this may increase the risk of myopathy-related adverse effects.

While many statin medications should be administered at bedtime for optimal effect, atorvastatin can be dosed at any time of day, as long as it is continually dosed once daily at the same time.

Specific Populations

- Geriatric: Plasma concentrations of atorvastatin in healthy elderly subjects are higher than those in young adults, and clinical data suggests a greater degree of LDL-lowering at any dose for patients in the population as compared to young adults.
- Pediatric: Pharmacokinetic data is not available for this population.
- Gender: Plasma concentrations are generally higher in women than in men, but there is no clinically significant difference in the extent of LDL reduction between men and women.
- Renal impairment: Renal disease has no influence on plasma concentrations of atorvastatin and dosing need not be adjusted in these patients.
- Hemodialysis: Hemodialysis will not significantly alter drug levels or change clinical effect of atorvastatin.
- Hepatic impairment: In patients with chronic alcoholic liver disease, levels of atorvastatin may be significantly increased depending upon the extent of liver disease.

Contraindications

- Active liver disease: cholestasis, hepatic encephalopathy, hepatitis, and jaundice
- Unexplained elevations in AST or ALT levels

- Pregnancy: Atorvastatin may cause fetal harm by affecting serum cholesterol and triglyceride levels, which are essential for fetal development.
- Breastfeeding: Small amounts of other statin drugs have been found to pass into breast milk, although atorvastatin has not been studied, specifically.
- Markedly elevated CPK levels or if a myopathy is suspected or diagnosed after dosing of atorvastatin has begun. Very rarely, atorvastatin may cause rhabdomyolysis, and it may be very serious leading to acute renal failure due to myoglobinuria. If rhabdomyolysis is suspected or diagnosed, atorvastatin therapy should be discontinued immediately. The likelihood of developing a myopathy is increased by the co-administration of cyclosporine, fibric acid derivatives, erythromycin, niacin, and azole antifungals.

Adverse Effects

Major

- Diabetes mellitus type 2, an uncommon class effect of all statins.
- Myopathy with elevation of creatinine kinase (CK) and rhabdomyolysis are the most serious side effects, occurring rarely at a rate of 2.3 to 9.1 per 10,000 person-years among patients taking atorvastatin. As mentioned previously, atorvastatin should be discontinued immediately if this occurs.
- Persistent liver enzyme abnormalities occurred in 0.7% of patients who received atorvastatin in clinical trials. It is recommended that hepatic function be assessed with laboratory tests before beginning atorvastatin treatment and repeated as clinically indicated thereafter. If evidence of serious liver injury occurs while a patient is taking atorvastatin, it should be discontinued and not restarted until the etiology of the patient's liver dysfunction is defined. If no other cause is found, atorvastatin should be discontinued permanently.

Common

The following have been shown to occur in 1–10% of patients taking atorvastatin in clinical trials.

- Arthralgia
- Diarrhea
- Dyspepsia
- Myalgia
- Nausea

High-dose atorvastatin have also been associated with worsening glycemic control.

Other

In 2014 the FDA reported memory loss, forgetfulness and confusion with all statin products including atorvastatin. The symptoms were not serious, and they were rare and reversible on cessation of drug treatment.

Interactions

Interactions with clofibrate, fenofibrate, gemfibrozil, which are fibrates used in accessory therapy in many forms of hypercholesterolemia, usually in combination with statins, increase the risk of myopathy and rhabdomyolysis.

Co-administration of atorvastatin with one of CYP3A4 inhibitors such as itraconazole, telithromycin, and voriconazole, may increase serum concentrations of atorvastatin, which may lead to adverse reactions. This is less likely to happen with other CYP3A4 inhibitors such as diltiazem, erythromycin, fluconazole, ketoconazole, clarithromycin, cyclosporine, protease inhibitors, or verapamil, and only rarely with other CYP3A4 inhibitors, such as amiodarone and aprepitant. Often, bosentan, fosphenytoin, and phenytoin, which are CYP3A4 inducers, can decrease the plasma concentrations of atorvastatin. Only rarely, though, barbiturates, carbamazepine, efavirenz, nevirapine, oxcarbazepine, rifampin, and rifamycin, which are also CYP3A4 inducers, can decrease the plasma concentrations of atorvastatin. Oral contraceptives increased AUC values for norethisterone and ethinyl estradiol; these increases should be considered when selecting an oral contraceptive for a woman taking atorvastatin.

Antacids can rarely decrease the plasma concentrations of statin drugs, but do not affect the LDL-C-lowering efficacy.

Niacin also is proved to increase the risk of myopathy or rhabdomyolysis.

Statins may also alter the concentrations of other drugs, such as warfarin or digoxin, leading to alterations in effect or a requirement for clinical monitoring.

Vitamin D supplementation lowers atorvastatin and active metabolite concentrations, yet synergistically reduces LDL and total cholesterol concentrations. Grapefruit juice components are known inhibitors of intestinal CYP3A4.

Co-administration of grapefruit juice with atorvastatin may cause an increase in C_{max} and AUC, which can lead to adverse reactions or overdose toxicity.

A few cases of myopathy have been reported when atorvastatin is given with colchicine.

Mechanism of Action

As with other statins, atorvastatin is a competitive inhibitor of HMG-CoA reductase. Unlike most others, however, it is a completely synthetic compound. HMG-CoA reductase catalyzes the reduction of 3-hydroxy-3-methylglutaryl-coenzyme A (HMG-CoA) to mevalonate, which is the rate-limiting step in hepatic cholesterol biosynthesis. Inhibition of the enzyme decreases *de novo* cholesterol synthesis, increasing expression of low-density lipoprotein receptors (LDL receptors) on hepatocytes. This increases LDL uptake by the hepatocytes, decreasing the amount of LDL-cholesterol in the blood. Like other statins, atorvastatin also reduces blood levels of triglycerides and slightly increases levels of HDL-cholesterol.

Recent studies have shown that in patients suffering from acute coronary syndrome, high-dose statin treatment may play a plaque-stabilizing role. At high doses, statins have anti-inflammatory effects, incite reduction of the necrotic plaque core, and improve endothelial function, leading to plaque stabilization and, sometimes, plaque regression. However, there is an increased risk of statin-associated adverse effects with such high-dose statin treatment. There is a similar thought process and risks associated with using high-dose statins to prevent recurrence of thrombotic stroke.

Pharmacodynamics

The liver is the primary site of action of atorvastatin, as this is the principal site of both cholesterol synthesis and LDL clearance. It is the dosage of atorvastatin, rather than systemic drug concentration, which correlates with extent of LDL-C reduction.

Pharmacokinetics

Absorption

Atorvastatin undergoes rapid absorption when taken orally, with an approximate time to maximum plasma concentration (T_{max}) of 1–2 h. The absolute bioavailability of the drug is about 14%, but the systemic availability for HMG-CoA reductase activity is approximately 30%. Atorvastatin undergoes high intestinal clearance and first-pass metabolism, which is the main cause for the low systemic availability. Administration of atorvastatin with food produces a 25% reduction in C_{max} (rate of absorption) and a 9% reduction in AUC (extent of absorption), although food does not affect the plasma LDL-C-lowering efficacy of atorvastatin. Evening dose administration is known to reduce the C_{max} and AUC by 30% each. However, time of administration does not affect the plasma LDL-C-lowering efficacy of atorvastatin.

Distribution

The mean volume of distribution of atorvastatin is approximately 381 L. It is highly protein bound (≥98%), and studies have shown it is likely secreted into human breastmilk.

Metabolism

Atorvastatin metabolism is primarily through cytochrome P450 3A4 hydroxylation to form active ortho- and parahydroxylated metabolites, as well as various beta-oxidation metabolites. The ortho- and parahydroxylated metabolites are responsible for 70% of systemic HMG-CoA reductase activity. The ortho-hydroxy metabolite undergoes further metabolism via glucuronidation. As a substrate for the CYP3A4 isozyme, it has shown susceptibility to inhibitors and inducers of CYP3A4 to produce increased or decreased plasma concentrations, respectively. This interaction was tested *in vitro* with concurrent administration of erythromycin, a known CYP3A4 isozyme inhibitor, which resulted in increased plasma concentrations of atorvastatin. It is also an inhibitor of cytochrome 3A4.

Excretion

Atorvastatin is primarily eliminated via hepatic biliary excretion, with less than 2% recovered in the urine. Bile elimination follows hepatic and/or extrahepatic metabolism. There does not appear to be any entero-hepatic recirculation. Atorvastatin has an approximate elimination half-life of 14 h. Noteworthy, the HMG-CoA reductase inhibitory activity appears to have a half-life of 20–30 h, which is thought to be due to the active metabolites. Atorvastatin is also a substrate of the intestinal P-glycoprotein efflux transporter, which pumps the drug back into the intestinal lumen during drug absorption.

In hepatic insufficiency, plasma drug concentrations are significantly affected by concurrent liver disease. Patients with A-stage liver disease show a four-fold increase in both C_{max} and AUC. Patients with B-stage liver disease show a 16-fold increase in C_{max} and an 11-fold increase in AUC.

Geriatric patients (>65 years old) exhibit altered pharmacokinetics of atorvastatin compared to young adults, with mean AUC and C_{max} values that are 40% and 30% higher, respectively. Additionally, healthy elderly patients show a greater pharmacodynamic response to atorvastatin at any dose; therefore, this population may have lower effective doses.

Pharmacogenetics

Several genetic polymorphisms have been found to be associated with a higher incidence of undesirable side effects of atorvastatin. This phenomenon is suspected to be related to increased plasma levels of pharmacologically active metabolites, such as atorvastatin lactone and p-hydroxy-atorvastatin. Atorvastatin and its active metabolites may be monitored in potentially susceptible patients using specific chromatographic techniques.

Chemical Synthesis

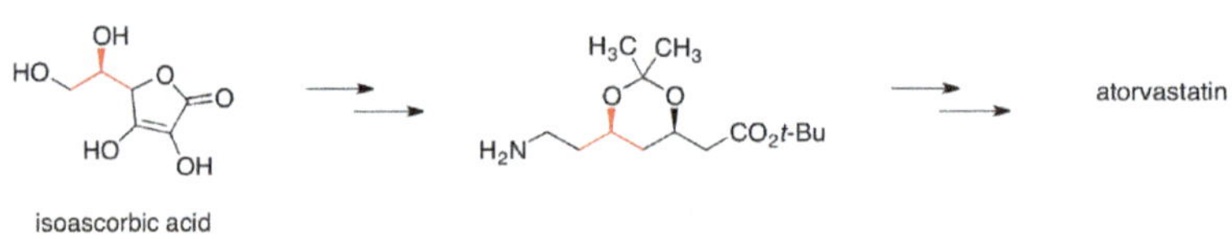

Atorvastatin synthesis in commercial production (process) chemistry. The key step of establishing this drug's stereocenters, through initial use of an inexpensive natural product (chiral pool approach).

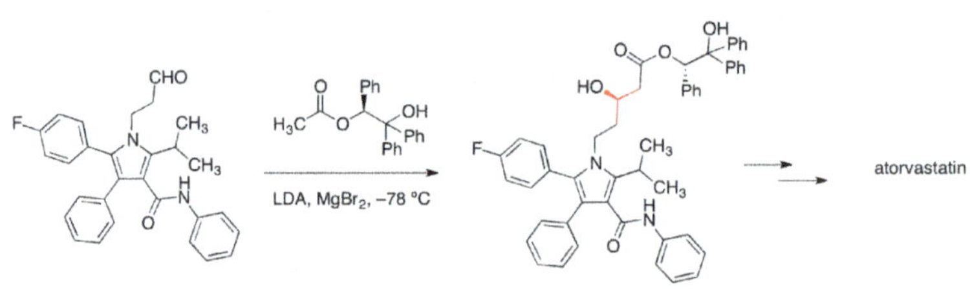

Atorvastatin synthesis during discovery chemistry. The key step of establishing stereocenters, using of a chiral ester auxiliary approach.

The first synthesis of atorvastatin at Parke-Davis that occurred during drug discovery was racemic followed by chiral chromatographic separation of the enantiomers. An early enantioselective route to atorvastatin made use of an ester chiral auxiliary to set the stereochemistry of the first of the two alcohol functional groups via a diastereoselective aldol reaction. Once the compound entered pre-clinical development, process chemistry developed a cost-effective and scalable synthesis. In atorvastatin's case, a key element of the overall synthesis was ensuring stereochemical purity in the final drug substance, and hence establishing the first stereocenter became a key aspect of the overall design. The final commercial production of atorvastatin relied on a chiral pool approach, where the stereochemistry of the first alcohol functional group was carried into the synthesis—through the choice of isoascorbic acid, an inexpensive and easily sourced plant-derived natural product.

History

Bruce Roth, who was hired by Warner-Lambert as a young chemist in 1982, had created an "experimental compound" codenamed CI 981 – later called atorvastatin – and moved it "from synthesis in 1985 into expensive clinical trials" when company executives began to balk at the cost. Roth's boss Roger Newton and Ronald Cresswell supported the young chemist's costly research but Warner-Lambert was concerned that atorvastatin was a me-too version of rival Merck & Co.'s orphan drug lovastatin (brand name Mevacor). Mevacor, which was first marketed in 1987, was the industry's first statin and Merck's synthetic version – Zocor – was in the advanced stages of development.

In 1994 the findings of a Merck-funded study were published in *The Lancet* concluding the efficacy of statins in lowering cholesterol in 4444 Scandinavian patients proving for the first time not only that a "statin reduced "bad" LDL cholesterol but also that it led to a sharp drop in fatal heart attacks among patients with heart disease." In 1996 Warner-Lambert entered into a co-marketing agreement with Pfizer to sell Lipitor and in 2000 Pfizer acquired Warner-Lambert 2000 for $90.2 billion. Lipitor was on the market by 1996. By 2003 Lipitor had become the best selling pharmaceutical in the United States. From 1996 to 2012 under the trade name Lipitor, atorvastatin became the world's best-selling drug of all time, with more than $125 billion in sales over approximately 14.5 years. Lipitor alone "provided up to a quarter of Pfizer Inc.'s annual revenue for years."

Pfizer's patent on atorvastatin expired in November 2011.

Formulations

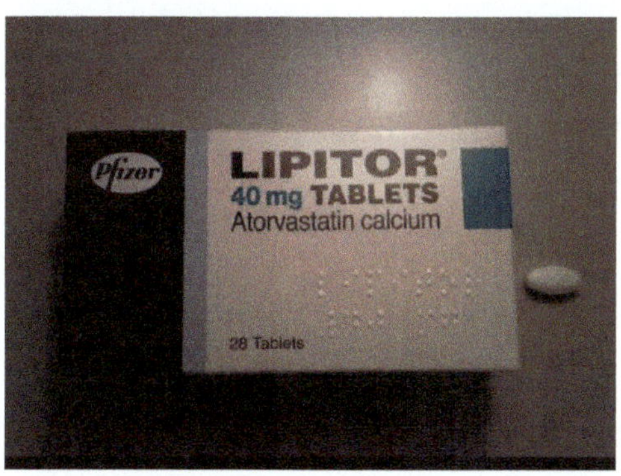

Pack and tablet of atorvastatin (Lipitor) 40mg

Atorvastatin calcium tablets are marketed by Pfizer under the trade name Lipitor for oral administration. Tablets are white, elliptical, and film-coated. Pfizer also packages the drug in combination with other drugs, such as with Caduet. Pfizer recommends that patients do not break tablets in half to take half-doses, even when this is recommended by their doctors.

Generic Availability

Pfizer's U.S. patent on Lipitor expired on 30 November 2011. Initially, generic atorvastatin was manufactured only by Watson Pharmaceuticals and India's Ranbaxy Laboratories. Prices for the generic version did not drop to the level of other generics—$10 or less for a month's supply—until other manufacturers began to supply the drug in May 2012.

In other countries, atorvastatin calcium is made in tablet form by generic drug makers under various brand names including Stator, Atorvastatin Teva, Litorva, Torid, Atoris, Atorlip, Mactor, Lipvas, Sortis, Torvast, Torvacard, Totalip, and Tulip. Pfizer also makes its own generic version under the name Zarator, which is the sole Pharmac-subsidised brand of atorvastatin in New Zealand.

Drug Recalls

On 9 November 2012, Indian drugmaker Ranbaxy Laboratories Ltd. voluntarily recalled 10-, 20- and 40-mg doses of its generic version of atorvastatin in the United States. The lots of atorvastatin, packaged in bottles of 90 and 500 tablets, were recalled due to possible contamination with very small glass particles similar to the size of a grain of sand (less than 1 mm in size). The FDA received no reports of injury from the contamination. Ranbaxy also issued recalls of bottles of 10-milligram tablets in August 2012 and March 2014, due to concerns that the bottles might contain larger, 20-milligram tablets and thus cause potential dosing errors.

ETFE

Ethylene tetrafluoroethylene (ETFE) a fluorine-based plastic. It was designed to have high corrosion resistance and strength over a wide temperature range. ETFE is a polymer and its source-based name is poly(ethene-co-tetrafluoroethene). ETFE has a relatively high melting temperature, excellent chemical, electrical and high-energy radiation resistance properties. When burned, ETFE releases hydrofluoric acid.

Chemical, Mechanical, Physical, and Thermal Properties

Useful comparison tables of PTFE against FEP, PFA and ETFE can be found on DuPont's website, listing the mechanical, thermal, chemical, electrical and vapour properties of each, side by side.

ETFE is effectively the high-strength version of the other three in this group, often featuring slightly diminished capacities in other fields by comparison.

Combustion of ETFE occurs in the same way as a number of other fluoropolymers, in terms of releasing hydrofluoric acid (HF). HF is extremely corrosive and toxic, and so appropriate caution must be exercised.

ETFE film is self-cleaning (due to its nonstick surface) and recyclable. It is prone to punctures by sharp edges and therefore mostly used for roofs. In sheet form as commonly employed for architecture, it is able to stretch to three times its length without loss of elasticity. Employing heat welding, tears can be repaired with a patch or multiple sheets assembled into larger panels.

ETFE has an approximate tensile strength of 42 N/mm² (6100 psi), with a working temperature range of 89 K to 423 K (−185 °C to +150 °C or −300 °F to +300 °F).

ETFE resins are resistant to ultraviolet light. An accelerated weathering test (comparable to 30 years' exposure) produced almost no signs of film deterioration.

Applications

The Eden Project, Cornwall, UK. Biomes are constructed with ETFE cushions.

ETFE roof at Manchester Piccadilly station, Manchester, UK

An example of its use is as pneumatic panels to cover the outside of the football stadium Allianz Arena or the Beijing National Aquatics Centre (a.k.a. the Water Cube of the 2008 Olympics) – the world's largest structure made of ETFE film (laminate). The panels of the Eden Project are also made from ETFE, and the Tropical Islands have a 20,000 m² window made from this translucent material.

Another key use of ETFE is for the covering of electrical and fiber-optic wiring used in high-stress, low-fume-toxicity and high-reliability situations. Aircraft and spacecraft wiring are primary examples. Some small cross-section wires like the wire used for the wire-wrap technique are coated with ETFE.

As a dual laminate, ETFE can be bonded with FRP as a thermoplastic liner and used in pipes, tanks, and vessels for additional corrosion protection.

ETFE is commonly used in the nuclear industry for tie or cable wraps and in the aviation and aerospace industries for wire coatings. This is because ETFE has better mechanical toughness than PTFE. In addition, ETFE exhibits a high-energy radiation resistance and can withstand moderately high temperatures for a long period of time. Commercially deployed brand names of ETFE include *Tefzel* by DuPont, *Fluon* by Asahi Glass Company, *Neoflon ETFE* by Daikin, and *Texlon* by Vector Foiltec.

Due to its high temperature resistance ETFE is also used in film mode as a mold-release film. ETFE film such as PMP film by Honeywell is used in aerospace applications such as carbon fiber pre-preg curing as a release film for molds or hot high-pressure plates.

ETFE cushions roof with integrated photovoltaic cells. Munich's municipal waste management department

Notable Buildings and Designs Using ETFE as a Significant Architectural Element

- Allianz Arena, Munich, Germany
- Beijing National Aquatics Centre, (the Water Cube) Beijing, China
- Eden Project, Cornwall, United Kingdom
- Khan Shatyr Entertainment Center, Astana, Kazakhstan

National Space Centre, Leicester UK

- U.S. Bank Stadium, Minneapolis, Minnesota
- National Space Centre, Leicester, United Kingdom
- Cuauhtémoc Stadium, Puebla, México.
- Midland Metropolitan Hospital, Smethwick, Birmingham, United Kingdom

Other Notable Constructions Using ETFE

- Avenues Phase-III, Al-Rai, Kuwait
- Dworzec Tramwajowy Centrum, tram station in Łódź, Poland.

- Solaris, Clamart, France
- Discovery College, Lantau Island, Hong Kong
- Green 18, Hong Kong Science Park, Hong Kong
- Pavilion, Alnwick Castle, Alnwick, United Kingdom
- BC Place, Vancouver, Canada

Detail of Beijing National Aquatics Centre showing ETFE exterior cushions

- River Culture Pavillon The ARC, Daegu, South Korea
- Munich's municipal waste management department, Munich, Germany
- Beijing National Stadium, Beijing, China
- FestiveWalk, Resorts World at Sentosa, Singapore
- Dolce Vita Tejo Shopping Centre, Amadora, Lisbon, Portugal
- roof, dedicated underground rail station at the Heathrow Airport Terminal 5, London, United Kingdom
- Manchester Victoria station concourse, Manchester, United Kingdom
- Forsyth Barr Stadium at University Plaza, Dunedin, New Zealand
- Islazul Shopping Centre, Madrid, Spain
- Kansas City Power & Light District, Kansas City, Missouri, United States
- South Campus skylight structures, Art Center College of Design, Pasadena, California, United States
- Tanaka Business School, London, United Kingdom
- Tropical Islands, Brandenburg, Germany
- Barnsley Interchange, Barnsley, United Kingdom
- The Mall Athens, Athens, Greece
- Newport railway station, Newport, United Kingdom

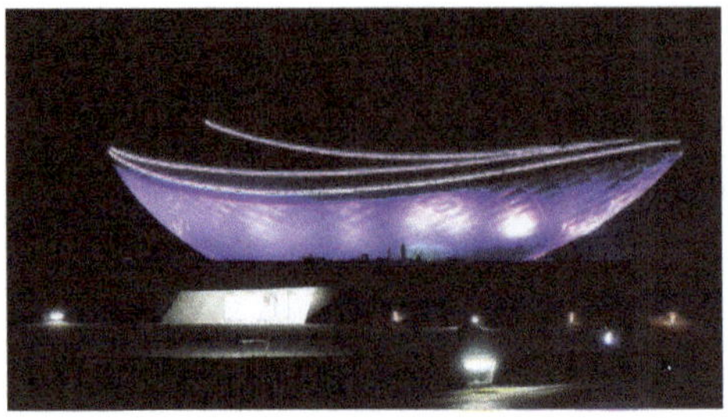

ETFE Facade with integrated LED Lights

- The Elements, Livingston, United Kingdom
- Experimental Media and Performing Arts Center, Rensselaer Polytechnic Institute, Troy, New York, United States
- Arena Pernambuco, Recife, Brazil
- Sandton City, Sandton, South Africa
- Key West Shopping Centre, Krugersdorp, South Africa
- Oceanus Casino, Macau, Special Administrative Region of China.
- Masdar city, Abu Dhabi, United Arab Emirates
- ISS Building Lancaster University
- Empire City Casino, Yonkers, New York
- The SSE Hydro, Glasgow, Scotland
- Anaheim Regional Transportation Intermodal Center, California (12-13-14)
- National Stadium, Singapore
- Orto Botanico di Padova Biodiversity Garden roof, Padua, Italy
- Guangzhou South Railway Station, China
- Yujiapu Railway Station, China

Under Construction

- Avenues Phase IV & IVB, Al-Rai, Kuwait
- Jungle Exhibit, Sedgwick County Zoo, Wichita, Kansas (2015)
- Dockside Pavilion, Sydney Australia 2014
- Baku Olympic Stadium, Baku, Azerbaijan (2015)
- Australian Embassy, Jakarta, 2014
- Wharf Retail, Bluewaters Island, DUBAI (2016)

Inorganic Compound

An inorganic compound is a compound that is not organic. The term is not well defined, but in its simplest definition refers simply to compounds that do not contain carbon, and not consisting of or deriving from living matter. Inorganic compounds are traditionally viewed as being synthesized by the agency of geological systems. In contrast, organic compounds are found in biological systems. The distinction between inorganic and organic compounds is not always clear. Organic chemists traditionally refer to any molecule containing carbon as an organic compound and by default this means that inorganic chemistry deals with molecules lacking carbon. As many minerals are of biological origin, biologists may distinguish organic from inorganic compounds in a different way that does not hinge on the presence of a carbon atom. Pools of organic matter, for example, that have been metabolically incorporated into living tissues persist in decomposing tissues, but as molecules become oxidized into the open environment, such as atmospheric CO_2, this creates a separate pool of inorganic compounds. The International Union of Pure and Applied Chemistry, an agency widely recognized for defining chemical terms, does not offer definitions of inorganic or organic. Hence, the definition for an inorganic versus an organic compound in a multidisciplinary context spans the division between organic life living (or animate) and inorganic non-living (or inanimate) matter.

Traditional Usage

The Wöhler synthesis is the conversion of ammonium cyanate into urea. This chemical reaction was discovered in 1828 by Friedrich Wöhler and is considered the starting point of modern organic chemistry.

The Wöhler synthesis is of great historical significance because for the first time an organic compound was produced from inorganic reactants. This finding went against the mainstream theory of that time called vitalism, which stated that organic matter possessed a special force or *vital force* inherent to all things living. For this reason a sharp boundary existed between organic and inorganic compounds. Urea was discovered in 1799 and could until 1828 only be obtained from biological sources such as urine. Wöhler reported to his mentor Berzelius:

"I cannot, so to say, hold my chemical water and must tell you that I can make urea without thereby needing to have kidneys, or anyhow, an animal, be it human or dog".

Modern Usage

Inorganic compounds can be defined as any compound that is not organic compound. Some simple compounds which contain carbon are usually considered inorganic. These include carbon monoxide, carbon dioxide, carbonates, cyanides, cyanates, carbides, and thiocyanates. In contrast, methane and formic acid are generally considered to be simple examples of organic compounds, although the Inorganic Crystal Structure Database (ICSD), in its definition of "inorganic" carbon compounds, states that such compounds may contain *either* C-H or C-C bonds, but not both.

Coordination Chemistry

A large class of compounds discussed in inorganic chemistry textbooks are coordination compounds.

Examples range from substances that are strictly inorganic, such as $[Co(NH_3)_6]Cl_3$, to organometallic compounds, such as $Fe(C_5H_5)_2$, and extending to bioinorganic compounds, such as the hydrogenase enzymes.

Mineralogy

Minerals are mainly oxides and sulfides, which are strictly inorganic, although they may be of biological origin. In fact, most of the Earth is inorganic. Although the components of Earth's crust are well-elucidated, the processes of mineralization and the composition of the deep mantle remain active areas of investigation, which are covered mainly in geology-oriented venues.

Lithium (Medication)

Lithium compounds, also known as lithium salts, are primarily used as a psychiatric medication. This includes in the treatment of major depressive disorder that does not improve following the use of other antidepressants and bipolar disorder. In these disorders, it reduces the risk of suicide. Lithium is taken by mouth.

Common side effects include increased urination, shakiness of the hands, and increased thirst. Serious side effects include hypothyroidism, diabetes insipidus, and lithium toxicity. Blood level monitoring is recommended to decrease the risk of potential toxicity. If levels become too high, diarrhea, vomiting, poor coordination, sleepiness, and ringing in the ears may occur. If used during pregnancy, lithium can cause problems for the baby. It appears to be safe to use while breastfeeding. Lithium salts are classified as mood stabilizers. How they work is not specifically known.

In the 1800s, lithium was used in people who had gout, epilepsy, and cancer. Its use in the treatment of mental illness began in 1948 by John Cade in Australia. It is on the World Health Organization's List of Essential Medicines, the most important medications needed in a basic health system. It is available as a generic medication. The wholesale cost in the developing world in 2014 was between 0.12 and 0.20 USD per day. In the United States at usual doses it costs about 0.90 to 1.20 USD per day.

Medical Uses

Lithium is used primarily for bipolar disorder. It is sometimes used when other treatments are not effective in a number of other conditions, including major depression, schizophrenia, and some psychiatric disorders in children. In mood disorders, of which bipolar is one, it decreases the risk of suicide. This benefit is not seen with other medications.

Bipolar

Lithium carbonate treatment was previously considered to be unsuitable for children; however, more recent studies show its effectiveness for treatment of early-onset bipolar disorder in children as young as eight. The required dosage is slightly less than the toxic level, requiring close monitoring of blood levels of lithium carbonate during treatment. A limited amount of evidence suggests lithium carbonate may contribute to treatment of substance abuse for some people with bipolar disorder.

Suicide

High levels of naturally occurring lithium in drinking water have been associated with lower rates of suicide.

Monitoring

Those who use lithium should receive regular serum level tests and should monitor thyroid and kidney function for abnormalities, as it interferes with the regulation of sodium and water levels in the body, and can cause dehydration. Dehydration, which is compounded by heat, can result in increasing lithium levels. The dehydration is due to lithium inhibition of the action of antidiuretic hormone, which normally enables the kidney to reabsorb water from urine. This causes an inability to concentrate urine, leading to consequent loss of body water and thirst.

Lithium concentrations in whole blood, plasma, serum or urine may be measured using instrumental techniques as a guide to therapy, to confirm the diagnosis in potential poisoning victims or to assist in the forensic investigation in a case of fatal overdosage. Serum lithium concentrations are usually in the 0.5–1.3 mmol/l range in well-controlled people, but may increase to 1.8–2.5 mmol/l in those who accumulate the drug over time and to 3–10 mmol/l in acute overdose.

Lithium salts have a narrow therapeutic/toxic ratio, so should not be prescribed unless facilities for monitoring plasma concentrations are available. Doses are adjusted to achieve plasma concentrations of 0.4 to 1.2 mmol Li+/l (lower end of the range for maintenance therapy and the elderly, higher end for children) on samples taken 12 hours after the preceding dose.

Side Effects

Sources for the Following Lists.

Very Common (>10% incidence) adverse effects of lithium include

- Confusion
- Constipation (usually transient, but can persist in some)
- Decreased memory
- Diarrhea (usually transient, but can persist in some)
- Dry mouth
- EKG changes — usually benign changes in T waves.
- Hand tremor (usually transient, but can persist in some)
- Headache
- Hyperreflexia — overresponsive reflexes.

- Leukocytosis — elevated white blood cell count
- Muscle weakness (usually transient, but can persist in some)
- Myoclonus — muscle twitching.
- Nausea (usually transient, but can persist in some)
- Polydypsia — increased thirst.
- Polyuria — increased urination.
- Renal (kidney) toxicity which may lead to chronic kidney failure
- Vomiting (usually transient, but can persist in some)
- Vertigo
- Weight gain

Common (1-10%) adverse effects include

- Acne
- Extrapyramidal side effects — movement-related problems such as muscle rigidity, parkinsonism, dystonia, etc.
- Euthyroid goitre — i.e. the formation of a goitre despite normal thyroid functioning.
- Hypothyroidism — a deficiency of thyroid hormone.
- Hair loss/hair thinning

Rare/Uncommon (<1%) adverse effects include

- Brugada syndrome — a potentially fatal abnormality in the electrical activity of the heart.
- Coma
- Erythema multiforme — a potentially fatal skin reaction
- Hallucinations
- Hypercalcaemia — elevated blood levels of calcium.
- Hypermagnesaemia — elevated blood levels of magnesium.
- Hyperparathyroidism — elevated blood levels of parathyroid hormone.
- Hyperthyroidism — elevated blood concentrations of thyroid hormones.
- Increased intracranial pressure and papilledema
- Myasthenia gravis — an autoimmune condition where the body's own defences attack the

neuromuscular junction — the gap across which the nerves communicate with the muscles — leading to muscle weakness.

- Oedema
- Pseudotumor cerebri
- Renal interstitial fibrosis
- Seizure
- Sinus node dysfunction
- Transient reduction in peripheral circulation as a whole

Unknown frequency adverse effects include

- Abdominal pain
- Albuminuria — protein in the urine, a sign of impaired kidney function.
- Bradycardia — low heart rate.
- Changes in taste
- Decreased creatinine clearance — another sign of impaired kidney function.
- Flatulence
- Gastritis
- Glycosuria — glucose (blood sugar) in the urine.
- Hyperthermia
- Hypotension — low blood pressure.
- Indigestion
- Nystagmus — involuntary eye movements that can interfere with vision.
- Oliguria — low urine output, although excess urine output is more likely.
- Sexual dysfunction including impotence, decreased libido, vaginal dryness, erectile dysfunction, etc.
- Slurred speech
- Somnolence
- Weight loss (gain is more common with prolonged treatment).

Lithium is known to be responsible for 1–2 kg of weight gain. Weight gain may be a source of low self-esteem for the clinically depressed.

In addition to tremors, lithium treatment appears to be a risk factor for development of parkinsonism symptoms, although the causal mechanism remains unknown.

Most side effects of lithium are dose-dependent. The lowest effective dose is used to limit the risk of side effects.

Hypothyroidism

Most patients treated with lithium carbonate show elevated thyroid stimulating hormone levels in response to injections of thyrotropin-releasing hormone. According to an Australian study, "The incidence of hypothyroidism is six-fold higher in people on lithium as compared to the general population. Hypothyroidism in turn increases the likelihood of developing clinical depression."

Because lithium competes with the receptors for the antidiuretic hormone in the kidney, it increases water output into the urine, a condition called nephrogenic diabetes insipidus. Clearance of lithium by the kidneys is usually successful with certain diuretic medications, including amiloride and triamterene. It increases the appetite and thirst ("polydypsia") and reduces the activity of thyroid hormone (hypothyroidism). The latter can be corrected by treatment with thyroxine and does not require the lithium dose to be adjusted. Lithium is also believed to permanently affect renal function, although this does not appear to be common.

Pregnancy and Breast Feeding

Lithium is a teratogen, causing birth defects in a small number of newborn babies. Case reports and several retrospective studies have demonstrated possible increases in the rate of a congenital heart defect known as Ebstein's anomaly, if taken during a woman's pregnancy. As a consequence, fetal echocardiography is routinely performed in pregnant women taking lithium to exclude the possibility of cardiac anomalies. Lamotrigine seems to be a possible alternative to lithium in pregnant women. Gabapentin and clonazepam are also indicated as antipanic medications during the childbearing years and during pregnancy. Valproic acid and carbamazepine also tend to be associated with teratogenicity.

While it appears to be safe to use while breastfeeding a number of guidelines list it as a contraindication including the British National Formulary.

Dehydration

Dehydration in people taking lithium salts can be very hazardous, especially when combined with lithium-induced nephrogenic diabetes insipidus with polyuria. Such situations include preoperative fluid restrictions, warm weather conditions, sporting events, hiking, or other periods of fluid inaccessibility. Dehydration can result in increased plasma lithium levels due to decreased glomerular filtration rate, which causes lithium retention. Further, in the case of diabetes insipidus, free water is lost in greater proportion to sodium and other electrolytes, artificially raising lithium's concentration in the blood.

Another danger is that if the period of dehydration and diuresis has been prolonged, total body stores of sodium may actually be depleted, despite elevated plasma levels. Thus, rapid hydration with a large volume of plain water may very quickly produce hyponatremia, as total stores of sodium may be insufficient to support normal concentrations at a normal blood volume. Rapid overcorrection of hypernatremia also increases the risk of developing cerebral edema. Hyponatremia can also promote lithium retention by increasing reabsorption in the distal nephron, thus increasing lithium levels.

Kidney Damage

Lithium has been associated with several forms of kidney injury. It is estimated that impaired urinary concentrating ability is present in at least 50% of individuals on chronic lithium therapy. Continued use of lithium can lead to more serious kidney damage, a common type of which is lithium-induced diabetes insipidus. Evidence for chronic kidney failure was found in about one third of people undergoing long-term lithium treatment, according to one study. Some forms of lithium-caused kidney damage may be progressive and lead to end-stage kidney failure.

Interactions

Lithium concentrations are known to be increased with concurrent use of diuretics — especially loop diuretics (such as furosemide) and thiazides — and non-steroidal anti-inflammatory drugs (NSAIDs) such as ibuprofen. Lithium concentrations can also be increased with concurrent use of ACE inhibitors such as captopril, enalapril, and lisinopril.

Lithium is mainly removed from the body through glomerular function, but some is then reabsorbed together with sodium through the proximal tubule. Its levels are therefore sensitive to water and electrolyte balance. Diuretics act by lowering water and sodium levels; this causes more reabsorption of lithium in the proximal tubules so that the removal of lithium from the body is less, leading to increased levels of lithium. ACE inhibitors have also been shown in a retrospective case-control study to increase lithium concentrations. This is likely due to constriction of the afferent arteriole of the glomerulus, resulting in decreased glomerular filtration rate and clearance. Another possible mechanism is that ACE inhibitors can lead to a decrease in sodium and water. This will increase lithium reabsorption and its concentrations in the body.

There are also drugs that can increase the clearance of lithium from the body, which can result in decreased lithium levels in the blood. These drugs include theophylline, caffeine, and acetazolamide. Additionally, increasing dietary sodium intake may also reduce lithium levels by prompting the kidneys to excrete more lithium.

Lithium is also known to be a potential precipitant of serotonin syndrome in people concurrently on serotonergic medications such as antidepressants, buspirone and certain opioids such as pethidine (meperidine), tramadol, oxycodone, fentanyl and others. Lithium co-treatment is also a risk factor for neuroleptic malignant syndrome in people on antipsychotics and other antidopaminergic medications.

High doses of haloperidol, fluphenazine, or flupenthixol may be hazardous when used with lithium; irreversible toxic encephalopathy has been reported. Indeed these and other antipsychotics have been associated with increased risk of lithium neurotoxicity, even with low therapeutic lithium doses.

Overdose

Lithium toxicity may occur on an acute basis, in persons taking excessive amounts either accidentally or intentionally, or on a chronic basis, in people who accumulate high levels during ongoing therapy. The manifestations include nausea, emesis, diarrhea, asthenia, ataxia, confusion, lethargy, polyuria, seizures and coma. Other toxic effects of lithium include coarse tremor, muscle twitching, convulsions and renal failure. People who survive a poisoning episode may develop persistent neurotoxicity. Several authors have described a "Syndrome of Irreversible Lithium-Effected Neurotoxicity" (SILENT), associated with episodes of acute lithium toxicity or long-term treatment within the appropriate dosage range. Symptoms are said to include cerebellar dysfunction.

Overdosage, usually with plasma concentrations over 1.5 mmol Li+/l, may be fatal, and toxic effects include tremor, ataxia, dysarthria, nystagmus, renal impairment, confusion, and convulsions. If these potentially hazardous signs occur, treatment should be stopped, plasma lithium concentrations redetermined, and steps taken to reverse lithium toxicity.

Lithium toxicity is compounded by sodium depletion. Concurrent use of diuretics that inhibit the uptake of sodium by the distal tubule (e.g. thiazides) is hazardous and should be avoided because this can cause increased resorption of lithium in the proximal convoluted tubule, leading to elevated, potentially toxic levels. In mild cases, withdrawal of lithium and administration of generous amounts of sodium and fluid will reverse the toxicity. Plasma concentrations in excess of 2.5 mmol Li+ /l are usually associated with serious toxicity requiring emergency treatment. When toxic concentrations are reached, there may be a delay of one or two days before maximum toxicity occurs.

In long-term use, therapeutic concentrations of lithium have been thought to cause histological and functional changes in the kidney. The significance of such changes is not clear, but is of sufficient concern to discourage long-term use of lithium unless it is definitely indicated. Doctors may change a bipolar patient's medication from lithium to another mood-stabilizing drug, such as valproate (Depakote), if problems with the kidneys arise. An important potential consequence of long-term lithium use is the development of renal diabetes insipidus (inability to concentrate urine). Patients should therefore be maintained on lithium treatment after three to five years only if, on assessment, benefit persists. Conventional and sustained-release tablets are available. Preparations vary widely in bioavailability, and a change in the formulation used requires the same precautions as initiation of treatment. There are few reasons to prefer any one simple salt of lithium; the carbonate has been the more widely used, but the citrate is also available.

Mechanism of Action

The specific biochemical mechanism of lithium action in stabilizing mood is unknown.

Upon ingestion, lithium becomes widely distributed in the central nervous system and interacts with a number of neurotransmitters and receptors, decreasing norepinephrine release and increasing serotonin synthesis.

Unlike many other psychoactive drugs, Li+typically produces no obvious psychotropic effects (such as euphoria) in normal individuals at therapeutic concentrations. Lithium may also increase the release of serotonin by neurons in the brain. In vitro studies performed on serotonergic neurons from rat raphe nuclei have shown that when these neurons are treated with lithium, serotonin release is enhanced during a depolarization compared to no lithium treatment and the same depolarization.

Several authors proposed that pAp-phosphatase could be one of the therapeutic targets of lithium. This hypothesis was supported by the low Ki of lithium for human pAp-phosphatase compatible within the range of therapeutic concentrations of lithium in the plasma of people (0.8–1 mM). Importantly, the Ki of human pAp-phosphatase is ten times lower than that of GSK3β (glycogen synthase kinase 3β). Inhibition of pAp-phosphatase by lithium leads to increased levels of pAp (3′-5′ phosphoadenosine phosphate), which was shown to inhibit PARP-1

Another mechanism proposed in 2007 is that lithium may interact with nitric oxide (NO) signalling pathway in the central nervous system, which plays a crucial role in the neural plasticity. The NO system could be involved in the antidepressant effect of lithium in the Porsolt forced swimming test in mice. It was also reported that NMDA receptor blockage augments antidepressant-like effects of lithium in the mouse forced swimming test, indicating the possible involvement of NMDA receptor/NO signaling in the action of lithium in this animal model of learned helplessness.

Lithium possesses neuroprotective properties by preventing apoptosis and increasing cell longevity.

Oxidative Metabolism

Evidence suggests that mitochondrial dysfunction is present in patients with bipolar disorder. Oxidative stress and reduced levels of anti-oxidants (such as glutathione) lead to cell death. Lithium may protect against oxidative stress by up-regulating complex I and II of the mitochondrial electron transport chain.

Dopamine and G-protein Coupling

During mania, there is an increase in neurotransmission of dopamine that causes a secondary homeostatic down-regulation, resulting in decreased neurotransmission of dopamine, which can cause depression. Additionally, the post-synaptic actions of dopamine are mediated through G-protein coupled receptors. Once dopamine is coupled to the G-protein receptors, it stimulates other secondary messenger systems that modulate neurotransmission. Studies found that in autopsies (which do not necessarily reflect living people), people with bipolar disorder had increased G-protein coupling compared to people without bipolar disorder. Lithium treatment alters the function of certain subunits of the dopamine associated G-protein, which may be part of its mechanism of action.

Glutamate and NMDA Receptors

Glutamate levels are observed to be elevated during mania. Lithium is thought to provide long-term mood stabilization and have anti-manic properties by modulating glutamate levels. It is proposed that lithium competes with magnesium for binding to NMDA glutamate receptor, increasing the availability of glutamate in post-synaptic neurons. The NMDA receptor is also affected by other neurotransmitters such as serotonin and dopamine. Effects observed appear exclusive to lithium and have not been observed by other monovalent ions such as rubidium and caesium.

GABA Receptors

GABA is an inhibitory neurotransmitter that plays an important role in regulating dopamine and glutamate neurotransmission. It was found that patients with bipolar disorder had lower GABA levels, which results in excitotoxicity and can cause apoptosis (cell loss). Lithium counteracts these degrading processes by decreasing pro-apoptotic proteins and stimulating release of neuroprotective proteins.

Cyclic AMP Secondary Messengers

The cyclic AMP secondary messenger system is shown to be modulated by lithium. Lithium was found to increase the basal levels of cyclic AMP but impair receptor coupled stimulation of cyclic AMP production. It is hypothesized that the dual effects of lithium are due the inhibition of G-proteins that then mediate cyclic AMP production. Over a long period of lithium treatment, cyclic AMP and adenylate cyclase levels are further changed by gene transcription factors.

Inositol Depletion Hypothesis

Lithium treatment has been found to inhibit the enzyme inositol monophosphatase, involved in degrading inositol monophosphate to inositol required in PIP_2 synthesis. This leads to lower levels of inositol triphosphate, created by decomposition of PIP_2. This effect has been suggested to be further enhanced with an inositol triphosphate reuptake inhibitor. Inositol disruptions have been linked to memory impairment and depression. It is known with good certainty that signals from the receptors coupled to the phosphoinositide signal transduction is effected by lithium. myo-inositol is also regulated by the high affinity sodium mI transport system (SMIT). Lithium is hypothesized to inhibit mI entering the cells and mitigating the function of SMIT. Reductions of cellular levels of myo-inositol results in the inhibition of the phosphoinositide cycle

Intracellular Calcium

In 2014, it was proposed that lithium treatment works by affecting calcium signaling by blocking excitotoxic processes such as antagonizing NMDA receptors and inhibiting inositol monophosphatase (IMPase).

History

Lithium was first used in the 19th century as a treatment for gout after scientists discovered that, at least in the laboratory, lithium could dissolve uric acid crystals isolated from the kidneys. The levels of lithium needed to dissolve urate in the body, however, were toxic. Because of prevalent theories linking excess uric acid to a range of disorders, including depressive and manic disorders, Carl Lange in Denmark and William Alexander Hammond in New York used lithium to treat mania from the 1870s onwards. By the turn of the 20th century, as theory regarding mood disorders evolved and so-called "brain gout" disappeared as a medical entity, the use of lithium in psychiatry was largely abandoned. Some suggest that the pharmaceutical industry was reluctant to invest in a drug that could not be patented, however a number of lithium preparations were yet produced for the control of renal calculi and "uric acid diathesis". As accumulating knowledge indicated a role for excess sodium intake in hypertension and heart disease, lithium salts were prescribed to patients for use as a replacement for dietary table salt (sodium chloride). This practice and the sale of lithium itself were both banned in 1949, following publication of reports detailing side effects and deaths.

Also in 1949, the Australian psychiatrist John Cade rediscovered the usefulness of lithium salts in treating mania. Cade was injecting rodents with urine extracts taken from schizophrenic patients in an attempt to isolate a metabolic compound which might be causing mental symptoms. Since uric acid in gout was known to be psychoactive, (adenosine receptors on neurons are stimulated by it; caffeine blocks them), Cade needed soluble urate for a control. He used lithium urate, already known to be the most soluble urate compound, and observed that it caused the rodents to become tranquil. Cade traced the effect to the lithium ion itself, and so proposed lithium salts as tranquilizers. He soon succeeded in controlling mania in chronically hospitalized patients with them. This was one of the first successful applications of a drug to treat mental illness, and it opened the door for the development of medicines for other mental problems in the next decades.

The rest of the world was slow to adopt this treatment, largely because of deaths which resulted from even relatively minor overdosing, including those reported from use of lithium chloride as a substitute for table salt. Largely through the research and other efforts of Denmark's Mogens Schou and Paul Baastrup in Europe, and Samuel Gershon and Baron Shopsin in the U.S., this resistance was slowly overcome. The application of lithium in manic illness was approved by the United States Food and Drug Administration in 1970. In 1974, this application was extended to its use as a preventive agent for manic-depressive illness.

Ronald R. Fieve, who had opened the first lithium clinic in North America in 1966, helped popularize the psychiatric use of lithium through his national TV appearances and his bestselling book, Moodswing. In addition, Fieve and David L. Dunner developed the concept of 'rapid cycling' Bipolar Disorder based on non-response to lithium.

Lithium has now become a part of Western popular culture. Characters in *Pi*, *Premonition*, *Stardust Memories*, *American Psycho*, *Garden State*, and *An Unmarried Woman* all take lithium. Sirius XM Satellite Radio in North America has a 1990s alternative rock station called Lithium, and several songs refer to the use of lithium as a mood stabilizer. These include: "Equilibrium met Lithium" by South African artist Koos Kombuis, "Lithium" by Evanescence, "Lithium" by Nirvana, "Lithium and a Lover" by Sirenia, "Lithium Sunset", from the album *Mercury Falling* by Sting, and "Lithium" by Thin White Rope.

7 Up

As with cocaine in Coca-Cola, lithium was widely marketed as one of a number of patent medicine products popular in the late-19th and early-20th centuries, and was the medicinal ingredient of a refreshment beverage. Charles Leiper Grigg, who launched his St. Louis-based company The Howdy Corporation, invented a formula for a lemon-lime soft drink in 1920. The product, originally named "Bib-Label Lithiated Lemon-Lime Soda", was launched two weeks before the Wall Street Crash of 1929. It contained the mood stabilizer lithium citrate, and was one of a number of patent medicine products popular in the late-19th and early-20th centuries. Its name was soon changed to 7 Up. All American beverage makers were forced to remove lithium in 1948. Despite the 1948 ban, in 1950 the *Painesville Telegraph* still carried an advertisement for a lithiated lemon beverage.

Names

Lithium carbonate (Li2CO3), sold under several trade names, is the most commonly prescribed, while lithium citrate (Li3C6H5O7) is also used in conventional pharmacological treatments. Lithium orotate (C5H3LiN2O4), has been presented as an alternative. Lithium bromide and lithium chloride have been used in the past as table salt; however, they fell out of use in the 1940s, when it was discovered they were toxic in those large doses. Many other lithium salts and compounds exist, such as lithium fluoride and lithium iodide, but they are presumed to be toxic as well and have never been evaluated for pharmacological effects.

Brandnames include Camcolit (UK), Eskalith (US), Li-Liquid (UK), Liskonum (UK), Lithicarb (Australia), Lithobid (US), Priadel (UK), Quilonum (Australia), Lithosun (India)

Research

Lithium may protect the brain and encourage the growth of gray matter in the cerebral cortex and prevent and slow the progression of Alzheimer's disease, senile dementia and Parkinson's disease.

However, for the amyotrophic lateral sclerosis (ALS), lithium has failed to produce positive results.

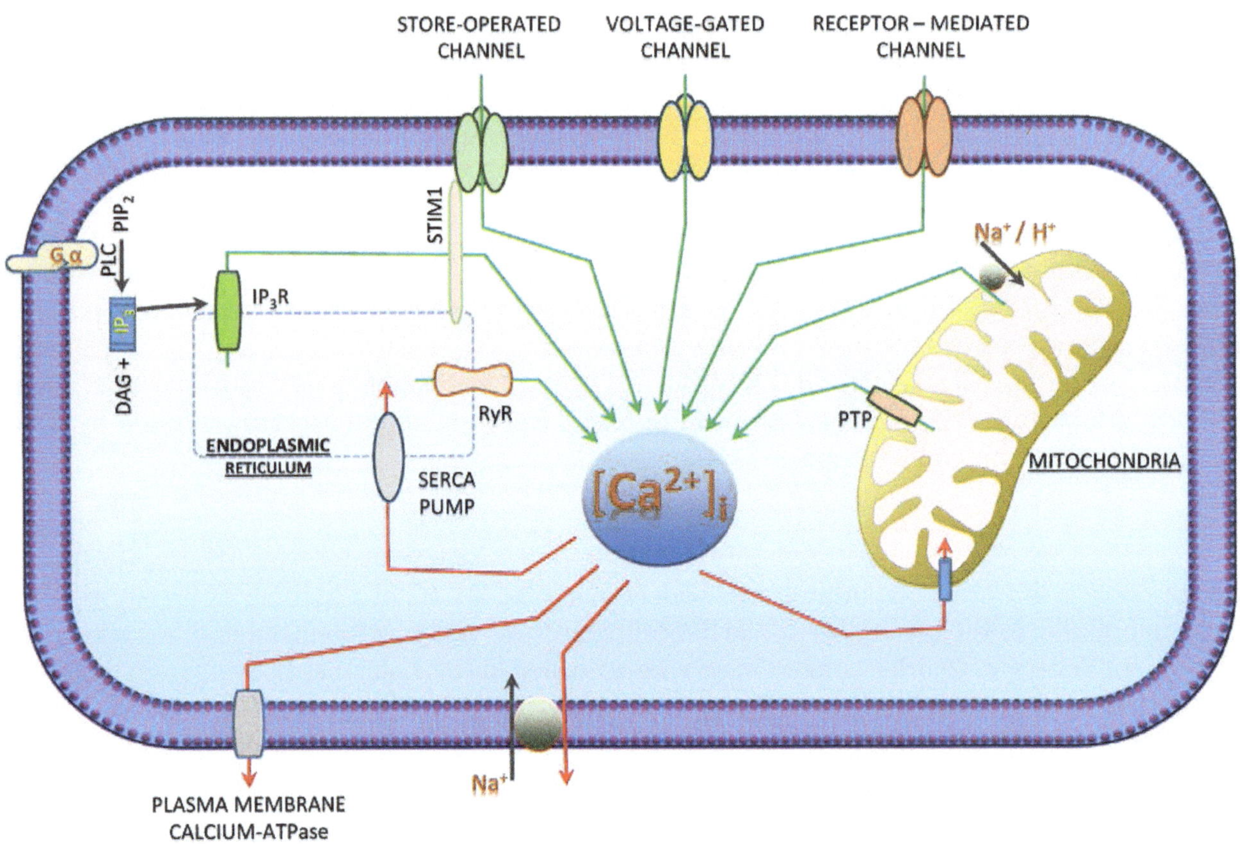

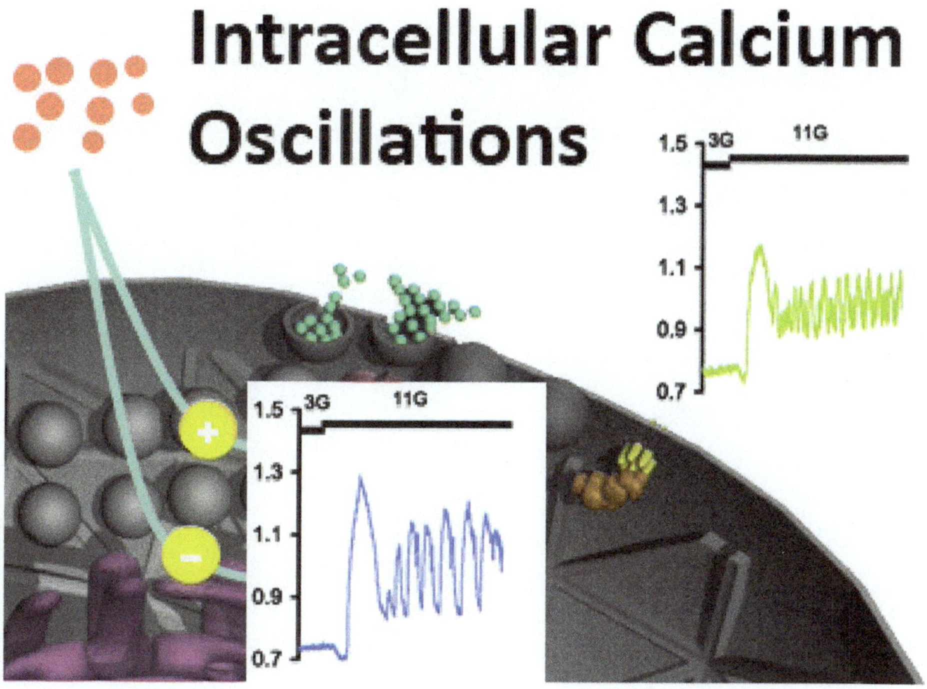

References

- Spencer L. Seager, Michael R. Slabaugh. Chemistry for Today: general, organic, and biochemistry. Thomson Brooks/Cole, 2004, p. 342. ISBN 0-534-39969-X
- Jie Jack Li, Douglas S. Johnson, Drago R. Sliskovic, Bruce D. Roth (2004). "Chapter 9. Atorvastatin Calcium (Lipitor)". Contemporary Drug Synthesis. John Wiley & Sons, Inc. pp. 113–125. ISBN 0-471-21480-9. CS1 maint: Uses authors parameter (link).
- Arnoldischen Buchhandlung, Dresden and Leipzig, 1827. ISBN 1-148-99953-1. Brief English commentary in English can be found in Bent Soren Jorgensen "More on Berzelius and the vital force" J. Chem. Educ., 1965, vol. 42, p 394. doi:10.1021/ed042p394
- Joint Formulary Committee (2013). British National Formulary (BNF) 65. London: Pharmaceutical Press. pp. 240–242. ISBN 9780857110848.
- Sperner-Unterweger, Barbara; W. Wolfgang Fleischhacker; Wolfgang P. Kaschka (2001). Psychoneuroimmunology. Karger Publishers. p. 22. ISBN 3-8055-7262-X.
- Silver, M; Factor, S (2015). "Chapter 12: VPA, lithium, amiodarone, and other non-DA". In Friedman, J. Medication-Induced Movement Disorders. Cambridge University Press. pp. 131–140. ISBN 978-1-107-06600-7.
- Keshavan, Matcheri S.; John S. Kennedy (2001). Drug-induced dysfunction in psychiatry. Taylor & Francis. p. 305. ISBN 0-89116-961-X.
- al.], edited by Brian K. Alldredge ... [et (2013). Applied therapeutics : the clinical use of drugs. (10th ed.). Baltimore: Wolters Kluwer Health/Lippincott Williams & Wilkins. p. 1991. ISBN 978-1609137137.
- Brunton, L; Chabner, B; Knollman, B (2010). Goodman and Gilman's The Pharmacological Basis of Therapeutics (12th ed.). New York: McGraw-Hill Professional. ISBN 978-0-07-162442-8.
- Winslow, Ron (24 January 2000). "The Birth of a Blockbuster: Lipitor's Route out of the Lab". The Wall Street Journal. Ann Arbor, Michigan. Retrieved 24 November 2015.
- "FDA Expands Advice on Statin Risks", FDA, 31 January 2014, retrieved 24 November 2015

- Macedo AF, Taylor FC, Casas JP, Adler A, Prieto-Merino D, Ebrahim S (2014). "Unintended effects of statins from observational studies in the general population: systematic review and meta-analysis". BMC Med. 12: 51. doi:10.1186/1741-7015-12-51. PMC 3998050. PMID 24655568.
- Cipriani, A.; Hawton, K.; Stockton, S.; Geddes, JR. (2013). "Lithium in the prevention of suicide in mood disorders: updated systematic review and meta-analysis.". BMJ. 346: f3646. doi:10.1136/bmj.f3646. PMID 23814104.

4

Medicinal Drugs: A Significant Aspect of Medicinal Chemistry

A drug that can be used for medicinal purposes is known as a medicinal drug. Medicinal drugs are either used for a prescribed duration or on a regular basis if the patient is diagnosed with a chronic disorder. The text will provide an integrated understanding of medicinal drugs.

Drug Class

For the UK meaning of Class A,B,C and D see Drugs controlled by the UK Misuse of Drugs Act in other countries this may be called schedules, e.g. USA drug schedules

A drug class is a set of medications that have similar chemical structures, the same mechanism of action (i.e., bind to the same biological target), a related mode of action, and/or are used to treat the same disease.

In several dominant drug classification systems, these four types of classifications form a hierarchy. For example, the fibrates are a chemical class of drugs (amphipathic carboxylic acids) that share the same mechanism of action (PPAR agonist), mode of action (reducing blood triglycerides), and are used to prevent and to treat the same disease (atherosclerosis). Conversely not all PPAR agonists are fibrates, not all triglyceride lowering agents are PPAR agonists, and not all drugs that are used to treat atherosclerosis are triglyceride lowering agents.

The most widely used drug classification system is the Anatomical Therapeutic Chemical Classification System (ATC). The Systematized Nomenclature of Medicine (SNOMED) also includes a section devoted to drug classification.

Chemical Class

Examples of drug classes that are based on chemical structures include:

- β-lactam antibiotic
- Benzodiazepine
- Cardiac glycoside
- Fibrate
- Thiazide diuretic

Mechanism of Action

Drug classes that share a common molecular mechanism of action by modulating the activity of a specific biological target. The definition of a mechanism of action also includes the type of activity at that biological target. For receptors, these activities include agonist, antagonist, inverse agonist, or modulator. Enzyme target mechanisms include activator or inhibitor. Ion channel modulators include opener or blocker. The following are specific examples of drug classes whose definition is based on a specific mechanism of action:

- 5-Alpha-reductase inhibitor
- Angiotensin II receptor antagonist
- ACE inhibitor
- Alpha-adrenergic agonist
- Beta blocker
- Dopamine agonist
- Dopamine antagonist
- Incretin mimetic
- Nonsteroidal anti-inflammatory drug – cyclooxygenase inhibitor
- Proton-pump inhibitor
- Renin inhibitor
- Selective glucocorticoid receptor modulator
- Selective serotonin reuptake inhibitor
- Statin – HMG-CoA reductase inhibitor

Mode of Action

Drug classes that are defined by common cellular mode of action include:

- Diuretic
- Cholinergic
- Dopaminergic
- GABAergic
- Serotonergic

Therapeutic Class

Drug classes that are defined by their therapeutic use include:

- Analgesic

- Antibiotic
- Anticoagulant
- Antidepressant
- Anticancer
- Antiepileptic
- Antipsychotic
- Antiviral
- Sedative

Amalgamated Classes

Some drug classes have been amalgamated from these three principles to meet practical needs. The class of non-steroidal anti-inflammatory drugs (NSAIDs) is one such example. Strictly speaking, and also historically, the wider class of anti-inflammatory drugs also comprises *steroidal* anti-inflammatory drugs. These drugs were in fact the predominant anti-inflammatories during the decade leading up to the introduction of the term "non-steroidal anti-inflammatory drugs". Because of the disastrous reputation that the corticosteroids had got in the 1950s, the new term, which offered to signal that an anti-inflammatory drug was not a steroid, rapidly gained currency. The drug class of "non-steroidal anti-inflammatory drugs" (NSAIDs) is thus composed by one element ("anti-inflammatory") that designates the mechanism of action, and one element ("non-steroidal") that separates it from other drugs with that same mechanism of action. Similarly, one might argue that the class of disease-modifying anti-rheumatic drugs (DMARD) is composed by one element ("disease-modifying") that albeit vaguely designates a mechanism of action, and one element ("anti-rheumatic drug") that indicates its therapeutic use.

- Nonsteroidal anti-inflammatory drug (NSAID)
- Disease-modifying antirheumatic drug (DMARD)

Pharmaceutical Drug

A pharmaceutical drug (also referred to as medication and drug) is a drug used to diagnose, cure, treat, or prevent disease. Drug therapy (pharmacotherapy) is an important part of the medical field and relies on the science of pharmacology for continual advancement and on pharmacy for appropriate management.

Drugs are classified in various ways. One of the key divisions is by level of control, which distinguishes prescription drugs (those that a pharmacist dispenses only on the order of a physician, physician assistant, or qualified nurse) from over-the-counter drugs (those that consumers can order for themselves). Another key distinction is between traditional small-molecule drugs, usually derived from chemical synthesis, and biopharmaceuticals, which include recombinant proteins, vaccines, blood products used therapeutically (such as IVIG), gene therapy, monoclonal antibodies and cell therapy (for instance, stem-cell therapies). Other ways to classify medicines are by mode of action, route of administration, biological system affected, or therapeutic effects. An elaborate and widely used classification system is the Anatomical Therapeutic Chemical Classification System (ATC system). The World Health Organization keeps a list of essential medicines.

Drug discovery and drug development are complex and expensive endeavors undertaken by pharmaceutical companies, academic scientists, and governments. Governments generally regulate what drugs can be marketed, how drugs are marketed, and in some jurisdictions, drug pricing. Controversies have arisen over drug pricing and disposal of used drugs.

Definition

In Europe, the term is "medicinal product", and it is defined by EU law as: "(a) Any substance or combination of substances presented as having properties for treating or preventing disease in human beings; or

(b) Any substance or combination of substances which may be used in or administered to human beings either with a view to restoring, correcting or modifying physiological functions by exerting a pharmacological, immunological or metabolic action, or to making a medical diagnosis."

In the US, a "drug" is:

- A substance recognized by an official pharmacopoeia or formulary.
- A substance intended for use in the diagnosis, cure, mitigation, treatment, or prevention of disease.
- A substance (other than food) intended to affect the structure or any function of the body.
- A substance intended for use as a component of a medicine but not a device or a component, part or accessory of a device.
- Biological products are included within this definition and are generally covered by the same laws and regulations, but differences exist regarding their manufacturing processes (chemical process versus biological process.)

Usage

Drug use among elderly Americans has been studied; in a group of 2377 people with average age of 71 surveyed between 2005 and 2006, 84% took at least one prescription drug, 44% took at least one over-the-counter (OTC) drug, and 52% took at least one dietary supplement; in a group of 2245 elderly Americans (average age of 71) surveyed over the period 2010 - 2011, those percentages were 88%, 38%, and 64%.

Classification

Pharmaceutical or a drug is classified on the basis of their origin.

1. Drug from natural origin: Herbal or plant or mineral origin, some drug substances are of marine origin.

2. Drug from chemical as well as natural origin: Derived from partial herbal and partial chemical synthesis Chemical, example steroidal drugs
3. Drug derived from chemical synthesis.
4. Drug derived from animal origin: For example, hormones, and enzymes.
5. Drug derived from microbial origin: Antibiotics
6. Drug derived by biotechnology genetic-engineering, hybridoma technique for example
7. Drug derived from radioactive substances.

One of the key classifications is between traditional small molecule drugs, usually derived from chemical synthesis, and biologic medical products, which include recombinant proteins, vaccines, blood products used therapeutically (such as IVIG), gene therapy, and cell therapy (for instance, stem cell therapies).

Pharmaceutical or drug or medicines are classified in various other groups besides their origin on the basis of pharmacological properties like mode of action and their pharmacological action or activity, such as by chemical properties, mode or route of administration, biological system affected, or therapeutic effects. An elaborate and widely used classification system is the Anatomical Therapeutic Chemical Classification System (ATC system). The World Health Organization keeps a list of essential medicines.

A sampling of classes of medicine includes:

1. Antipyretics: reducing fever (pyrexia/pyresis)
2. Analgesics: reducing pain (painkillers)
3. Antimalarial drugs: treating malaria
4. Antibiotics: inhibiting germ growth
5. Antiseptics: prevention of germ growth near burns, cuts and wounds
6. Mood stabilizers: lithium and valpromide
7. Hormone replacements: Premarin
8. Oral contraceptives: Enovid, “biphasic” pill, and “triphasic” pill
9. Stimulants: methylphenidate, amphetamine
10. Tranquilizers: meprobamate, chlorpromazine, reserpine, chlordiazepoxide, diazepam, and alprazolam
11. Statins: lovastatin, pravastatin, and simvastatin

Pharmaceuticals may also be described as “specialty”, independent of other classifications, which is an ill defined class of drugs that might be difficult to administer, require special handling during administration, require patient monitoring during and immediately after administration, have particular regulatory requirements restricting their use, and are generally expensive relative to other drugs.

Types of Medicines

For the Digestive System

- Upper digestive tract: antacids, reflux suppressants, antiflatulents, antidopaminergics, proton pump inhibitors (PPIs), H_2-receptor antagonists, cytoprotectants, prostaglandin analogues
- Lower digestive tract: laxatives, antispasmodics, antidiarrhoeals, bile acid sequestrants, opioid

For the Cardiovascular System

- General: β-receptor blockers ("beta blockers"), calcium channel blockers, diuretics, cardiac glycosides, antiarrhythmics, nitrate, antianginals, vasoconstrictors, vasodilators.
- Affecting blood pressure/(antihypertensive drugs): ACE inhibitors, angiotensin receptor blockers, beta-blockers, α blockers, calcium channel blockers, thiazide diuretics, loop diuretics, aldosterone inhibitors
- Coagulation: anticoagulants, heparin, antiplatelet drugs, fibrinolytics, anti-hemophilic factors, haemostatic drugs
- HMG-CoA reductase inhibitors (statins) for lowering LDL cholesterol inhibitors: hypolipidaemic agents.

For the Central Nervous System

Drugs affecting the central nervous system include: Psychedelics, hypnotics, anaesthetics, antipsychotics, eugeroics, antidepressants (including tricyclic antidepressants, monoamine oxidase inhibitors, lithium salts, and selective serotonin reuptake inhibitors (SSRIs)), antiemetics, Anticonvulsants/antiepileptics, anxiolytics, barbiturates, movement disorder (e.g., Parkinson's disease) drugs, stimulants (including amphetamines), benzodiazepines, cyclopyrrolones, dopamine antagonists, antihistamines, cholinergics, anticholinergics, emetics, cannabinoids, and 5-HT (serotonin) antagonists.

For Pain

The main classes of painkillers are NSAIDs, opioids and Local anesthetics.

For Consciousness (Anesthetic Drugs)

Some anesthetics include Benzodiazepines ande Barbiturates

For Musculo-skeletal Disorders

The main categories of drugs for musculoskeletal disorders are: NSAIDs (including COX-2 selective inhibitors), muscle relaxants, neuromuscular drugs, and anticholinesterases.

For the Eye

- General: adrenergic neurone blocker, astringent, ocular lubricant

- Diagnostic: topical anesthetics, sympathomimetics, parasympatholytics, mydriatics, cycloplegics
- Antibacterial: antibiotics, topical antibiotics, sulfa drugs, aminoglycosides, fluoroquinolones
- Antiviral drug
- Anti-fungal: imidazoles, polyenes
- Anti-inflammatory: NSAIDs, corticosteroids
- Anti-allergy: mast cell inhibitors
- Anti-glaucoma: adrenergic agonists, beta-blockers, carbonic anhydrase inhibitors/hyperosmotics, cholinergics, miotics, parasympathomimetics, prostaglandin agonists/prostaglandin inhibitors. nitroglycerin

For the Ear, Nose and Oropharynx

Antibiotics, sympathomimetics, antihistamines, anticholinergics, NSAIDs, corticosteroids, antiseptics, local anesthetics, antifungals, cerumenolytic

For the Respiratory System

bronchodilators, antitussives, mucolytics, decongestants inhaled and systemic corticosteroids, Beta2-adrenergic agonists, anticholinergics, Mast cell stabilizers. Leukotriene antagonists

For Endocrine Problems

androgens, antiandrogens, estrogens, gonadotropin, corticosteroids, human growth hormone, insulin, antidiabetics (sulfonylureas, biguanides/metformin, thiazolidinediones, insulin), thyroid hormones, antithyroid drugs, calcitonin, diphosponate, vasopressin analogues

For the Reproductive System or Urinary System

antifungal, alkalinizing agents, quinolones, antibiotics, cholinergics, anticholinergics, antispasmodics, 5-alpha reductase inhibitor, selective alpha-1 blockers, sildenafils, fertility medications

For Contraception

- Hormonal contraception
- Ormeloxifene
- Spermicide

For Obstetrics and Gynecology

NSAIDs, anticholinergics, haemostatic drugs, antifibrinolytics, Hormone Replacement Therapy

(HRT), bone regulators, beta-receptor agonists, follicle stimulating hormone, luteinising hormone, LHRH gamolenic acid, gonadotropin release inhibitor, progestogen, dopamine agonists, oestrogen, prostaglandins, gonadorelin, clomiphene, tamoxifen, Diethylstilbestrol

For the Skin

emollients, anti-pruritics, antifungals, disinfectants, scabicides, pediculicides, tar products, vitamin A derivatives, vitamin D analogues, keratolytics, abrasives, systemic antibiotics, topical antibiotics, hormones, desloughing agents, exudate absorbents, fibrinolytics, proteolytics, sunscreens, antiperspirants, corticosteroids, immune modulators

For Infections and Infestations

antibiotics, antifungals, antileprotics, antituberculous drugs, antimalarials, anthelmintics, amoebicides, antivirals, antiprotozoals, probiotics, prebiotics, antitoxins and antivenoms.

For the Immune System

vaccines, immunoglobulins, immunosuppressants, interferons, monoclonal antibodies

For Allergic Disorders

anti-allergics, antihistamines, NSAIDs, Corticosteroids

For Nutrition

Tonics, electrolytes and mineral preparations (including iron preparations and magnesium preparations), parenteral nutritions, vitamins, anti-obesity drugs, anabolic drugs, haematopoietic drugs, food product drugs

For Neoplastic Disorders

cytotoxic drugs, therapeutic antibodies, sex hormones, aromatase inhibitors, somatostatin inhibitors, recombinant interleukins, G-CSF, erythropoietin

For Diagnostics

contrast media

For Euthanasia

A euthanaticum is used for euthanasia and physician-assisted suicide. Euthanasia is not permitted by law in many countries, and consequently medicines will not be licensed for this use in those countries.

Administration

Administration is the process by which a patient takes a medicine. There are three major categories of drug administration; enteral (by mouth), parenteral (into the blood stream), and other (which includes giving a drug through intranasal, topical, inhalation, and rectal means).

It can be performed in various dosage forms such as pills, tablets, or capsules.

There are many variations in the routes of administration, including intravenous (into the blood through a vein) and oral administration (through the mouth).

They can be administered all at once as a bolus, at frequent intervals or continuously. Frequencies are often abbreviated from Latin, such as *every 8 hours* reading Q8H from *Quaque VIII Hora.*

Drug Discovery

In the fields of medicine, biotechnology and pharmacology, drug discovery is the process by which new candidate drugs are discovered.

Historically, drugs were discovered through identifying the active ingredient from traditional remedies or by serendipitous discovery. Later chemical libraries of synthetic small molecules, natural products or extracts were screened in intact cells or whole organisms to identify substances that have a desirable therapeutic effect in a process known as classical pharmacology. Since sequencing of the human genome which allowed rapid cloning and synthesis of large quantities of purified proteins, it has become common practice to use high throughput screening of large compounds libraries against isolated biological targets which are hypothesized to be disease modifying in a process known as reverse pharmacology. Hits from these screens are then tested in cells and then in animals for efficacy. Even more recently, scientists have been able to understand the shape of biological molecules at the atomic level, and to use that knowledge to design drug candidates.

Modern drug discovery involves the identification of screening hits, medicinal chemistry and optimization of those hits to increase the affinity, selectivity (to reduce the potential of side effects), efficacy/potency, metabolic stability (to increase the half-life), and oral bioavailability. Once a compound that fulfills all of these requirements has been identified, it will begin the process of drug development prior to clinical trials. One or more of these steps may, but not necessarily, involve computer-aided drug design.

Despite advances in technology and understanding of biological systems, drug discovery is still a lengthy, "expensive, difficult, and inefficient process" with low rate of new therapeutic discovery. In 2010, the research and development cost of each new molecular entity (NME) was approximately US$1.8 billion. Drug discovery is done by pharmaceutical companies, with research assistance from universities. The "final product" of drug discovery is a patent on the potential drug. The drug requires very expensive Phase I, II and III clinical trials, and most of them fail. Small companies have a critical role, often then selling the rights to larger companies that have the resources to run the clinical trials.

Development

Drug development is a blanket term used to define the process of bringing a new drug to the market once a lead compound has been identified through the process of drug discovery. It includes pre-clinical research (microorganisms/animals) and clinical trials (on humans) and may include the step of obtaining regulatory approval to market the drug.

Regulation

The regulation of drugs varies by jurisdiction. In some countries, such as the United States, they are regulated at the national level by a single agency. In other jurisdictions they are regulated at the state level, or at both state and national levels by various bodies, as is the case in Australia. The role of therapeutic goods regulation is designed mainly to protect the health and safety of the population. Regulation is aimed at ensuring the safety, quality, and efficacy of the therapeutic goods which are covered under the scope of the regulation. In most jurisdictions, therapeutic goods must be registered before they are allowed to be marketed. There is usually some degree of restriction of the availability of certain therapeutic goods depending on their risk to consumers.

Depending upon the jurisdiction, drugs may be divided into over-the-counter drugs (OTC) which may be available without special restrictions, and prescription drugs, which must be prescribed by a licensed medical practitioner. The precise distinction between OTC and prescription depends on the legal jurisdiction. A third category, "behind-the-counter" drugs, is implemented in some jurisdictions. These do not require a prescription, but must be kept in the dispensary, not visible to the public, and only be sold by a pharmacist or pharmacy technician. Doctors may also prescribe prescription drugs for off-label use - purposes which the drugs were not originally approved for by the regulatory agency. The Classification of Pharmaco-Therapeutic Referrals helps guide the referral process between pharmacists and doctors.

The International Narcotics Control Board of the United Nations imposes a world law of prohibition of certain drugs. They publish a lengthy list of chemicals and plants whose trade and consumption (where applicable) is forbidden. OTC drugs are sold without restriction as they are considered safe enough that most people will not hurt themselves accidentally by taking it as instructed. Many countries, such as the United Kingdom have a third category of "pharmacy medicines", which can only be sold in registered pharmacies by or under the supervision of a pharmacist.

Drug Pricing

In many jurisdictions drug prices are regulated.

United Kingdom

In the UK the Pharmaceutical Price Regulation Scheme is intended to ensure that the National Health Service is able to purchase drugs at reasonable prices.

Canada

In Canada, the Patented Medicine Prices Review Board examines drug pricing, compares the proposed Canadian price to that of seven other countries and determines if a price is excessive or not. In these circumstances, drug manufacturers must submit a proposed price to the appropriate regulatory agency.

Brazil

In Brazil, the prices are regulated through a legislation under the name of *Medicamento Genérico* (generic drugs) since 1999.

India

In India, drug prices are regulated by the National Pharmaceutical Pricing Authority.

United States

In the United States, drug costs are unregulated, but instead are the result of negotiations between drug companies and insurance companies. High prices have been attributed to monopolies given to manufacturers by the government and a lack of ability for organizations to negotiate prices.

Blockbuster Drug

A blockbuster drug is a drug generating more than $1 billion of revenue for the pharmaceutical company that sells it each year. Cimetidine was the first drug ever to reach more than $1 billion a year in sales, thus making it the first blockbuster drug.

> "In the pharmaceutical industry, a blockbuster drug is one that achieves acceptance by prescribing physicians as a therapeutic standard for, most commonly, a highly prevalent chronic (rather than acute) condition. Patients often take the medicines for long periods."

Leading Blockbuster Drugs as of 2011

Drug	Trade name	Type	Indication	Company	Sales ($billion/year)*
Atorvastatin	Lipitor	Small molecule	Hypercholesterolemia	Pfizer	12.5
Clopidogrel	Plavix	Small molecule	Atherosclerosis	Bristol-Myers Squibb Sanofi	9.1
Fluticasone/salmeterol	Advair	Small molecule	Asthma	GlaxoSmithKline	8.7
Esomeprazole	Nexium	Small molecule	Gastroesophageal reflux disease	AstraZeneca	8.3
Rosuvastatin	Crestor	Small molecule	Hypercholesterolemia	AstraZeneca	7.4
Quetiapine	Seroquel	Small molecule	Bipolar disorder Schizophrenia Major Depressive Disorder	AstraZeneca	7.2
Adalimumab	Humira	Biologic	Rheumatoid Arthritis	AbbVie	6.6
Etanercept	Enbrel	Biologic	Rheumatoid Arthritis	Amgen Pfizer	6.5

Drug	Trade name	Type	Indication	Company	Sales ($billion/year)*
Infliximab	Remicade	Biologic	Crohn's Disease Rheumatoid Arthritis	Johnson & Johnson	6.4
Olanzapine	Zyprexa	Small molecule	Schizophrenia	Eli Lilly and Company	6.2

*Sales are for the 12 months preceding 30 June 2011.

History

Prescription Drug History

Antibiotics first arrived on the medical scene in 1932 thanks to Gerhard Domagk; and coined the "wonder drugs". The introduction of the sulfa drugs led to a decline in the U.S. mortality rate from pneumonia to drop from 0.2% each year to 0.05% by 1939. Antibiotics inhibit the growth or the metabolic activities of bacteria and other microorganisms by a chemical substance of microbial origin. Penicillin, introduced a few years later, provided a broader spectrum of activity compared to sulfa drugs and reduced side effects. Streptomycin, found in 1942, proved to be the first drug effective against the cause of tuberculosis and also came to be the best known of a long series of important antibiotics. A second generation of antibiotics was introduced in the 1940s: aureomycin and chloramphenicol. Aureomycin was the best known of the second generation.

Lithium was discovered in the 19th century for nervous disorders and its possible mood-stabilizing or prophylactic effect; it was cheap and easily produced. As lithium fell out of favor in France, valpromide came into play. This antibiotic was the origin of the drug that eventually created the mood stabilizer category. Valpromide had distinct psychotrophic effects that were of benefit in both the treatment of acute manic states and in the maintenance treatment of manic depression illness. Psychotropics can either be sedative or stimulant; sedatives aim at damping down the extremes of behavior. Stimulants aim at restoring normality by increasing tone. Soon arose the notion of a tranquilizer which was quite different from any sedative or stimulant. The term tranquilizer took over the notions of sedatives and became the dominant term in the West through the 1980s. In Japan, during this time, the term tranquilizer produced the notion of a psyche-stabilizer and the term mood stabilizer vanished.

Premarin (conjugated estrogens, introduced in 1942) and Prempro (a combination estrogen-progestin pill, introduced in 1995) dominated the hormone replacement therapy (HRT) during the 1990s. HRT is not a life-saving drug, nor does it cure any disease. HRT has been prescribed to improve one's quality of life. Doctors prescribe estrogen for their older female patients both to treat short-term menopausal symptoms and to prevent long-term diseases. In the 1960s and early 1970s more and more physicians began to prescribe estrogen for their female patients. between 1991 and 1999, Premarin was listed as the most popular prescription and best-selling drug in America.

The first oral contraceptive, Enovid, was approved by FDA in 1960. Oral contraceptives inhibit ovulation and so prevent conception. Enovid was known to be much more effective than alternatives including the condom and the diaphragm. As early as 1960, oral contraceptives were available in several different strengths by every manufacturer. In the 1980s and 1990s an increasing number of options arose including, most recently, a new delivery system for the oral contraceptive via a transdermal patch. In 1982, a new version of the Pill was introduced, known as the "biphasic" pill. By 1985, a new triphasic pill was approved. Physicians began to think of the Pill as an excellent means of birth control for young women.

Stimulants such as Ritalin (methylphenidate) came to be pervasive tools for behavior management and modification in young children. Ritalin was first marketed in 1955 for narcolepsy; its potential users were middle-aged and the elderly. It wasn't until some time in the 1980s along with hyperactivity in children that Ritalin came onto the market. Medical use of methlyphanidate is predominately for symptoms of attention deficit/hyperactivity disorder (ADHD). Consumption of methylphenidate in the U.S. out-paced all other countries between 1991 and 1999. Significant growth in consumption was also evident in Canada, New Zealand, Australia, and Norway. Currently, 85% of the world's methylphanidate is consumed in America.

The first minor tranquilizer was Meprobamate. Only fourteen months after it was made available, meprobamate had become the country's largest-selling prescription drug. By 1957, meprobamate had become the fastest-growing drug in history. The popularity of meprobamate paved the way for Librium and Valium, two minor tranquilizers that belonged to a new chemical class of drugs called the benzodiazepines. These were drugs that worked chiefly as anti-anxiety agents and muscle relaxants. The first benzodiazepine was Librium. Three months after it was approved, Librium had become the most prescribed tranquilizer in the nation. Three years later, Valium hit the shelves and was ten times more effective as a muscle relaxant and anti-convulsant. Valium was the most versatile of the minor tranquilizers. Later came the widespread adoption of major tranquilizers such as chlorpromazine and the drug reserpine. In 1970 sales began to decline for Valium and Librium, but sales of new and improved tranquilizers, such as Xanax, introduced in 1981 for the newly created diagnosis of panic disorder, soared.

Mevacor (lovastatin) is the first and most influential statin in the American market. The 1991 launch of Pravachol (pravastatin), the second available in the United States, and the release of Zocor (simvastatin) made Mevacor no longer the only statin on the market. In 1998, Viagra was released as a treatment for erectile dysfunction.

Ancient Pharmacology

Using plants and plant substances to treat all kinds of diseases and medical conditions is believed to date back to prehistoric medicine.

The Kahun Gynaecological Papyrus, the oldest known medical text of any kind, dates to about 1800 BC and represents the first documented use of any kind of drug. It and other medical papyri describe Ancient Egyptian medical practices, such as using honey to treat infections and the legs of bee-eaters to treat neck pains.

Ancient Babylonian medicine demonstrate the use of prescriptions in the first half of the 2nd millennium BC. Medicinal creams and pills were employed as treatments.

On the Indian subcontinent, the Atharvaveda, a sacred text of Hinduism whose core dates from the 2nd millennium BC, although the hymns recorded in it are believed to be older, is the first Indic text dealing with medicine. It describes plant-based drugs to counter diseases. The earliest foundations of ayurveda were built on a synthesis of selected ancient herbal practices, together with a massive addition of theoretical conceptualizations, new nosologies and new therapies dating from about 400 BC onwards. The student of Āyurveda was expected to know ten arts that were indispensable in the preparation and application of his medicines: distillation, operative skills, cooking, horticulture, metallurgy, sugar manufacture, pharmacy, analysis and separation of minerals, compounding of metals, and preparation of alkalis.

The Hippocratic Oath for physicians, attributed to 5th century BC Greece, refers to the existence of "deadly drugs", and ancient Greek physicians imported drugs from Egypt and elsewhere.

Medieval Pharmacology

Al-Kindi's 9th century AD book, *De Gradibus* and Ibn Sina (Avicenna)'s *The Canon of Medicine* cover a range of drugs known to Medicine in the medieval Islamic world.

Medieval medicine saw advances in surgery, but few truly effective drugs existed, beyond opium (found in such extremely popular drugs as the "Great Rest" of the Antidotarium Nicolai at the time) and quinine. Folklore cures and potentially poisonous metal-based compounds were popular treatments. Theodoric Borgognoni, (1205–1296), one of the most significant surgeons of the medieval period, responsible for introducing and promoting important surgical advances including basic antiseptic practice and the use of anaesthetics. Garcia de Orta described some herbal treatments that were used.

Modern Pharmacology

For most of the 19th century, drugs were not highly effective, leading Oliver Wendell Holmes, Sr. to famously comment in 1842 that "if all medicines in the world were thrown into the sea, it would be all the better for mankind and all the worse for the fishes".

During the First World War, Alexis Carrel and Henry Dakin developed the Carrel-Dakin method of treating wounds with an irrigation, Dakin's solution, a germicide which helped prevent gangrene.

In the inter-war period, the first anti-bacterial agents such as the sulpha antibiotics were developed. The Second World War saw the introduction of widespread and effective antimicrobial therapy with the development and mass production of penicillin antibiotics, made possible by the pressures of the war and the collaboration of British scientists with the American pharmaceutical industry.

Medicines commonly used by the late 1920s included aspirin, codeine, and morphine for pain; digitalis, nitroglycerin, and quinine for heart disorders, and insulin for diabetes. Other drugs included antitoxins, a few biological vaccines, and a few synthetic drugs. In the 1930s antibiotics emerged: first sulfa drugs, then penicillin and other antibiotics. Drugs increasingly became "the center of medical practice". In the 1950s other drugs emerged including corticosteroids for inflammation, rauwolfia alkaloids as tranqulizers and antihypertensives, antihistamines for nasal allergies, xanthines for asthma, and typical antipsychotics for psychosis. As of 2007, thousands of approved drugs have been developed. Increasingly, biotechnology is used to discover biopharmaceuticals. Recently, multi-disciplinary approaches have yielded a wealth of new data on the development of novel antibiotics and antibacterials and on the use of biological agents for antibacterial therapy.

In the 1950s new psychiatric drugs, notably the antipsychotic chlorpromazine, were designed in laboratories and slowly came into preferred use. Although often accepted as an advance in some ways, there was some opposition, due to serious adverse effects such as tardive dyskinesia. Patients often opposed psychiatry and refused or stopped taking the drugs when not subject to psychiatric control.

Governments have been heavily involved in the regulation of drug development and drug sales. In the U.S., the Elixir Sulfanilamide disaster led to the establishment of the Food and Drug Administration, and the 1938 Federal Food, Drug, and Cosmetic Act required manufacturers to file new drugs with the FDA. The 1951 Humphrey-Durham Amendment required certain drugs to be sold by prescription. In 1962 a subsequent amendment required new drugs to be tested for efficacy and safety in clinical trials.

Until the 1970s, drug prices were not a major concern for doctors and patients. As more drugs became prescribed for chronic illnesses, however, costs became burdensome, and by the 1970s nearly every U.S. state required or encouraged the substitution of generic drugs for higher-priced brand names. This also led to the 2006 U.S. law, Medicare Part D, which offers Medicare coverage for drugs.

As of 2008, the United States is the leader in medical research, including pharmaceutical development. U.S. drug prices are among the highest in the world, and drug innovation is correspondingly high. In 2000 U.S. based firms developed 29 of the 75 top-selling drugs; firms from the second-largest market, Japan, developed eight, and the United Kingdom contributed 10. France, which imposes price controls, developed three. Throughout the 1990s outcomes were similar.

Controversies

Controversies concerning pharmaceutical drugs include patient access to drugs under development and not yet approved, pricing, and environmental issues.

Access to Unapproved Drugs

Governments worldwide have created provisions for granting access to drugs prior to approval for patients who have exhausted all alternative treatment options and do not match clinical trial entry criteria. Often grouped under the labels of compassionate use, expanded access, or named patient supply, these programs are governed by rules which vary by country defining access criteria, data collection, promotion, and control of drug distribution.

Within the United States, pre-approval demand is generally met through treatment IND (investigational new drug) applications (INDs), or single-patient INDs. These mechanisms, which fall under the label of expanded access programs, provide access to drugs for groups of patients or individuals residing in the US. Outside the US, Named Patient Programs provide controlled, pre-approval access to drugs in response to requests by physicians on behalf of specific, or "named", patients before those medicines are licensed in the patient's home country. Through these programs, patients are able to access drugs in late-stage clinical trials or approved in other countries for a genuine, unmet medical need, before those drugs have been licensed in the patient's home country.

Patients who have not been able to get access to drugs in development have organized and advocated for greater access. In the United States, ACT UP formed in the 1980s, and eventually formed its Treatment Action Group in part to pressure the US government to put more resources into discovering treatments for AIDS and then to speed release of drugs that were under development.

The Abigail Alliance was established in November 2001 by Frank Burroughs in memory of his daughter, Abigail. The Alliance seeks broader availability of investigational drugs on behalf of terminally ill patients.

In 2013, BioMarin Pharmaceutical was at the center of a high-profile debate regarding expanded access of cancer patients to experimental drugs.

Access to Medicines and Drug Pricing

Essential medicines as defined by the World Health Organization (WHO) are "those drugs that satisfy the health care needs of the majority of the population; they should therefore be available at all times in adequate amounts and in appropriate dosage forms, at a price the community can afford." Recent studies have found that most of the medicines on the WHO essential medicines list, outside of the field of HIV drugs, are not patented in the developing world, and that lack of widespread access to these medicines arise from issues fundamental to economic development - lack of infrastructure and poverty. Médecins Sans Frontières also runs a Campaign for Access to Essential Medicines campaign, which includes advocacy for greater resources to be devoted to currently untreatable diseases that primarily occur in the developing world. The Access to Medicine Index tracks how well pharmaceutical companies make their products available in the developing world.

World Trade Organization negotiations in the 1990s, including the TRIPS Agreement and the Doha Declaration, have centered on issues at the intersection of international trade in pharmaceuticals and intellectual property rights, with developed world nations seeking strong intellectual property rights to protect investments made to develop new drugs, and developing world nations seeking to promote their generic pharmaceuticals industries and their ability to make medicine available to their people via compulsory licenses.

Some have raised ethical objections specifically with respect to pharmaceutical patents and the high prices for drugs that they enable their proprietors to charge, which poor people in the developed world, and developing world, cannot afford. Critics also question the rationale that exclusive patent rights and the resulting high prices are required for pharmaceutical companies to recoup the large investments needed for research and development. One study concluded that marketing expenditures for new drugs often doubled the amount that was allocated for research and development. Other critics claim that patent settlements would be costly for consumers, the health care system, and state and federal governments because it would result in delaying access to lower cost generic medicines.

Novartis fought a protracted battle with the government of India over the patenting of its drug, Gleevec, in India, which ended up in India's Supreme Court in a case known as Novartis v. Union of India & Others. The Supreme Court ruled narrowly against Novartis, but opponents of patenting drugs claimed it as a major victory.

Environmental Issues

The environmental impact of pharmaceuticals and personal care products is controversial. PPCPs are substances used by individuals for personal health or cosmetic reasons and the products used by agribusiness to boost growth or health of livestock. PPCPs comprise a diverse collection of thousands of chemical substances, including prescription and over-the-counter therapeutic drugs, veterinary drugs, fragrances, and cosmetics. PPCPs have been detected in water bodies throughout the world and ones that persist in the environment are called Environmental Persistent Pharmaceutical Pollutants. The effects of these chemicals on humans and the environment are not yet known, but to date there is no scientific evidence that they affect human health.

Biopharmaceutical

A biopharmaceutical, also known as a biologic(al) medical product, biological, or biologic, is any pharmaceutical drug product manufactured in, extracted from, or semisynthesized from biological sources. Different from chemically synthesized pharmaceuticals, they include vaccines, blood, blood components, allergenics, somatic cells, gene therapies, tissues, recombinant therapeutic protein, and living cells used in cell therapy. Biologics can be composed of sugars, proteins, or nucleic acids or complex combinations of these substances, or may be living cells or tissues. They are isolated from natural sources—human, animal, or microorganism.

Terminology surrounding biopharmaceuticals varies between groups and entities, with different terms referring to different subsets of therapeutics within the general biopharmaceutical category. Some regulatory agencies use the terms *biological medicinal products* or *therapeutic biological product* to refer specifically to engineered macromolecular products like protein- and nucleic acid-based drugs, distinguishing them from products like blood, blood components, or vaccines, which are usually extracted directly from a biological source. Specialty drugs, a recent classification of pharmaceuticals, are high-cost drugs that are often biologics.

Gene-based and cellular biologics, for example, often are at the forefront of biomedical research, and may be used to treat a variety of medical conditions for which no other treatments are available.

In some jurisdictions, biologics are regulated via different pathways than other small molecule drugs and medical devices.

The term biopharmacology is sometimes used to describe the branch of pharmacology that studies biopharmaceuticals.

Major Classes

Blood plasma is a type of biopharmaceutical directly extracted from living systems.

Extracted from Living Systems

Some of the oldest forms of biologics are extracted from the bodies of animals, and other humans especially. Important biologics include:

- Whole blood and other blood components
- Organs and tissue transplants
- Stem cell therapy
- Antibodies for passive immunization (e.g., to treat a virus infection)

Some biologics that were previously extracted from animals, such as insulin, are now more commonly produced by recombinant DNA.

Produced by Recombinant DNA

As indicated the term "biologics" can be used to refer to a wide range of biological products in medicine. However, in most cases, the term "biologics" is used more restrictively for a class of therapeutics (either approved or in development) that are produced by means of biological processes involving recombinant DNA technology. These medications are usually one of three types:

1. Substances that are (nearly) identical to the body's own key signalling proteins. Examples are the blood-production stimulating protein erythropoetin, or the growth-stimulating hormone named (simply) "growth hormone" or biosynthetic human insulin and its analogues.

2. Monoclonal antibodies. These are similar to the antibodies that the human immune system uses to fight off bacteria and viruses, but they are "custom-designed" (using hybridoma technology or other methods) and can therefore be made specifically to counteract or block any given substance in the body, or to target any specific cell type; examples of such monoclonal antibodies for use in various diseases are given in the table below.

3. Receptor constructs (fusion proteins), usually based on a naturally-occurring receptor linked to the immunoglobulin frame. In this case, the receptor provides the construct with detailed specificity, whereas the immunoglobulin-structure imparts stability and other useful features in terms of pharmacology. Some examples are listed in the table below.

Biologics as a class of medications in this narrower sense have had a profound impact on many medical fields, primarily rheumatology and oncology, but also cardiology, dermatology, gastroenterology, neurology, and others. In most of these disciplines, biologics have added major therapeutic options for the treatment of many diseases, including some for which no effective therapies were available, and others where previously existing therapies were clearly inadequate. However, the advent of biologic therapeutics has also raised complex regulatory issues and significant pharmacoeconomic concerns, because the cost for biologic therapies has been dramatically higher than for conventional (pharmacological) medications. This factor has been particularly relevant since many biological medications are used for the treatment of chronic diseases, such as rheumatoid arthritis or inflammatory bowel disease, or for the treatment of otherwise untreatable cancer during the remainder of life. The cost of treatment with a typical monoclonal antibody therapy for relatively common indications is generally in the range of €7,000–14,000 per patient per year.

Older patients who receive biologic therapy for diseases such as rheumatoid arthritis, psoriatic arthritis, or ankylosing spondylitis are at increased risk for life-threatening infection, adverse cardiovascular events, and malignancy.

The first such substance approved for therapeutic use was biosynthetic "human" insulin made via recombinant DNA. Sometimes referred to as rHI, under the trade name Humulin, was developed by Genentech, but licensed to Eli Lilly and Company, who manufactured and marketed it starting in 1982.

Major kinds of biopharmaceuticals include:

- Blood factors (Factor VIII and Factor IX)
- Thrombolytic agents (tissue plasminogen activator)
- Hormones (insulin, glucagon, growth hormone, gonadotrophins)
- Haematopoietic growth factors (Erythropoietin, colony stimulating factors)
- Interferons (Interferons-α, -β, -γ)
- Interleukin-based products (Interleukin-2)
- Vaccines (Hepatitis B surface antigen)
- Monoclonal antibodies (Various)
- Additional products (tumour necrosis factor, therapeutic enzymes)

Research and development investment in new medicines by the biopharmaceutical industry stood at $65.2 billion in 2008. A few examples of biologics made with recombinant DNA technology include:

USAN/INN	Trade name	Indication	Technology	Mechanism of action
abatacept	Orencia	rheumatoid arthritis	immunoglobin CTLA-4 fusion protein	T-cell deactivation
adalimumab	Humira	rheumatoid arthritis, ankylosing spondylitis, psoriatic arthritis, psoriasis, Ulcerative Colitis, Crohn's disease	monoclonal antibody	TNF antagonist
alefacept	Amevive	chronic plaque psoriasis	immunoglobin G1 fusion protein	incompletely characterized
erythropoietin	Epogen	anemia arising from cancer chemotherapy, chronic renal failure, etc.	recombinant protein	stimulation of red blood cell production
etanercept	Enbrel	rheumatoid arthritis, ankylosing spondylitis, psoriatic arthritis, psoriasis	recombinant human TNF-receptor fusion protein	TNF antagonist

infliximab	Remicade	rheumatoid arthritis, ankylosing spondylitis, psoriatic arthritis, psoriasis, Ulcerative Colitus, Crohn's disease	monoclonal antibody	TNF antagonist
trastuzumab	Herceptin	breast cancer	humanized monoclonal antibody	HER2/neu (erbB2) antagonist
ustekinumab	Stelara	psoriasis	humanized monoclonal antibody	IL-12 and IL-23 antagonist
denileukin diftitox	Ontak	cutaneous T-cell lymphoma (CTCL)	Diphtheria toxin engineered protein combining Interleukin-2 and Diphtheria toxin	Interleukin-2 receptor binder
golimumab	Simponi	rheumatoid arthritis, psoriatic arthritis, ankylosing spondylitis, Ulcerative colitis	monoclonal antibody	TNF antagonist

Vaccines

Many vaccines are grown in tissue cultures.

Gene Therapy

Viral gene therapy involves artificially manipulating a virus to include a desirable piece of genetic material.

Biosimilars

With the expiration of numerous patents for blockbuster biologics between 2012 and 2019, the interest in biosimilar production, i.e., follow-on biologics, has increased. Compared to small molecules that consist of chemically identical active ingredients, biologics are vastly more complex and consist of a multitude of subspecies. Due to their heterogeneity and the high process sensitivity, neither originators nor follow-on manufacturers produce reliably constant quality profiles over time. The process variations are monitored by modern analytical tools (e.g., liquid chromatography, immunoassays, mass spectrometry, etc.) and describe a unique design space for each biologic.

Thus, biosimilars require a different regulatory framework compared to small-molecule generics. Legislation in the 21st century has addressed this by recognizing an intermediate ground of testing for biosimilars. The filing pathway requires more testing than for small-molecule generics, but less testing than for registering completely new therapeutics.

In 2003, the European Medicines Agency introduced an adapted pathway for biosimilars, termed *similar biological medicinal products*. This pathway is based on a thorough demonstration of "comparability" of the "similar" product to an existing approved product. Within the United States, the Patient Protection and Affordable Care Act of 2010 created an abbreviated approval pathway for biological products shown to be biosimilar to, or interchangeable with, an FDA-licensed reference biological product. A major hope linked to the introduction of biosimilars is a reduction of costs to the patients and the healthcare system.

Commercialization

When a new biopharmaceutical is developed, the company will typically apply for a patent, which is a grant for exclusive manufacturing rights. This is the primary means by which the developer of the drug can recover the investment cost for development of the biopharmaceutical. The patent laws in the United States and Europe differ somewhat on the requirements for a patent, with the European requirements are perceived as more difficult to satisfy. The total number of patents granted for biopharmaceuticals has risen significantly since the 1970s. In 1978 the total patents granted was 30. This had climbed to 15,600 in 1995, and by 2001 there were 34,527 patent applications.

Large-scale Production

Biopharmaceuticals may be produced from microbial cells (e.g., recombinant *E. coli* or yeast cultures), mammalian cell lines and plant cell cultures (plant tissue culture) and moss plants in bioreactors of various configurations, including photo-bioreactors. Import-ant issues of concern are cost of production (low-volume, high-purity products are desirable) and microbial contamination (by bacteria, viruses, mycoplasma). Alternative platforms of production which are being tested include whole plants (plant-made pharmaceuticals).

Transgenics

A potentially controversial method of producing biopharmaceuticals involves transgenic organisms, particularly plants and animals that have been genetically modified to produce drugs. This production is a significant risk for the investor, due to production failure or scrituny from regulatory bodies based on perceived risks and ethical issues. Biopharmaceutical crops also represent a risk of cross-contamination with non-engineered crops, or crops engineered for non-medical purposes.

One potential approach to this technology is the creation of a transgenic mammal that can produce the biopharmaceutical in its milk, blood, or urine. Once an animal is produced, typically using the pronuclear microinjection method, it becomes efficacious to use cloning technology to create additional offspring that carry the favorable modified genome. The first such drug manufactured from the milk of a genetically modified goat was ATryn, but marketing permission was blocked by the European Medicines Agency in February 2006. This decision was reversed in June 2006 and approval was given August 2006.

Regulation

European Union

In the European Union, a biological medicinal product is one of the active substance(s) produced from or extracted from a biological (living) system, and requires, in addition to physico-chemical testing, biological testing for full characterisation. The characterisation of a biological medicinal product is a combination of testing the active substance and the final medicinal product together with the production process and its control. For example:

- Production process – it can be derived from biotechnology or from other technologies. It may be prepared using more conventional techniques as is the case for blood or plasma-derived products and a number of vaccines.

- Active substance – consisting of entire microorganisms, mammalian cells, nucleic acids, proteinaceous, or polysaccharide components originating from a microbial, animal, human, or plant source.
- Mode of action – therapeutic and immunological medicinal products, gene transfer materials, or cell therapy materials.

United States

In the United States, biologics are regulated by the FDA's Center for Biologics Evaluation and Research (CBER) whereas drugs are regulated by the Center for Drug Evaluation and Research. Approval may require several years of clinical trials, including trials with human volunteers. Even after the drug is released, it will still be monitored for performance and safety risks. The manufacture process must satisfy the FDA's "Good Manufacturing Practices", which are typically manufactured in a clean room environment with strict limits on the amount of airborne particles.

Drug Design

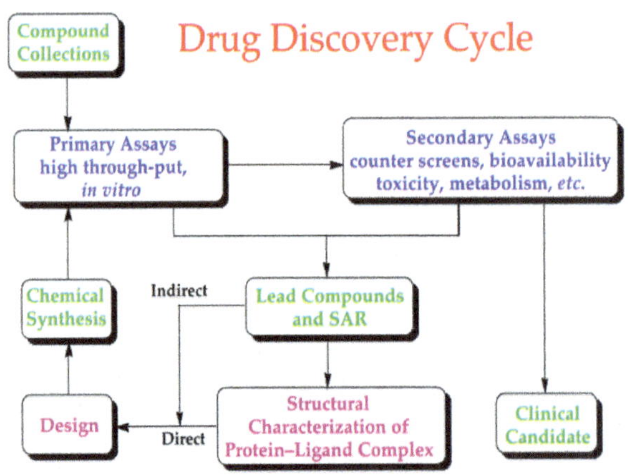

Drug design, often referred to as rational drug design or simply rational design, is the inventive process of finding new medications based on the knowledge of a biological target. The drug is most commonly an organic small molecule that activates or inhibits the function of a biomolecule such as a protein, which in turn results in a therapeutic benefit to the patient. In the most basic sense, drug design involves the design of molecules that are complementary in shape and charge to the biomolecular target with which they interact and therefore will bind to it. Drug design frequently but not necessarily relies on computer modeling techniques. This type of modeling is sometimes referred to as computer-aided drug design. Finally, drug design that relies on the knowledge of the three-dimensional structure of the biomolecular target is known as structure-based drug design. In addition to small molecules, biopharmaceuticals and especially therapeutic antibodies are an increasingly important class of drugs and computational methods for improving the affinity, selectivity, and stability of these protein-based therapeutics have also been developed.

The phrase "drug design" is to some extent a misnomer. A more accurate term is ligand design (i.e., design of a molecule that will bind tightly to its target). Although design techniques for prediction of binding affinity are reasonably successful, there are many other properties, such as bioavailability, metabolic half-life, side effects, etc., that first must be optimized before a ligand can become a safe and efficacious drug. These other characteristics are often difficult to predict with rational design techniques. Nevertheless, due to high attrition rates, especially during clinical phases of drug development, more attention is being focused early in the drug design process on selecting candidate drugs whose physicochemical properties are predicted to result in fewer complications during development and hence more likely to lead to an approved, marketed drug. Furthermore, in vitro experiments complemented with computation methods are increasingly used in early drug discovery to select compounds with more favorable ADME (absorption, distribution, metabolism, and excretion) and toxicological profiles.

Drug Targets

A biomolecular target (most commonly a protein or nucleic acid) is a key molecule involved in a particular metabolic or signaling pathway that is associated with a specific disease condition or pathology or to the infectivity or survival of a microbial pathogen. Potential drug targets are not necessarily disease causing but must by definition be disease modifying. In some cases, small molecules will be designed to enhance or inhibit the target function in the specific disease modifying pathway. Small molecules (for example receptor agonists, antagonists, inverse agonists, or modulators; enzyme activators or inhibitors; or ion channel openers or blockers) will be designed that are complementary to the binding site of target. Small molecules (drugs) can be designed so as not to affect any other important "off-target" molecules (often referred to as antitargets) since drug interactions with off-target molecules may lead to undesirable side effects. Due to similarities in binding sites, closely related targets identified through sequence homology have the highest chance of cross reactivity and hence highest side effect potential.

Most commonly, drugs are organic small molecules produced through chemical synthesis, but biopolymer-based drugs (also known as biopharmaceuticals) produced through biological processes are becoming increasingly more common. In addition, mRNA-based gene silencing technologies may have therapeutic applications.

Rational Drug Discovery

In contrast to traditional methods of drug discovery (known as forward pharmacology), which rely on trial-and-error testing of chemical substances on cultured cells or animals, and matching the apparent effects to treatments, rational drug design (also called reverse pharmacology) begins with a hypothesis that modulation of a specific biological target may have therapeutic value. In order for a biomolecule to be selected as a drug target, two essential pieces of information are required. The first is evidence that modulation of the target will be disease modifying. This knowledge may come from, for example, disease linkage studies that show an association between mutations in the biological target and certain disease states. The second is that the target is "druggable". This means that it is capable of binding to a small molecule and that its activity can be modulated by the small molecule.

Once a suitable target has been identified, the target is normally cloned and produced and purified.

The purified protein is then used to establish a screening assay. In addition, the three-dimensional structure of the target may be determined.

The search for small molecules that bind to the target is begun by screening libraries of potential drug compounds. This may be done by using the screening assay (a "wet screen"). In addition, if the structure of the target is available, a virtual screen may be performed of candidate drugs. Ideally the candidate drug compounds should be "drug-like", that is they should possess properties that are predicted to lead to oral bioavailability, adequate chemical and metabolic stability, and minimal toxic effects. Several methods are available to estimate druglikeness such as Lipinski's Rule of Five and a range of scoring methods such as lipophilic efficiency. Several methods for predicting drug metabolism have also been proposed in the scientific literature.

Due to the large number of drug properties that must be simultaneously optimized during the design process, multi-objective optimization techniques are sometimes employed. Finally because of the limitations in the current methods for prediction of activity, drug design is still very much reliant on serendipity and bounded rationality.

Computer-aided Drug Design

The most fundamental goal in drug design is to predict whether a given molecule will bind to a target and if so how strongly. Molecular mechanics or molecular dynamics is most often used to estimate the strength of the intermolecular interaction between the small molecule and its biological target. These methods are also used to predict the conformation of the small molecule and to model conformational changes in the target that may occur when the small molecule binds to it. Semi-empirical, ab initio quantum chemistry methods, or density functional theory are often used to provide optimized parameters for the molecular mechanics calculations and also provide an estimate of the electronic properties (electrostatic potential, polarizability, etc.) of the drug candidate that will influence binding affinity.

Molecular mechanics methods may also be used to provide semi-quantitative prediction of the binding affinity. Also, knowledge-based scoring function may be used to provide binding affinity estimates. These methods use linear regression, machine learning, neural nets or other statistical techniques to derive predictive binding affinity equations by fitting experimental affinities to computationally derived interaction energies between the small molecule and the target.

Ideally, the computational method will be able to predict affinity before a compound is synthesized and hence in theory only one compound needs to be synthesized, saving enormous time and cost. The reality is that present computational methods are imperfect and provide, at best, only qualitatively accurate estimates of affinity. In practice it still takes several iterations of design, synthesis, and testing before an optimal drug is discovered. Computational methods have accelerated discovery by reducing the number of iterations required and have often provided novel structures.

Drug design with the help of computers may be used at any of the following stages of drug discovery:

1. hit identification using virtual screening (structure- or ligand-based design)
2. hit-to-lead optimization of affinity and selectivity (structure-based design, QSAR, etc.)
3. lead optimization of other pharmaceutical properties while maintaining affinity

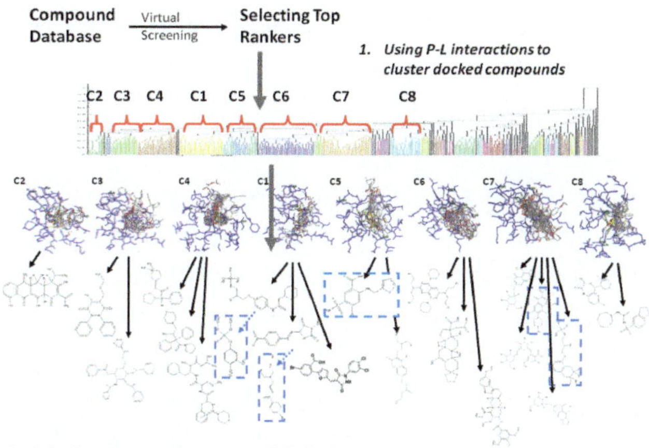

Flowchart of a Usual Clustering Analysis for Structure-Based Drug Design

In order to overcome the insufficient prediction of binding affinity calculated by recent scoring functions, the protein-ligand interaction and compound 3D structure information are used for analysis. For structure-based drug design, several post-screening analyses focusing on protein-ligand interaction have been developed for improving enrichment and effectively mining potential candidates:

- Consensus scoring
  - Selecting candidates by voting of multiple scoring functions
  - May lose the relationship between protein-ligand structural information and scoring criterion
- Cluster analysis
  - Represent and cluster candidates according to protein-ligand 3D information
  - Needs meaningful representation of protein-ligand interactions.

Types

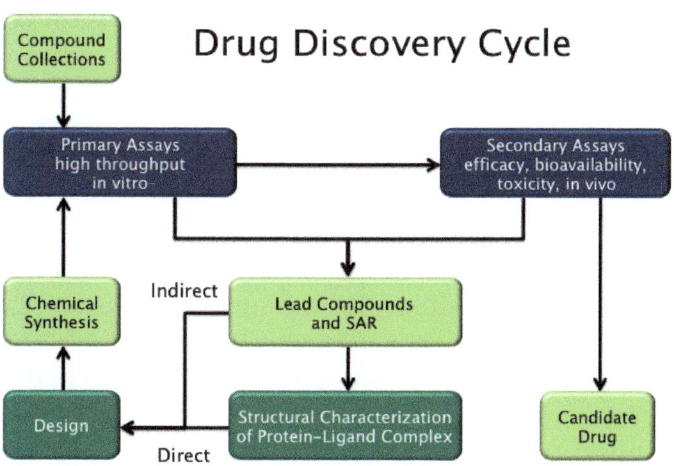

Drug discovery cycle highlighting both ligand-based (indirect) and structure-based (direct) drug design strategies.

There are two major types of drug design. The first is referred to as ligand-based drug design and the second, structure-based drug design.

Ligand-based

Ligand-based drug design (or indirect drug design) relies on knowledge of other molecules that bind to the biological target of interest. These other molecules may be used to derive a pharmacophore model that defines the minimum necessary structural characteristics a molecule must possess in order to bind to the target. In other words, a model of the biological target may be built based on the knowledge of what binds to it, and this model in turn may be used to design new molecular entities that interact with the target. Alternatively, a quantitative structure-activity relationship (QSAR), in which a correlation between calculated properties of molecules and their experimentally determined biological activity, may be derived. These QSAR relationships in turn may be used to predict the activity of new analogs.

Structure-based

Structure-based drug design (or direct drug design) relies on knowledge of the three dimensional structure of the biological target obtained through methods such as x-ray crystallography or NMR spectroscopy. If an experimental structure of a target is not available, it may be possible to create a homology model of the target based on the experimental structure of a related protein. Using the structure of the biological target, candidate drugs that are predicted to bind with high affinity and selectivity to the target may be designed using interactive graphics and the intuition of a medicinal chemist. Alternatively various automated computational procedures may be used to suggest new drug candidates.

Current methods for structure-based drug design can be divided roughly into three main categories. The first method is identification of new ligands for a given receptor by searching large databases of 3D structures of small molecules to find those fitting the binding pocket of the receptor using fast approximate docking programs. This method is known as virtual screening. A second category is de novo design of new ligands. In this method, ligand molecules are built up within the constraints of the binding pocket by assembling small pieces in a stepwise manner. These pieces can be either individual atoms or molecular fragments. The key advantage of such a method is that novel structures, not contained in any database, can be suggested. A third method is the optimization of known ligands by evaluating proposed analogs within the binding cavity.

Binding Site Identification

Binding site identification is the first step in structure based design. If the structure of the target or a sufficiently similar homolog is determined in the presence of a bound ligand, then the ligand should be observable in the structure in which case location of the binding site is trivial. However, there may be unoccupied allosteric binding sites that may be of interest. Furthermore, it may be that only apoprotein (protein without ligand) structures are available and the reliable identification of unoccupied sites that have the potential to bind ligands with high affinity is non-trivial. In brief, binding site identification usually relies on identification of concave surfaces on the protein that can accommodate drug sized molecules that also possess appropriate "hot spots" (hydrophobic surfaces, hydrogen bonding sites, etc.) that drive ligand binding.

Scoring Functions

Structure-based drug design attempts to use the structure of proteins as a basis for designing new ligands by applying the principles of molecular recognition. Selective high affinity binding to the target is generally desirable since it leads to more efficacious drugs with fewer side effects. Thus, one of the most important principles for designing or obtaining potential new ligands is to predict the binding affinity of a certain ligand to its target (and known antitargets) and use the predicted affinity as a criterion for selection.

One early general-purposed empirical scoring function to describe the binding energy of ligands to receptors was developed by Böhm. This empirical scoring function took the form:

where:

- ΔG_0 – empirically derived offset that in part corresponds to the overall loss of translational and rotational entropy of the ligand upon binding.
- ΔG_{hb} – contribution from hydrogen bonding
- ΔG_{ionic} – contribution from ionic interactions
- ΔG_{lip} – contribution from lipophilic interactions where $|A_{lipo}|$ is surface area of lipophilic contact between the ligand and receptor
- ΔG_{rot} – entropy penalty due to freezing a rotatable in the ligand bond upon binding

A more general thermodynamic "master" equation is as follows:

where:

- desolvation – enthalpic penalty for removing the ligand from solvent
- motion – entropic penalty for reducing the degrees of freedom when a ligand binds to its receptor
- configuration – conformational strain energy required to put the ligand in its "active" conformation
- interaction – enthalpic gain for "resolvating" the ligand with its receptor

The basic idea is that the overall binding free energy can be decomposed into independent components that are known to be important for the binding process. Each component reflects a certain kind of free energy alteration during the binding process between a ligand and its target receptor. The Master Equation is the linear combination of these components. According to Gibbs free energy equation, the relation between dissociation equilibrium constant, K_d, and the components of free energy was built.

Various computational methods are used to estimate each of the components of the master equation. For example, the change in polar surface area upon ligand binding can be used to estimate the desolvation energy. The number of rotatable bonds frozen upon ligand binding is proportional to the motion term. The configurational or strain energy can be estimated using molecular mechanics calculations. Finally the interaction energy can be estimated using methods such as the change in non polar surface, statistically derived potentials of mean force, the number of hydrogen bonds formed, etc. In practice, the components of the master equation are fit to experimental data using multiple linear regression. This can be done with a diverse training set including many types of ligands and receptors to produce a less accurate but more general "global" model or a more restricted set of ligands and receptors to produce a more accurate but less general "local" model.

Examples

A particular example of rational drug design involves the use of three-dimensional information about biomolecules obtained from such techniques as X-ray crystallography and NMR spectroscopy. Computer-aided drug design in particular becomes much more tractable when there is a high-resolution structure of a target protein bound to a potent ligand. This approach to drug discovery is sometimes referred to as structure-based drug design. The first unequivocal example of the application of structure-based drug design leading to an approved drug is the carbonic anhydrase inhibitor dorzolamide, which was approved in 1995.

Another important case study in rational drug design is imatinib, a tyrosine kinase inhibitor designed specifically for the *bcr-abl* fusion protein that is characteristic for Philadelphia chromosome-positive leukemias (chronic myelogenous leukemia and occasionally acute lymphocytic leukemia). Imatinib is substantially different from previous drugs for cancer, as most agents of chemotherapy simply target rapidly dividing cells, not differentiating between cancer cells and other tissues.

Additional examples include:

- Many of the atypical antipsychotics
- Cimetidine, the prototypical H_2-receptor antagonist from which the later members of the class were developed
- Selective COX-2 inhibitor NSAIDs
- Enfuvirtide, a peptide HIV entry inhibitor
- Nonbenzodiazepines like zolpidem and zopiclone
- Raltegravir, an HIV integrase inhibitor
- SSRIs (selective serotonin reuptake inhibitors), a class of antidepressants
- Zanamivir, an antiviral drug

Case Studies

- 5-HT3 antagonists
- Acetylcholine receptor agonists
- Angiotensin receptor antagonists

- Bcr-Abl tyrosine-kinase inhibitors
- Cannabinoid receptor antagonists
- CCR5 receptor antagonists
- Cyclooxygenase 2 inhibitors
- Dipeptidyl peptidase-4 inhibitors
- HIV protease inhibitors
- NK1 receptor antagonists
- Non-nucleoside reverse transcriptase inhibitors
- Nucleoside and nucleotide reverse transcriptase inhibitors
- PDE5 inhibitors
- Proton pump inibitors
- Renin inhibitors
- Triptans
- TRPV1 antagonists
- c-Met inhibitors

Criticism

It has been argued that the highly rigid and focused nature of rational drug design suppresses serendipity in drug discovery. Because many of the most significant medical discoveries have been inadvertent, the recent focus on rational drug design may limit the progress of drug discovery. Furthermore, the rational design of a drug may be limited by a crude or incomplete understanding of the underlying molecular processes of the disease it is intended to treat.

Drug Metabolism

Drug metabolism is the metabolic breakdown of drugs by living organisms, usually through specialized enzymatic systems. More generally, xenobiotic metabolism (from the Greek xenos "stranger" and biotic "related to living beings") is the set of metabolic pathways that modify the chemical structure of xenobiotics, which are compounds foreign to an organism's normal biochemistry, such any drug or poison. These pathways are a form of biotransformation present in all major groups of organisms, and are considered to be of ancient origin. These reactions often act to detoxify poisonous compounds (although in some cases the intermediates in xenobiotic metabolism can themselves cause toxic effects). The study of drug metabolism is called pharmacokinetics.

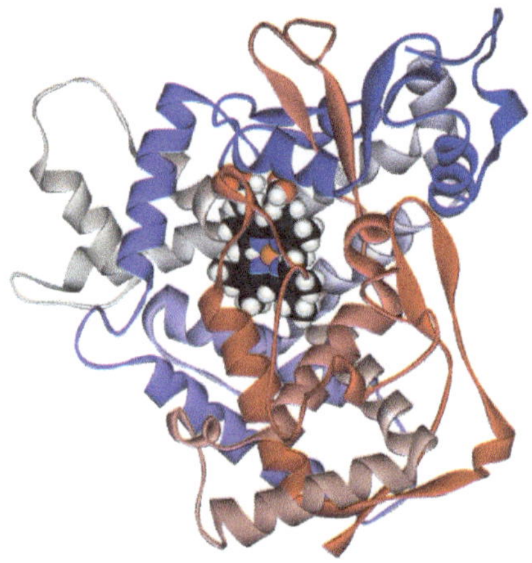

Cytochrome P450 oxidases are important enzymes in xenobiotic metabolism.

The metabolism of pharmaceutical drugs is an important aspect of pharmacology and medicine. For example, the rate of metabolism determines the duration and intensity of a drug's pharmacologic action. Drug metabolism also affects multidrug resistance in infectious diseases and in chemotherapy for cancer, and the actions of some drugs as substrates or inhibitors of enzymes involved in xenobiotic metabolism are a common reason for hazardous drug interactions. These pathways are also important in environmental science, with the xenobiotic metabolism of microorganisms determining whether a pollutant will be broken down during bioremediation, or persist in the environment. The enzymes of xenobiotic metabolism, particularly the glutathione S-transferases are also important in agriculture, since they may produce resistance to pesticides and herbicides.

Drug metabolism is divided into three phases. In phase I, enzymes such as cytochrome P450 oxidases introduce reactive or polar groups into xenobiotics. These modified compounds are then conjugated to polar compounds in phase II reactions. These reactions are catalysed by transferase enzymes such as glutathione S-transferases. Finally, in phase III, the conjugated xenobiotics may be further processed, before being recognised by efflux transporters and pumped out of cells. Drug metabolism often converts lipophilic compounds into hydrophilic products that are more readily excreted.

Permeability Barriers and Detoxification

The exact compounds an organism is exposed to will be largely unpredictable, and may differ widely over time; these are major characteristics of xenobiotic toxic stress. The major challenge faced by xenobiotic detoxification systems is that they must be able to remove the almost-limitless number of xenobiotic compounds from the complex mixture of chemicals involved in normal metabolism. The solution that has evolved to address this problem is an elegant combination of physical barriers and low-specificity enzymatic systems.

All organisms use cell membranes as hydrophobic permeability barriers to control access to their internal environment. Polar compounds cannot diffuse across these cell membranes, and the uptake of useful molecules is mediated through transport proteins that specifically select substrates from the extracellular mixture. This selective uptake means that most hydrophilic molecules cannot enter cells, since they are not recognised by any specific transporters. In contrast, the diffusion of hydrophobic compounds across these barriers cannot be controlled, and organisms, therefore, cannot exclude lipid-soluble xenobiotics using membrane barriers.

However, the existence of a permeability barrier means that organisms were able to evolve detoxification systems that exploit the hydrophobicity common to membrane-permeable xenobiotics. These systems therefore solve the specificity problem by possessing such broad substrate specificities that they metabolise almost any non-polar compound. Useful metabolites are excluded since they are polar, and in general contain one or more charged groups.

The detoxification of the reactive by-products of normal metabolism cannot be achieved by the systems outlined above, because these species are derived from normal cellular constituents and usually share their polar characteristics. However, since these compounds are few in number, specific enzymes can recognize and remove them. Examples of these specific detoxification systems are the glyoxalase system, which removes the reactive aldehyde methylglyoxal, and the various antioxidant systems that eliminate reactive oxygen species.

Phases of Detoxification

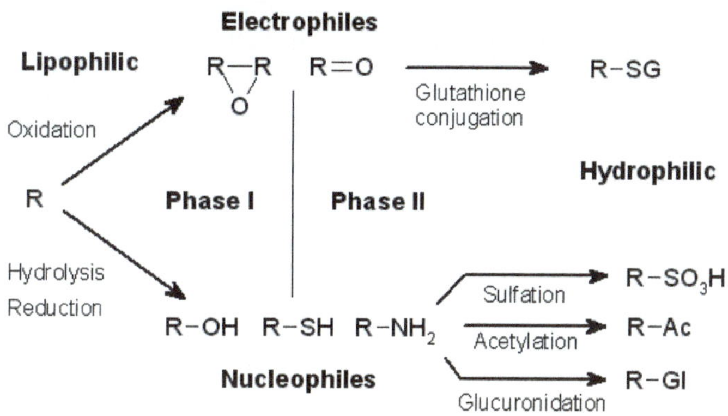

Phases I and II of the metabolism of a lipophilic xenobiotic.

The metabolism of xenobiotics is often divided into three phases:- modification, conjugation, and excretion. These reactions act in concert to detoxify xenobiotics and remove them from cells.

Phase I – Modification

In phase I, a variety of enzymes act to introduce reactive and polar groups into their substrates. One of the most common modifications is hydroxylation catalysed by the cytochrome P-450-dependent mixed-function oxidase system. These enzyme complexes act to incorporate an atom of oxygen into nonactivated hydrocarbons, which can result in either the introduction of hydroxyl groups or N-, O- and S-dealkylation of substrates. The reaction mechanism of the P-450 oxidases proceeds through the reduction of cytochrome-bound oxygen and the generation of a highly-reactive oxyferryl species, according to the following scheme:

$$O_2 + NADPH + H^+ + RH \rightarrow NADP^+ + H_2O + ROH$$

Phase I reactions (also termed nonsynthetic reactions) may occur by oxidation, reduction, hydrolysis, cyclization, decyclization, and addition of oxygen or removal of hydrogen, carried out by mixed function oxidases, often in the liver. These oxidative reactions typically involve a cytochrome

P450 monooxygenase (often abbreviated CYP), NADPH and oxygen. The classes of pharmaceutical drugs that utilize this method for their metabolism include phenothiazines, paracetamol, and steroids. If the metabolites of phase I reactions are sufficiently polar, they may be readily excreted at this point. However, many phase I products are not eliminated rapidly and undergo a subsequent reaction in which an endogenous substrate combines with the newly incorporated functional group to form a highly polar conjugate.

A common Phase I oxidation involves conversion of a C-H bond to a C-OH. This reaction sometimes converts a pharmacologically inactive compound (a prodrug) to a pharmacologically active one. By the same token, Phase I can turn a nontoxic molecule into a poisonous one (toxification). Simple hydrolysis in the stomach is normally an innocuous reaction, however there are exceptions. For example, phase I metabolism converts acetonitrile to $HOCH_2CN$, which rapidly dissociates into formaldehyde and hydrogen cyanide, both of which are toxic.

Phase I metabolism of drug candidates can be simulated in the laboratory using non-enzyme catalysts. This example of a biomimetic reaction tends to give products that often contains the Phase I metabolites. As an example, the major metabolite of the pharmaceutical trimebutine, desmethyltrimebutine (nor-trimebutine), can be efficiently produced by in vitro oxidation of the commercially available drug. Hydroxylation of an N-methyl group leads to expulsion of a molecule of formaldehyde, while oxidation of the O-methyl groups takes place to a lesser extent.

Oxidation

- Cytochrome P450 monooxygenase system
- Flavin-containing monooxygenase system
- Alcohol dehydrogenase and aldehyde dehydrogenase
- Monoamine oxidase
- Co-oxidation by peroxidases

Reduction

- NADPH-cytochrome P450 reductase

Cytochrome P450 reductase, also known as NADPH:ferrihemoprotein oxidoreductase, NADPH:hemoprotein oxidoreductase, NADPH:P450 oxidoreductase, P450 reductase, POR, CPR, CYPOR, is a membrane-bound enzyme required for electron transfer to cytochrome P450 in the microsome of the eukaryotic cell from a FAD- and FMN-containing enzyme NADPH:cytochrome P450 reductase The general scheme of electron flow in the POR/P450 system is: NADPH → FAD → FMN → P450 → O_2

- Reduced (ferrous) cytochrome P450

During reduction reactions, a chemical can enter *futile cycling*, in which it gains a free-radical electron, then promptly loses it to oxygen (to form a superoxide anion).

Hydrolysis

- Esterases and amidase
- Epoxide hydrolase

Phase II – Conjugation

In subsequent phase II reactions, these activated xenobiotic metabolites are conjugated with charged species such as glutathione (GSH), sulfate, glycine, or glucuronic acid. Sites on drugs where conjugation reactions occur include carboxyl (-COOH), hydroxyl (-OH), amino (NH_2), and sulfhydryl (-SH) groups. Products of conjugation reactions have increased molecular weight and tend to be less active than their substrates, unlike Phase I reactions which often produce active metabolites. The addition of large anionic groups (such as GSH) detoxifies reactive electrophiles and produces more polar metabolites that cannot diffuse across membranes, and may, therefore, be actively transported.

These reactions are catalysed by a large group of broad-specificity transferases, which in combination can metabolise almost any hydrophobic compound that contains nucleophilic or electrophilic groups. One of the most important classes of this group is that of the glutathione S-transferases (GSTs).

Mechanism	Involved enzyme	Co-factor	Location
methylation	methyltransferase	S-adenosyl-L-methionine	liver, kidney, lung, CNS
sulphation	sulfotransferases	3'-phosphoadenosine-5'-phosphosulfate	liver, kidney, intestine
acetylation	• N-acetyltransferases • bile acid-CoA:amino acid N-acyltransferases	acetyl coenzyme A	liver, lung, spleen, gastric mucosa, RBCs, lymphocytes
glucuronidation	UDP-glucuronosyltransferases	UDP-glucuronic acid	liver, kidney, intestine, lung, skin, prostate, brain
glutathione conjugation	glutathione S-transferases	glutathione	liver, kidney
glycine conjugation	acetyl Co-enzyme As	glycine	liver, kidney

Phase III – Further Modification and Excretion

After phase II reactions, the xenobiotic conjugates may be further metabolised. A common example is the processing of glutathione conjugates to acetylcysteine (mercapturic acid) conjugates. Here, the γ-glutamate and glycine residues in the glutathione molecule are removed by Gamma-glutamyl transpeptidase and dipeptidases. In the final step, the cystine residue in the conjugate is acetylated.

Conjugates and their metabolites can be excreted from cells in phase III of their metabolism, with the anionic groups acting as affinity tags for a variety of membrane transporters of the multidrug resistance protein (MRP) family. These proteins are members of the family of ATP-binding cassette transporters and can catalyse the ATP-dependent transport of a huge variety of hydrophobic anions, and thus act to remove phase II products to the extracellular medium, where they may be further metabolised or excreted.

Endogenous Toxins

The detoxification of endogenous reactive metabolites such as peroxides and reactive aldehydes often cannot be achieved by the system described above. This is the result of these species' being derived from normal cellular constituents and usually sharing their polar characteristics. However, since these compounds are few in number, it is possible for enzymatic systems to utilize specific molecular recognition to recognize and remove them. The similarity of these molecules to useful metabolites therefore means that different detoxification enzymes are usually required for the metabolism of each group of endogenous toxins. Examples of these specific detoxification systems are the glyoxalase system, which acts to dispose of the reactive aldehyde methylglyoxal, and the various antioxidant systems that remove reactive oxygen species.

Sites

Quantitatively, the smooth endoplasmic reticulum of the liver cell is the principal organ of drug metabolism, although every biological tissue has some ability to metabolize drugs. Factors responsible for the liver's contribution to drug metabolism include that it is a large organ, that it is the first organ perfused by chemicals absorbed in the gut, and that there are very high concentrations of most drug-metabolizing enzyme systems relative to other organs. If a drug is taken into the GI tract, where it enters hepatic circulation through the portal vein, it becomes well-metabolized and is said to show the *first pass effect*.

Other sites of drug metabolism include epithelial cells of the gastrointestinal tract, lungs, kidneys, and the skin. These sites are usually responsible for localized toxicity reactions.

Factors that Affect Drug Metabolism

The duration and intensity of pharmacological action of most lipophilic drugs are determined by the rate they are metabolized to inactive products. The Cytochrome P450 monooxygenase system is the most important pathway in this regard. In general, anything that *increases* the rate of metabolism (*e.g.*, enzyme induction) of a pharmacologically active metabolite will *decrease* the duration and intensity of the drug action. The opposite is also true (*e.g.*, enzyme inhibition). However, in cases where an enzyme is responsible for metabolizing a pro-drug into a drug, enzyme induction can speed up this conversion and increase drug levels, potentially causing toxicity.

Various *physiological* and *pathological* factors can also affect drug metabolism. Physiological factors that can influence drug metabolism include age, individual variation (*e.g.*, pharmacogenetics), enterohepatic circulation, nutrition, intestinal flora, or sex differences.

In general, drugs are metabolized more slowly in fetal, neonatal and elderly humans and animals than in adults.

Genetic variation (polymorphism) accounts for some of the variability in the effect of drugs. With N-acetyltransferases (involved in *Phase II* reactions), individual variation creates a group of people who acetylate slowly (*slow acetylators*) and those who acetylate quickly, split roughly 50:50 in the population of Canada. This variation may have dramatic consequences, as the slow acetylators are more prone to dose-dependent toxicity.

Cytochrome P450 monooxygenase system enzymes can also vary across individuals, with deficiencies occurring in 1 – 30% of people, depending on their ethnic background.

Pathological factors can also influence drug metabolism, including liver, kidney, or heart diseases.

In silico modelling and simulation methods allow drug metabolism to be predicted in virtual patient populations prior to performing clinical studies in human subjects. This can be used to identify individuals most at risk from adverse reaction.

History

Studies on how people transform the substances that they ingest began in the mid-nineteenth century, with chemists discovering that organic chemicals such as benzaldehyde could be oxidized and conjugated to amino acids in the human body. During the remainder of the nineteenth century, several other basic detoxification reactions were discovered, such as methylation, acetylation, and sulfonation.

In the early twentieth century, work moved on to the investigation of the enzymes and pathways that were responsible for the production of these metabolites. This field became defined as a separate area of study with the publication by Richard Williams of the book *Detoxication mechanisms* in 1947. This modern biochemical research resulted in the identification of glutathione *S*-transferases in 1961, followed by the discovery of cytochrome P450s in 1962, and the realization of their central role in xenobiotic metabolism in 1963.

Factors Affecting Drug Metabolism

Enterohepatic Circulation

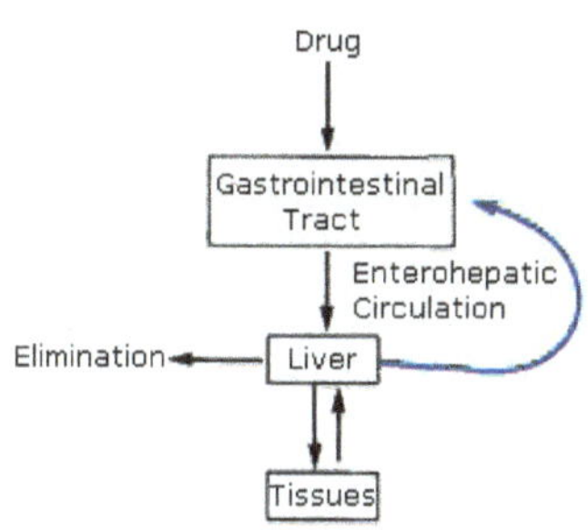

Enterohepatic circulation of drugs.

Enterohepatic circulation refers to the circulation of biliary acids, bilirubin, drugs, or other substances from the liver to the bile, followed by entry into the small intestine, absorption by the enterocyte and transport back to the liver.

Biliary Acids

The Circuit

Hepatocytes metabolize cholesterol to cholic acid and chenodeoxycholic acid. These lipid-soluble bile acids are conjugated (reversibly attached) mainly to glycine or taurine molecules to form water soluble primary conjugated bile acids, sometimes called "bile salts". These bile acids travel to the gall bladder during the interdigestive phase for storage and to the descending part of the duodenum via the common bile duct through the major duodenal papilla during digestion. 95% of the bile acids which are delivered to the duodenum will be recycled by the enterohepatic circulation.

Due to the pH of the small intestine, most of the bile acids are ionized and mostly occur as their sodium salts which are then called "primary conjugated bile salts." In the lower small intestine and colon, bacteria dehydroxylate some of the primary bile salts to form secondary conjugated bile salts (which are still water-soluble). Along the proximal and distal ileum, these conjugated primary bile salts are reabsorbed actively into hepatic portal circulation. Bacteria deconjugate some of the primary and secondary conjugated bile salts back to lipid-soluble bile acids, which are passively absorbed into hepatic portal circulation. Finally, the conjugated bile acids which remained un-ionized conjugated bile acids are passively absorbed.

Venous blood from the ileum goes straight into the portal vein and then into the liver sinusoids. There, hepatocytes extract bile acids very efficiently, and little escapes the healthy liver into systemic circulation.

The net effect of enterohepatic recirculation is that each bile salt molecule is reused about 20 times, often multiple times during a single digestive phase.

Function

The presence of biliary acids in the intestines helps in digestion of fats and other substances.

Bilirubin

Bilirubin is conjugated with glucuronic acid in the liver by the enzyme glucuronyltransferase, making it soluble in water. Much of it goes into the bile and thus out into the small intestine. Although 95% of the secreted bile is reabsorbed by the small intestine, conjugated bilirubin is not reabsorbed in small intestine. All conjugated bilirubin in the large intestine is metabolised by colonic bacteria to urobilinogen, which is then further oxidized to urobilin and stercobilin. Urobilin, stercobilin and their degradation products give feces its brown color. However, just like bile, some of the urobilinogen reabsorbed is resecreted in the bile which is also part of enterohepatic circulation. The rest of the reabsorbed urobilinogen is excreted in the urine where it is converted to an oxidized form, urobilin, which gives urine its characteristic yellow color.

Drugs

Enterohepatic circulation also means that some molecules which would not otherwise be very toxic can become extremely hepatotoxic as they reach unexpectedly high hepatic concentrations. Drugs may remain in the enterohepatic circulation for a prolonged period of time as a result of this recycling process.

Gut Flora

Gut flora (gut microbiota, or gastrointestinal microbiota) is the complex community of microorganisms that live in the digestive tracts of humans and other animals, including insects. The gut metagenome is the aggregate of all the genomes of gut microbiota. The gut is one niche that human microbiota inhabit.

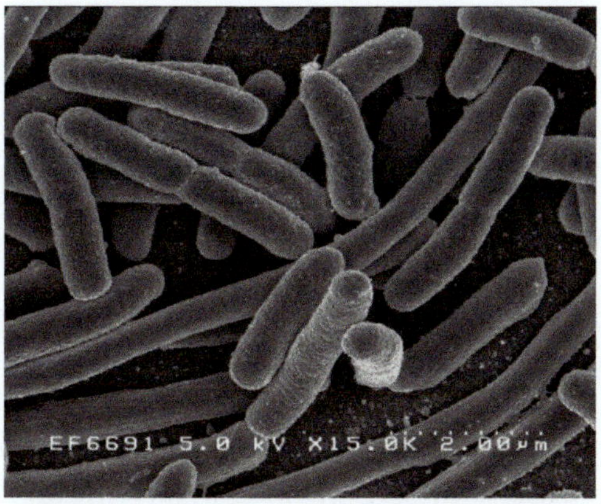

Escherichia coli, one of the many species of bacteria present in the human gut

In humans, the gut microbiota has the largest numbers of bacteria and the greatest number of species compared to other areas of the body. In humans the gut flora is established at one to two years after birth, and by that time the intestinal epithelium and the intestinal mucosal barrier that it secretes have co-developed in a way that is tolerant to, and even supportive of, the gut flora and that also provides a barrier to pathogenic organisms.

The relationship between some gut flora and humans is not merely commensal (a non-harmful coexistence), but rather a mutualistic relationship. Some human gut microorganisms benefit the host by fermenting dietary fiber into short-chain fatty acids (SCFAs), such as acetic acid and butyric acid, which are then absorbed by the host. Intestinal bacteria also play a role in synthesizing vitamin B and vitamin K as well as metabolizing bile acids, sterols, and xenobiotics. The systemic importance of the SCFAs and other compounds they produce are like hormones and the gut flora itself appears to function like an endocrine organ, and dysregulation of the gut flora has been correlated with a host of inflammatory and autoimmune conditions.

The composition of human gut flora changes over time, when the diet changes, and as overall health changes. A systematic review from July 2016 found that certain probiotics possess treatment efficacy for anxiety, depression, autism spectrum disorder, and obsessive–compulsive disorder.

Types

The microbial composition of the gut flora varies across the digestive tract. In the stomach and small intestine, relatively few species of bacteria are generally present. The colon, in contrast, contains a densely-populated microbial ecosystem with up to 10^{12} cells per gram of intestinal content. These bacteria represent between 300 and 1000 different species. However, 99% of the bacteria come from about 30 or 40 species. As a consequence of their abundance in the intestine, bacteria also make up to 60% of the dry mass of feces. Fungi, archaea, and viruses are also present in the gut flora, but less is known about their activities.

Over 99% of the bacteria in the gut are anaerobes, but in the cecum, aerobic bacteria reach high densities. It is estimated that these gut flora have around a hundred times as many genes in aggregate as there are in the human genome.

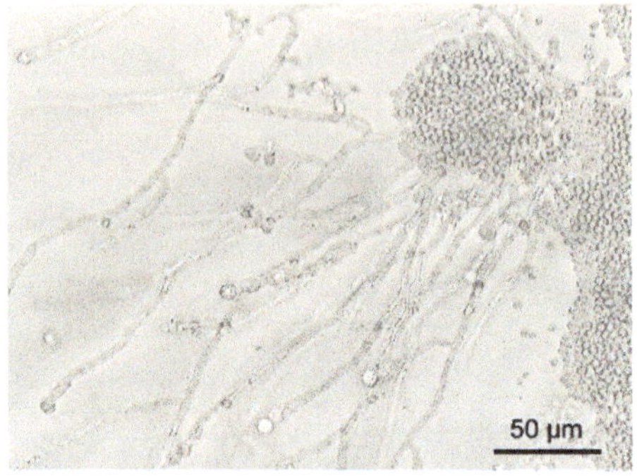

Candida albicans, a dimorphic fungus that grows as a yeast in the gut

Many species in the gut have not been studied outside of their hosts because most cannot be cultured. While there are a small number of core species of microbes shared by most individuals, populations of microbes can vary widely among different individuals. Within an individual, microbe populations stay fairly constant over time, even though some alterations may occur with changes in lifestyle, diet and age. The Human microbiome project has set out to better describe the microflora of the human gut and other body locations.

The four dominant bacterial phyla in the human gut are Firmicutes, Bacteroidetes, Actinobacteria, and Proteobacteria. Most bacteria belong to the genera *Bacteroides, Clostridium, Faecalibacterium, Eubacterium, Ruminococcus, Peptococcus, Peptostreptococcus*, and *Bifidobacterium*. Other genera, such as *Escherichia* and *Lactobacillus*, are present to a lesser extent. Species from the genus *Bacteroides* alone constitute about 30% of all bacteria in the gut, suggesting that this genus is especially important in the functioning of the host.

Fungal genera that have been detected in the gut include *Candida, Saccharomyces, Aspergillus, Penicillium, Rhodotorula, Trametes, Pleospora, Sclerotinia, Bullera*, and *Galactomyces*, among others. *Rhodotorula* is most frequently found in individuals with inflammatory bowel disease while *Candida* is most frequently found in individuals with hepatitis B cirrhosis and chronic hepatitis B.

Archaea constitute another large class of gut flora which are important in the metabolism of the bacterial products of fermentation.

Enterotype

An enterotype is a classification of living organisms based on its bacteriological ecosystem in the human gut microbiome not dictated by age, gender, body weight, or national divisions. There are indications that long-term diet influences enterotype. Three human enterotypes have been discovered.

Flora Composition

Anatomy

Stomach Flora

Due to the high acidity of the stomach, most microorganisms cannot survive. The main bacterial inhabitants of the stomach include: *Streptococcus, Staphylococcus, Lactobacillus, Peptostreptococcus*, and types of yeast. *Helicobacter pylori* is a Gram-negative spiral organism that establishes on gastric mucosa causing chronic gastritis and peptic ulcer disease and is a carcinogen for gastric cancer.

Intestinal Flora

Bacteria commonly found in the human colon	
Bacterium	**Incidence (%)**
Bacteroides fragilis	100
Bacteroides melaninogenicus	100
Bacteroides oralis	100
Enterococcus faecalis	100
Escherichia coli	100
Enterobacter sp.	40–80
Klebsiella sp.	40–80
Bifidobacterium bifidum	30–70
Staphylococcus aureus	30–50
Lactobacillus	20–60
Clostridium perfringens	25–35
Proteus mirabilis	5–55
Clostridium tetani	1–35
Clostridium septicum	5–25
Pseudomonas aeruginosa	3–11
Salmonella enteritidis	3–7
Faecalibacterium prausnitzii	?common
Peptostreptococcus sp.	?common
Peptococcus sp.	?common

The small intestine contains a trace amount of microorganisms due to the proximity and influence of the stomach. Gram positive cocci and rod shaped bacteria are the predominant microorganisms found in the small intestine. However, in the distal portion of the small intestine alkaline conditions support gram-positive bacteria of the *Enterobacteriaceae*. The bacterial flora of the small intestine aid in a wide range of intestinal functions. The bacterial flora provide regulatory signals that enable the development and utility of the gut. Overgrowth of bacteria in the small intestine can lead to intestinal failure. In addition the large intestine contains the largest bacterial ecosystem in the human body. Factors that disrupt the microorganism population of the large intestine include antibiotics, stress, and parasites.

Bacteria make up most of the flora in the colon and 60% of the dry mass of feces. This fact makes feces an ideal source to test for gut flora for any tests and experiments by extracting the nucleic acid from fecal specimens, and bacterial 16S rRNA gene sequences are generated with bacterial primers. This form of testing is also often preferable to more invasive techniques, such as biopsies. Somewhere between 300 and 1000 different species live in the gut, with most estimates at about 500., However, it is probable that 99% of the bacteria come from about 30 or 40 species, with *Faecalibacterium prausnitzii* being the most common species in healthy adults. Fungi and protozoa also make up a part of the gut flora, but little is known about their activities.

Research suggests that the relationship between gut flora and humans is not merely commensal (a non-harmful coexistence), but rather is a mutualistic, symbiotic relationship. Though people can (barely) survive with no gut flora, the microorganisms perform a host of useful functions, such as fermenting unused energy substrates, training the immune system via end products of metabolism like propionate and acetate, preventing growth of harmful species, regulating the development of the gut, producing vitamins for the host (such as biotin and vitamin K), and producing hormones to direct the host to store fats. Extensive modification and imbalances of the gut microbiota and its microbiome or gene collection are associated with obesity. However, in certain conditions, some species are thought to be capable of causing disease by causing infection or increasing cancer risk for the host.

Age

It has been demonstrated that there are common patterns of microbiome composition evolution during life. In general, the diversity of microbiota composition of fecal samples is significantly higher in adults than in children, although interpersonal differences are higher in children than in adults. Much of the maturation of microbiota into an adult-like configuration happens during the three first years of life.

As the microbiome composition changes, so does the composition of bacterial proteins produced in the gut. In adult microbiomes, a high prevalence of enzymes involved in fermentation, methanogenesis and the metabolism of arginine, glutamate, aspartate and lysine have been found. In contrast, in infant microbiomes the dominant enzymes are involved in cysteine metabolism and fermentation pathways.

Diet

Studies and statistical analyses have identified the different bacterial genera in gut microbiota and their associations with nutrient intake. Gut microflora is mainly composed of three enterotypes: *Prevotella*, *Bacteroides*, and *Ruminococcus*. There is an association between the concentration of each microbial community and diet. For example, *Prevotella* is related to carbohydrates and simple sugars, while *Bacteroides* is associated with proteins, amino acids, and saturated fats. One enterotype will dominate depending on the diet. Altering the diet will result in a corresponding change in the numbers of species.

Malnourished human children have less mature and less diverse gut microbiota than healthy children, and changes in the microbiome associated with nutrient scarcity can in turn be a pathophysiological cause of malnutrition. Malnourished children also typically have more potentially pathogenic gut flora, and more yeast in their mouths and throats.

Geography

Gut microbiome composition depends on the geographic origin of populations. Variations in trade off of *Prevotella*, the representation of the urease gene, and the representation of genes encoding glutamate synthase/degradation or other enzymes involved in amino acids degradation or vitamin biosynthesis show significant differences between populations from USA, Malawi or Amerindian origin.

The US population has a high representation of enzymes encoding the degradation of glutamine and enzymes involved in vitamin and lipoic acid biosynthesis; whereas Malawi and Amerindian populations have a high representation of enzymes encoding glutamate synthase and they also have an overrepresentation of α-amylase in their microbiomes. As the US population has a diet richer in fats than Amerindian or Malawian populations which have a corn-rich diet, the diet is probably a main determinant of gut bacterial composition.

Further studies have indicated a large difference in the composition of microbiota between European and rural African children. The fecal bacteria of children from Florence were compared to that of children from the small rural village of Boulpon in Burkina Faso. The diet of a typical child living in this village is largely lacking in fats and animal proteins and rich in polysaccharides and plant proteins. The fecal bacteria of European children was dominated by *Firmicutes* and showed a marked reduction in biodiversity, while the fecal bacteria of the Boulpon children was dominated by *Bacteroidetes*. The increased biodiversity and different composition of gut flora in African populations may aid in the digestion of normally indigestible plant polysaccharides and also may result in a reduced incidence of non-infectious colonic diseases.

On a smaller scale, it has been shown that sharing numerous common environmental exposures in a family is a strong determinant of individual microbiome composition. This effect has no genetic influence and it is consistently observed in culturally different populations.

Acquisition of Gut Flora in Human Infants

In humans, a gut flora similar to an adult's is formed within one to two years of birth. The gastrointestinal tract of a normal fetus has been considered to be sterile, however recently it has been acknowledged that microbial colonisation may occur in the fetus. During birth and rapidly thereafter, bacteria from the mother and the surrounding environment colonize the infant's gut. As of 2013, it was unclear whether most of colonizing arise from the mother or not. Infants born by caesarean section may also be exposed to their mothers' microflora, but the initial exposure is most likely to be from the surrounding environment such as the air, other infants, and the nursing staff, which serve as vectors for transfer. During the first year of life, the composition of the gut flora is generally simple and it changes a great deal with time and is not the same across individuals.

The initial bacterial population are generally facultative anaerobic organisms; investigators believe that these initial colonizers decrease the oxygen concentration in the gut, which in turn allows purely aneorobic bacteria like *Bacteroides*, *Actinobacteria*, and *Firmicutes* to become established and thrive. Breast-fed babies become dominated by bifidobacteria, possibly due to the contents of bifidobacterial growth factors in breast milk. In contrast, the microbiota of formula-fed infants is more diverse, with high numbers of *Enterobacteriaceae*, enterococci, bifidobacteria, *Bacteroides*, and clostridia.

Functions

When the gut flora first started to be studied, it was thought to have three key roles: directly defending against pathogens, fortifying host defense by its role in developing and maintaining the intestinal epithelium and inducing antibody production there, and metabolizing otherwise undigestable compounds in food; subsequent work discovered its role in training the developing immune system, and yet further work focused on its role in the gut-brain axis.

Direct Inhibition of Pathogens

The gut flora community plays a direct role in defensing against pathogens by fully colonizing the space, making use of all available nutrients, and by secreting compounds that kill or inhibit unwelcome organisms that would compete for nutrients with it. Disruption of the gut flora allows competing organisms like *Clostridium difficile* to become established that otherwise are kept in abeyance.

Development of Enteric Protection and Immune System

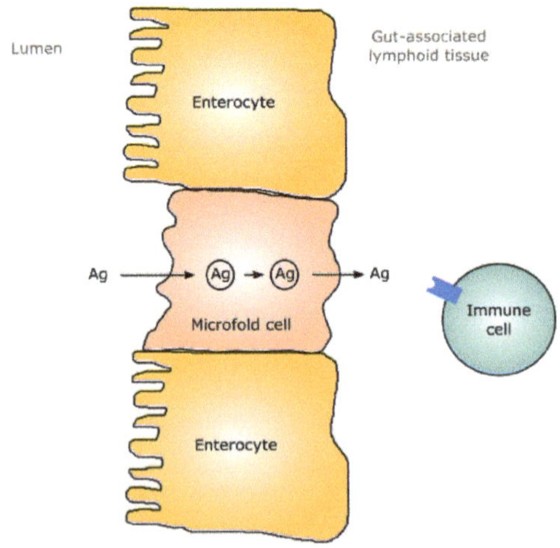

Microfold cells transfer antigens (Ag) from the lumen of the gut to gut-associated lymphoid tissue (GALT) via transcytosis and present them to different innate and adaptive immune cells.

In humans, a gut flora similar to an adult's is formed within one to two years of birth. As the gut flora gets established, the lining of the intestines – the intestinal epithelium and the intestinal mucosal barrier that it secretes – develop as well, in a way that is tolerant to, and even supportive of, commensurate microorganisms to a certain extent and also provides a barrier to pathogenic ones. Specifically, goblet cells that produce the muscosa proliferate, and the mucosa layer thickens, providing an outside mucosal layer in which "friendly" microorganisms can anchor and feed, and an inner layer that even these organisms cannot penetrate. Additionally, the development of gut-associated lymphoid tissue (GALT), which forms part of the intestinal epithelium and which detects and reacts to pathogens, appears and develops during the time that the gut flora develops and established. The GALT that develops is tolerant to gut flora species, but not to other microorganisms. GALT also normally becomes tolerant to food to which the infant is exposed, as well as digestive products of food, and gut flora's metabolites produced from food.

The human immune system creates cytokines that can drive the immune system to produce inflammation in order to protect itself, and that can tamp down the immune response to maintain homeostasis and allow healing after insult or injury. Different bacterial species that appear in gut flora have been shown to be able to drive the immune system to create cytokines selectively; for example *Bacteroides fragilis* and some *Clostridia* species appear to drive an anti-inflammatory response, while some segmented filamentous bacteria drive the production of inflammatory cytokines. Gut flora can also regulate the production of antibodies by the immune system. These cytokines and antibodies can have effects outside the gut, in the lungs and other tissues.

Metabolism

Tryptophan Metabolism by Human Gastrointestinal Microbiota (V • T • E)

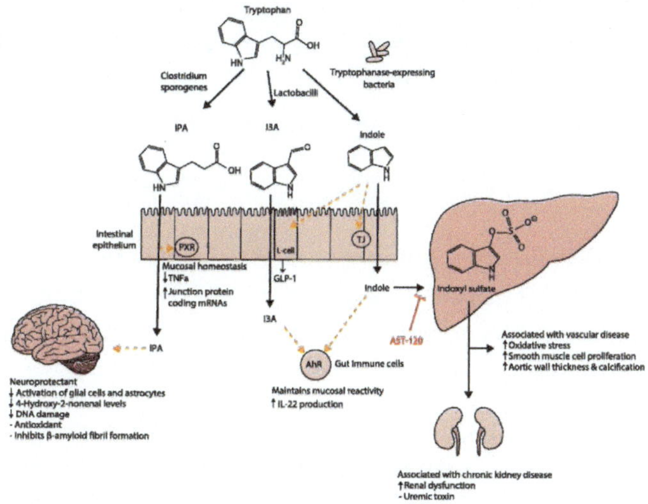

This diagram shows the biosynthesis of bioactive compounds (indole and certain derivatives) from tryptophan by bacteria in the gut. Indole is produced from tryptophan by bacteria that express tryptophanase. *Clostridium sporogenes* metabolizes indole into 3-indolepropionic acid (IPA), a highly potent neuroprotective antioxidant that scavenges hydroxyl radicals. In the intestine, IPA binds to pregnane X receptors (PXR) in intestinal cells, thereby facilitating mucosal homeostasis and barrier function. Following absorption from the intestine and distribution to the brain, IPA confers a neuroprotective effect against cerebral ischemia and Alzheimer's disease. *Lactobacillus* species metabolize tryptophan into indole-3-aldehyde (I3A) which acts on the aryl hydrocarbon receptor (AhR) in intestinal immune cells, in turn increasing interleukin-22 (IL-22) production. Indole itself acts as a glucagon-like peptide-1 (GLP-1) secretagogue in intestinal L cells and as a ligand for AhR. Indole can also be metabolized by the liver into indoxyl sulfate, a compound that is toxic in high concentrations and associated with vascular disease and renal dysfunction. AST-120 (activated charcoal), an intestinal sorbent that is taken by mouth, adsorbs indole, in turn decreasing the concentration of indoxyl sulfate in blood plasma.

Without gut flora, the human body would be unable to utilize some of the undigested carbohydrates it consumes, because some types of gut flora have enzymes that human cells lack for breaking down certain polysaccharides. Rodents raised in a sterile environment and lacking in gut flora need to eat 30% more calories just to remain the same weight as their normal counterparts. Carbohydrates that humans cannot digest without bacterial help include certain starches, fiber, oligosaccharides, and sugars that the body failed to digest and absorb like lactose in the case of lactose intolerance and sugar alcohols, mucus produced by the gut, and proteins.

Bacteria turn carbohydrates they ferment into short-chain fatty acids (SCFAs) by a form of fermentation called saccharolytic fermentation. Products include acetic acid, propionic acid and butyric acid. These materials can be used by host cells, providing a major source of useful energy and nutrients for humans, as well as helping the body to absorb essential dietary minerals such as calcium, magnesium and iron. Gases and organic acids, such as lactic acid, are also produced by saccharolytic fermentation. Acetic acid is used by muscle, propionic acid helps the liver produce

ATP, and butyric acid provides energy to gut cells and may prevent cancer. Evidence also indicates that bacteria enhance the absorption and storage of lipids and produce and then facilitate the body to absorb needed vitamins like vitamin K.

Gut flora also synthesize vitamins like biotin and folate, and help with absorption of dietary elements including magnesium, calcium and iron. Methanogenic archae such as *Methanobrevibacter smithii* are involved in the removal of end products of bacterial fermentation such as hydrogen.

Gut-brain Axis

The gut–brain axis is the biochemical signaling that takes place between the gastrointestinal tract and the central nervous system. That term has been expanded to include the role of the gut flora in the interplay; the term "microbiome-gut-brain axis" is sometimes used to describe paradigms explicitly including the gut flora.

Broadly defined, the gut-brain axis includes the central nervous system, neuroendocrine and neuroimmune systems including the hypothalamic–pituitary–adrenal axis (HPA axis), sympathetic and parasympathetic arms of the autonomic nervous system including the enteric nervous system, the vagus nerve, and the gut microbiota.

Interest in the field was sparked by a 2004 study showing that germ-free mice showed an exaggerated HPA axis response to stress compared to non-GF laboratory mice. As of January 2016, most of the work that has been done on the role of gut flora in the gut-brain axis had been conducted in animals, or characterizing the various neuroactive compounds that gut flora can produce, and studies with humans measuring differences between people with various psychiatric and neurological differences, or changes to gut flora in response to stress, or measuring effects of various probiotics (dubbed "psychobiotics in this context), had generally been small and could not be generalized; whether changes to gut flora are a result of disease, a cause of disease, or both in any number of possible feedback loops in the gut-brain axis, remained unclear.

A systematic review of 15 human randomized controlled trials from July 2016 found that certain commercially available strains of probiotic bacteria from the *Bifidobacterium* and *Lactobacillus* genera (*B. longum*, *B. breve*, *B. infantis*, *L. helveticus*, *L. rhamnosus*, *L. plantarum*, and *L. casei*), when taken by mouth in daily doses of 10^9–10^{10} colony forming units (CFU) for 1–2 months, possess treatment efficacy (i.e., improved behavioral outcomes) in certain central nervous system disorders – including anxiety, depression, autism spectrum disorder, and obsessive–compulsive disorder – and improved certain aspects of memory.

Alterations in Flora Balance

Effects of Antibiotic Use

Altering the numbers of gut bacteria, for example by taking broad-spectrum antibiotics, may affect the host's health and ability to digest food. Antibiotics can cause antibiotic-associated diarrhea (AAD) by irritating the bowel directly, changing the levels of gut flora, or allowing pathogenic bacteria to grow. Another harmful effect of antibiotics is the increase in numbers of antibiotic-resistant bacteria found after their use, which, when they invade the host, cause illnesses that are difficult to treat with antibiotics.

Changing the numbers and species of gut flora can reduce the body's ability to ferment carbohydrates and metabolize bile acids and may cause diarrhea. Carbohydrates that are not broken down may absorb too much water and cause runny stools, or lack of SCFAs produced by gut flora could cause the diarrhea.

A reduction in levels of native bacterial species also disrupts their ability to inhibit the growth of harmful species such as *C. difficile* and *Salmonella kedougou*, and these species can get out of hand, though their overgrowth may be incidental and not be the true cause of diarrhea. Emerging treatment protocols for C. difficile infections involve fecal microbiota transplantation of donor feces. Initial reports of treatment describe success rates of 90%, with few side effects. Efficacy is speculated to result from restoring bacterial balances of bacteroides and firmicutes classes of bacteria.

Gut flora composition also changes in severe illnesses, due not only to antibiotic use but also to such factors as ischemia of the gut, failure to eat, and immune compromise. Negative effects from this have led to interest in selective digestive tract decontamination (SDD), a treatment to kill only pathogenic bacteria and allow the re-establishment of healthy ones.

Antibiotics alter the population of the gastrointestinal (GI) tract microbiota, may change the intra-community metabolic interactions, modify caloric intake by using carbohydrates, and globally affects host metabolic, hormonal and immune homeostasis.

There is reasonable evidence that taking probiotics containing *Lactobacillus* species may help prevent antibiotic-associated diarrhea and that taking probiotics with *Saccharomyces* (e.g., *Saccharomyces boulardii*) may help to prevent *Clostridium difficile* infection following systemic antibiotic treatment.

Pregnancy

Women's gut microbiota change as pregnancy advances, with the changes similar to those seen in metabolic syndromes such as diabetes. The change in gut flora causes no ill effects. The newborn's gut biota resemble the mother's first-trimester samples. The diversity of the flora decreases from the first to third trimester, as the numbers of certain species go up.

Probiotics, Prebiotics, Synbiotics, and Pharmabiotics

Probiotics are microorganisms that are believed to provide health benefits when consumed. With regard to gut flora, prebiotics are typically non-digestible, fiber compounds that pass undigested through the upper part of the gastrointestinal tract and stimulate the growth or activity of advantageous gut flora by acting as substrate for them.

Synbiotics refers to food ingredients or dietary supplements combining probiotics and prebiotics in a form of synergism.

The term "pharmabiotics" is used in various ways, to mean: pharmaceutical formulations (standardized manufacturing that can obtain regulatory approval as a drug) of probiotics, prebiotics, or synbiotics; probiotics that have been genetically engineered or otherwise optimized for best performance (shelf life, survival in the digestive tract, etc.); and the natural products of gut flora metabolism (vitamins, etc.).

There is some evidence that treatment with some probiotic strains of bacteria may be effective in irritable bowel syndrome and chronic idiopathic constipation. Those organisms most likely to result in a decrease of symptoms have included:

- *Streptococcus faecium*
- *Lactobacillus plantarum*
- *Lactobacillus rhamnosus*
- *Propionibacterium freudenreichii*
- *Bifidobacterium breve*
- *Lactobacillus reuteri*
- *Lactobacillus salivarius*
- *Bifidobacterium infantis*
- *Streptococcus thermophilus*

Gram positive bacteria present in the lumen may be associated with extending the duration of relapse for ulcerative colitis.

Role in Disease

Bacteria in the digestive tract can contribute to and be affected by disease in various ways. The presence or overabundance of some kinds of bacteria may contribute to inflammatory disorders such as inflammatory bowel disease. Additionally, metabolites from certain members of the gut flora may influence host signaling pathways, contributing to disorders such as obesity and colon cancer. Alternatively, in the event of a breakdown of the gut epithelium, the intrusion of gut flora components into other host compartments can lead to sepsis.

Ulcers

Helicobacter pylori can cause stomach ulcers by crossing the epithelial lining of the stomach. Here the body produces an immune response. During this response parietal cells are stimulated and release extra hydrochloric acid (HCl^+) into the stomach. However, the response does not stimulate the mucus-secreting cells that protect and line the epithelium of the stomach. The extra acid sears holes into the epithelial lining of the stomach, resulting in stomach ulcers.

Inflammatory Bowel Diseases

The two main types of inflammatory bowel diseases, Crohn's disease and ulcerative colitis, are chronic inflammatory disorders of the gut; the causes of these disease are unknown and issues with the gut flora and its relationship with the host have been implicated in these conditions. Additionally, it appears that interactions of gut flora with the gut-brain axis have a role in IBD, with physiological stress mediated through the hypothalamic–pituitary–adrenal axis driving changes to intestinal epithelium and the gut flora in turn releasing factors and metabolites that trigger sigalling in the enteric nervous system and the vagus nerve.

The diversity of gut flora appears to be significantly diminished in people with inflammatory bowel diseases compared to healthy people; additionally, in people with ulcerative colitis, Proteobacteria and Actinobacteria appear to dominate; in people with Crohn's, *Enterococcus faecium* and several Proteobacteria appear to be over-represented.

There is reasonable evidence that correcting gut flora imbalances by taking probiotics with *Lactobacilli* and *Bifidobacteria* can reduce visceral pain and gut inflammation in IBD.

Irritable Bowel Syndrome

Irritable bowel syndrome is a result of stress and chronic activation of the HPA axis; its symptoms include abdominal pain, changes in bowel movements, and an increase in proinflammatory cytokines. Overall, studies have found that the luminal and mucosal microbiota are changed in irritable bowel syndrome individuals, and these changes can relate to the type of irritation such as diarrhea or constipation. Also, there is a decrease in the diversity of the microbiome with low levels of fecal Lactobacilli and Bifidobacteria, high levels of facultative anaerobic bacteria such as Escherichia coli, and increased ratios of Firmicutes:Bacteroidetes.

Other Inflammatory or Autoimmune Conditions

Allergy, asthma and Diabetes mellitus type 1 are autoimmune and inflammatory disorders' the causes of these disease are unknown and issues with the gut flora and its relationship with the host have been implicated in these conditions.

Two hypotheses have been posed to explain the rising prevalence of these diseases in the developed world: the hygiene hypothesis, which posits that children in the developed world are not exposed to a wide enough range of pathogens and end up with an overreactive immune system, and the role of the Western pattern diet which lacks whole grains and fiber and has an overabundance of simple sugars. Both hypotheses converge on the changes in the gut flora and its role in modulating the immune system, and as of 2016 this was an active area of research.

Similar hypotheses have been posited for the rise of food and other allergies.

As of 2016 it was not clear if changes to the gut flora cause these auto-immune and inflammatory disorders or are a product of them or adaptation to them.

Obesity and Metabolic Syndrome

The gut flora has also been implicated in obesity and metabolic syndrome due to the key role it plays in the digestive process; the Western pattern diet appears to drive and maintain changes in the gut flora that in turn change how much energy is derived from food and how that energy is used. One aspect of a healthy diet that is often lacking in the Western-pattern diet is fiber and other complex carbohydrates that a healthy gut flora require to flourish; changes to gut flora in response to a Western-pattern diet appear to increase the amount of energy generated by the gut flora which may contribute to obesity and metabolic syndrome. There is also evidence that microbiota influence eating behaviors based on the preferences of the microbiota, which can lead to the host consuming more food eventually resulting in obesity. It has generally been observed that with higher gut microbiome diversity, the microbiota will spend energy and resources on competing with other microbiota and less on manipulating the host. The opposite is seen with lower gut microbiome diversity, and these microbiotas may work together for to create host food cravings.

Additionally, the liver plays a dominant role in blood glucose homeostasis by maintaining a balance between the uptake and storage of glucose through the metabolic pathways of glycogenesis and gluconeogenesis. In recent studies, it is illustrated that intestinal lipids regulate glucose homeostasis involving a gut-brain-liver axis. The direct administration of lipids into the upper intestine increases the long chain fatty acyl-coenzyme A (LCFA-CoA) levels in the upper intestines and suppresses glucose production even under sub diaphragmatic vagotomy or gut vagal deafferentation. This interrupts the neural connection between the brain and the gut and blocks the upper intestinal lipids' ability to inhibit glucose production. The gut-brain-liver axis and gut microbiota composition can regulate the glucose homeostasis in the liver and provide potential therapeutic methods to treat obesity and diabetes.

Just as gut flora can function in a feedback loop that can drive the development of obesity, there is evidence that restricting intake of calories (i.e. dieting) can drive changes to the composition of the gut flora.

Liver Disease

As the liver is fed directly by the portal vein, whatever crosses the intestinal epithelium and the intestinal mucosal barrier enters the liver, as do cytokines generated there. Dysbiosis in the gut flora has been linked with the development of cirrhosis and non-alcoholic fatty liver disease.

Systemic Infections

Normally-commensal bacteria can be very harmful to the host if they get outside of the intestinal tract. Translocation, which occurs when bacteria leave the gut through its mucosal lining, the border between the lumen of the gut and the inside of the body, can occur in a number of different diseases, and can be caused by too much growth of bacteria in the small intestine, reduced immunity of the host, or increased gut lining permeability.

If the gut is perforated, bacteria can invade the body, causing a potentially fatal infection. Aerobic bacteria can make an infection worse by using up all available oxygen and creating an environment favorable to anaerobes.

Cancer

Some genera of bacteria, such as *Bacteroides* and *Clostridium*, have been associated with an increase in tumor growth rate, while other genera, such as *Lactobacillus* and *Bifidobacteria*, are known to prevent tumor formation.

Other Animals

Aside from mammals, some insects also possess complex and diverse gut microbiota that play key nutritional roles. Microbial communities associated termites can constitute a majority of the weight of the individuals and perform important roles in the digestion of lignocellulose and nitrogen fixation. These communities are host-specific, and closely related insect species share comparable similarities in gut microbiota composition. In cockroaches, gut microbiota have been shown to assemble in a deterministic fashion, irrespective of the inoculum; the reason for this host-specific assembly remains unclear. Bacterial communities associated with insects like termites and cockroaches are determined by a combination of forces, primarily diet, but there is some indication that host phylogeny may also be playing a role in the selection of lineages.

For more than 51 years we have known that the administration of low doses of antibacterial agents promotes the growth of farm animals to increase weight gain.

In a study performed on mice by Ilseung Cho, the ratio of *Firmicutes* and *Lachnospiraceae* was significantly elevated in animals treated with subtherapeutic doses of different antibiotics. By analyzing the caloric content of faeces and the concentration of small chain fatty acids (SCFAs) in the GI tract, they concluded that the changes in the composition of microbiota lead to an increased capacity to extract calories from otherwise indigestible constituents, and to an increased production of SCFAs. These findings provide evidence that antibiotics perturb not only the composition of the GI microbiome but also its metabolic capabilities, specifically with respect to SCFAs.

Prescription Drug

A prescription drug (also prescription medication or prescription medicine) is a pharmaceutical drug that legally requires a medical prescription to be dispensed. In contrast, over-the-counter drugs can be obtained without a prescription. The reason for this difference in substance control is the potential scope of misuse, from drug abuse to practicing medicine without a license and without sufficient education. Different jurisdictions have different definitions of what constitutes a prescription drug.

"Rx" (℞) is often used as a short form for prescription drug in North America- a contraction of the Latin word "*recipe*" (an imperative form of "recipere") meaning "take". Prescription drugs are often dispensed together with a monograph (in Europe, a Patient Information Leaflet or PIL) that gives detailed information about the drug.

The use of prescription drugs has been increasing since the 1960s. In the U.S., 88% of older adults (62–85 years) use at least 1 prescription drug, while 36% take at least 5 prescription medicines concurrently.

Regulation in Australia

In Australia the Standard for the Uniform Scheduling of Medicines and Poisons, abbreviated SUSMP, governs the manufacture and supply of drugs:

The categories defined by the SUSMP are:

- Schedule 1 - Defunct
- Schedule 2 - Pharmacy Medicine
- Schedule 3 - Pharmacist-Only Medicine

- Schedule 4 - Prescription-Only Medicine/Prescription Animal Remedy
- Schedule 5 - Caution
- Schedule 6 - Poison
- Schedule 7 - Dangerous Poison
- Schedule 8 - Controlled Drug (Possession without authority illegal)
- Schedule 9 - Prohibited Substance
- Unscheduled Substances

Similar to the UK, the patient visits a health practitioner, such as a doctor, nurse, dentist, podiatrist, etc., who is able to prescribe the drug.

Many prescriptions issued by health practitioners in Australia are covered by the Pharmaceutical Benefits Scheme, a scheme that provides subsidised prescription drugs to residents of Australia to ensure that all Australians have affordable and reliable access to a wide range of necessary medicines. When purchasing a drug under the PBS the maximum price a consumer pays is the patient co-payment contribution, which, as of January 1, 2014, is A$36.90 for general patients. Those covered by government entitlements (low-income earners, welfare recipients, Health Care Card holders, etc.) and those covered under the Repatriation Pharmaceutical Benefits Scheme (RPBS) have a reduced co-payment, which is $6.00 in 2014. The table below indicates the changes in co-payments over the years. These co-payments are compulsory and cannot be discounted by pharmacies under any circumstances.

Private prescriptions are issued for medicines not covered on the PBS, or being used off-label, for indications other than those covered by the PBS. The patient pays the pharmacy for medicines privately prescribed.

Regulation in United Kingdom

In the United Kingdom the Medicines Act 1968 and Prescription Only Medicines (Human Use) Order 1997 contain regulations that cover the supply of sale, use and production of medicines. Prescribing is also covered by this legislation. There are three categories of medicine:

- Prescription-only medicines (POM), which can be sold by a pharmacist if prescribed by a prescriber
- Pharmacy medicines (P), which may be sold by a pharmacist without prescription
- General sales list (GSL) medicines that may be sold without a prescription in any shop

Possession of prescription-only medicines without a prescription is not a criminal offence unless it falls under the Misuse of Drugs Act 1971.

A patient visits a medical practitioner authorised to prescribe drugs and certain other medical items, such as blood glucose-testing equipment for diabetics. Also, suitably qualified and experienced nurses and pharmacists may be independent prescribers. Both can prescribe all POMs but pharmacists are not allowed to prescribe schedule 1 controlled drugs. District nurses and health visitors have had limited prescribing rights since the mid-nineties, before which prescriptions for dressings and simple medicines would have had to have been signed by a doctor. Once issued, a prescription is taken by the patient to a pharmacy, which dispenses the medicine.

Most prescriptions in the UK are NHS prescriptions, subject to a standard charge unrelated to what is dispensed. The NHS prescription fee was increased to £8.40 per item in England on 1 April 2016; prescriptions are free of charge if prescribed and dispensed in Scotland, Wales or Northern Ireland.

The pharmacy charges the NHS the actual cost of the medicine; the patient pays a set prescription charge in England to the pharmacy. Most of the prescriptions dispensed on the NHS are exempt from charges even in England, as there are many exceptions on grounds such as age and chronic disease, and "season tickets" may be purchased to cover an unlimited amount of medication during a set period. The fee for each item does not depend upon the quantity prescribed by the doctor (which is subject to a maximum of three months' supply), or the cost of the medicine to the NHS, which may vary from a few pence to hundreds of pounds.

Outside the NHS, private prescriptions are issued by private medical practitioners, and sometimes under the NHS for medicines not covered by the NHS. NHS supply beyond three months' worth is not covered, and must be purchased privately. A patient pays the pharmacy the normal price for medicine prescribed outside the NHS.

Survey results published by Ipsos MORI in 2008 found that around 800,000 people in England were not collecting prescriptions or getting them dispensed due to the cost, a figure which was the same as in 2001.

Regulation in United States

In the United States, the Federal Food, Drug, and Cosmetic Act defines what requires a prescription. In general, prescription drugs are authorized by physicians, physician assistants, nurse practitioners, and other APRNs, veterinarians, dentists, and optometrists. In general, it is required that an MD, DO, PA, OD, DPM, NMD, ND, DVM, DDS, or DMD, some psychologists, clinical pharmacists, nurse practitioners, and other APRNs write the prescription; basic-level registered nurses, medical assistants, emergency medical technicians, most psychologists, and social workers as examples, do not have the authority to prescribe drugs.

The Controlled Substances Act (CSA) was enacted into law by the Congress of the United States in 1970. The CSA is the federal U.S. drug policy under which the manufacture, importation, possession, use and distribution of certain substances is regulated. The legislation created five Schedules (classifications), with varying qualifications for a substance to be included in each.

The safety and effectiveness of prescription drugs in the US is regulated by the federal Prescription Drug Marketing Act of 1987. The Food and Drug Administration (FDA) is charged with implementing this law. Misuse or abuse of prescription drugs can lead to adverse drug events, including those due to dangerous drug interactions. Fifteen percent of older American adults are at risk of potential major drug-drug interactions.

The package insert for a prescription drug contains information about the intended effect of the drug and how it works in the body. It also contains information about side-effects, how a patient should take the drug, and cautions for its use, including warnings about allergies.

As a general rule, *over-the-counter drugs* (OTC) are used to treat conditions not necessarily requiring care from a healthcare professional and have been proven to meet higher safety standards for self-medication by patients. Often a lower strength of a drug will be approved for OTC use, whereas higher strengths require a prescription to be obtained; a notable case is ibuprofen, which has been widely available as an OTC pain killer since the mid-1980s but is still available by prescription in doses up to four times the OTC dose for use in cases of severe pain not adequately controlled by the lower, OTC strength.

Herbal preparations, amino acids, vitamins, minerals, and other food supplements are regulated by the FDA as dietary supplements. Because specific health claims cannot be made, the consumer must make informed decisions when purchasing such products.

In the United States, the term "prescription drug" is most commonly used, but they are also called Rx-only drugs or legend drugs, after the Federal and State laws that mandate that all such drugs bear a "legend" prohibiting sale without a prescription; though more complex legends have been used, on most original drug packaging today the legend simply states "Rx only".

Also, pharmacies operated by membership clubs, such as Costco and Sam's Club, by law must allow non-members to use their pharmacy services and must charge the same prices as to members.

Physicians may legally prescribe drugs for uses other than those specified in the FDA approval; this is known as off-label use. Drug companies may not promote or market drugs for off-label uses.

Large U.S. retailers that operate pharmacies and pharmacy chains use inexpensive generic drugs as a way to attract customers into stores. Several chains, including Walmart, Kroger (including subsidiaries such as Dillons), Target, and others, offer $4 monthly prescriptions on select generic drugs as a customer draw. Publix Supermarkets, which has pharmacies in many of their stores, offers free prescriptions on a few older but still effective medications to their customers. The maximum supply is for 30 days.

A number of prescription drugs are commonly abused, including fentanyl (Duragesic), hydrocodone (Vicodin), oxycodone (OxyContin), oxymorphone (Opana), propoxyphene (Darvon), hydromorphone (Dilaudid), meperidine (Demerol), and diphenoxylate (Lomotil).

Regulation in European Union

In European Union (EU), there are two main routes for authorising medicines:

- Centralised Route
- National Route

Expiration Date

The expiration date, required in several countries, specifies the date up to which the manufacturer guarantees the full potency and safety of a drug. In the United States, expiration dates are determined by regulations established by the FDA. The FDA advises consumers not to use products after their expiration dates.

A study conducted by the U.S. Food and Drug Administration covered over 100 drugs, prescription and over-the-counter. The results showed that about 85% of them were safe and effective as far as 15 years past their expiration date. Joel Davis, a former FDA expiration-date compliance chief, said that with a handful of exceptions—notably nitroglycerin, insulin, some liquid antibiotics; outdated tetracyclines can cause Fanconi syndrome—most expired drugs are probably effective.

The American Medical Association (AMA) issued a report and statement on Pharmaceutical Expiration Dates. The Harvard Medical School Family Health Guide notes that, with rare exceptions, "it's true the effectiveness of a drug may decrease over time, but much of the original potency still remains even a decade after the expiration date".

The expiration date is the final day that the manufacturer guarantees the full potency and safety of a medication. Drug expiration dates exist on most medication labels, including prescription, over-the-counter (OTC) and dietary (herbal) supplements. U.S. pharmaceutical manufacturers are required by law to place expiration dates on prescription products prior to marketing. For legal and liability reasons, manufacturers will not make recommendations about the stability of drugs past the original expiration date.

Cost

Prices for prescription drugs vary widely around the world. Prescription costs for biosimilar and generic drugs are usually less than brand names, but the cost is different from one pharmacy to another.

Prescription drug prices including generic prices are rising faster than the average rate of inflation

Environment

Traces of prescription drugs — including antibiotics, anti-convulsants, mood stabilizers and sex hormones — have been detected in drinking water.

Drug Discovery

In the fields of medicine, biotechnology and pharmacology, drug discovery is the process by which new candidate medications are discovered. Historically, drugs were discovered through identifying the active ingredient from traditional remedies or by serendipitous discovery. Later chemical libraries of synthetic small molecules, natural products or extracts were screened in intact cells or whole organisms to identify substances that have a desirable therapeutic effect in a process known as classical pharmacology. Since sequencing of the human genome which allowed rapid cloning and synthesis of large quantities of purified proteins, it has become common practice to use high throughput screening of large compounds libraries against isolated biological targets which are hypothesized to be disease modifying in a process known as reverse pharmacology. Hits from these screens are then tested in cells and then in animals for efficacy.

Modern drug discovery involves the identification of screening hits, medicinal chemistry and optimization of those hits to increase the affinity, selectivity (to reduce the potential of side effects), efficacy/potency, metabolic stability (to increase the half-life), and oral bioavailability. Once a compound that fulfills all of these requirements has been identified, it will begin the process of drug development prior to clinical trials. One or more of these steps may, but not necessarily, involve computer-aided drug design. Modern drug discovery is thus usually a capital-intensive process that involves large investments by pharmaceutical industry corporations as well as national governments (who provide grants and loan guarantees). Despite advances in technology and understanding of biological systems, drug discovery is still a lengthy, "expensive, difficult, and inefficient process" with low rate of new therapeutic discovery. In 2010, the research and development cost of each new molecular entity (NME) was approximately US$1.8 billion. Drug discovery is done by pharmaceutical companies, with research assistance from universities. The "final product" of drug discovery is a patent on the potential drug. The drug requires very expensive Phase I, II and III clinical trials, and most of them fail. Small companies have a critical role, often then selling the rights to larger companies that have the resources to run the clinical trials.

Discovering drugs that may be a commercial success, or a public health success, involves a complex interaction between investors, industry, academia, patent laws, regulatory exclusivity, marketing and the need to balance secrecy with communication. Meanwhile, for disorders whose rarity means that no large commercial success or public health effect can be expected, the orphan drug funding process ensures that people who experience those disorders can have some hope of pharmacotherapeutic advances.

Historical Background

The idea that the effect of a drug in the human body is mediated by specific interactions of the drug molecule with biological macromolecules, (proteins or nucleic acids in most cases) led scientists to the conclusion that individual chemicals are required for the biological activity of the drug. This made for the beginning of the modern era in pharmacology, as pure chemicals, instead of crude extracts, became the standard drugs. Examples of drug compounds isolated from crude preparations are morphine, the active agent in opium, and digoxin, a heart stimulant originating from Digitalis lanata. Organic chemistry also led to the synthesis of many of the natural products isolated from biological sources.

Historically substances, whether crude extracts or purified chemicals were screened for biological activity without knowledge of the biological target. Only after an active substance was identified was an effort made to identify the target. This approach is known as classical pharmacology, forward pharmacology, or phenotypic drug discovery.

Later, small molecules were synthesized to specifically target a known physiological/pathological pathway, rather than adopt the mass screening of banks of stored compounds. This led to great success, such as the work of Gertrude Elion and George H. Hitchings on purine metabolism, the work of James Black on beta blockers and cimetidine, and the discovery of statins by Akira Endo. Another champion of the approach of developing chemical analogues of known active substances was Sir David Jack at Allen and Hanbury's, later Glaxo, who pioneered the first inhaled selective beta2-adrenergic agonist for asthma, the first inhaled steroid for asthma, ranitidine as a successor to cimetidine, and supported the development of the triptans.

Gertrude Elion, working mostly with a group of fewer than 50 people on purine analogues, contributed to the discovery of the first anti-viral; the first immunosuppressant (azathioprine) that allowed human organ transplantation; the first drug to induce remission of childhood leukaemia; pivotal anti-cancer treatments; an anti-malarial; an anti-bacterial; and a treatment for gout.

Cloning of human proteins made possible the screening of large libraries of compounds against specific targets thought to be linked to specific diseases. This approach is known as reverse pharmacology and is the most frequently used approach today.

Drug Targets

The definition of "target" itself is something argued within the pharmaceutical industry. Generally, the "target" is the naturally existing cellular or molecular structure involved in the pathology of interest that the drug-in-development is meant to act on. However, the distinction between a "new" and "established" target can be made without a full understanding of just what a "target" is. This distinction is typically made by pharmaceutical companies engaged in discovery and development of therapeutics. In an estimate from 2011, 435 human genome products were identified as therapeutic drug targets of FDA-approved drugs.

"Established targets" are those for which there is a good scientific understanding, supported by a lengthy publication history, of both how the target functions in normal physiology and how it is involved in human pathology. This does not imply that the mechanism of action of drugs that are thought to act through a particular established target is fully understood. Rather, "established" relates directly to the amount of background information available on a target, in particular functional information. The more such information is available, the less investment is (generally) required to develop a therapeutic directed against the target. The process of gathering such functional information is called "target validation" in pharmaceutical industry parlance. Established targets also include those that the pharmaceutical industry has had experience mounting drug discovery campaigns against in the past; such a history provides information on the chemical feasibility of developing a small molecular therapeutic against the target and can provide licensing opportunities and freedom-to-operate indicators with respect to small-molecule therapeutic candidates.

In general, "new targets" are all those targets that are not "established targets" but which have been or are the subject of drug discovery campaigns. These typically include newly discovered proteins, or proteins whose function has now become clear as a result of basic scientific research.

The majority of targets currently selected for drug discovery efforts are proteins. Two classes predominate: G-protein-coupled receptors (or GPCRs) and protein kinases.

Screening and Design

The process of finding a new drug against a chosen target for a particular disease usually involves high-throughput screening (HTS), wherein large libraries of chemicals are tested for their ability to modify the target. For example, if the target is a novel GPCR, compounds will be screened for their ability to inhibit or stimulate that receptor (antagonist and agonist): if the target is a protein kinase, the chemicals will be tested for their ability to inhibit that kinase.

Another important function of HTS is to show how selective the compounds are for the chosen target. The ideal is to find a molecule which will interfere with only the chosen target, but not other, related targets. To this end, other screening runs will be made to see whether the "hits" against the chosen target will interfere with other related targets - this is the process of cross-screening. Cross-screening is important, because the more unrelated targets a compound hits, the more likely that off-target toxicity will occur with that compound once it reaches the clinic.

It is very unlikely that a perfect drug candidate will emerge from these early screening runs. It is more often observed that several compounds are found to have some degree of activity, and if these compounds share common chemical features, one or more pharmacophores can then be developed. At this point, medicinal chemists will attempt to use structure-activity relationships (SAR) to improve certain features of the lead compound:

- increase activity against the chosen target
- reduce activity against unrelated targets
- improve the druglikeness or ADME properties of the molecule.

This process will require several iterative screening runs, during which, it is hoped, the properties of the new molecular entities will improve, and allow the favoured compounds to go forward to in vitro and in vivo testing for activity in the disease model of choice.

Amongst the physico-chemical properties associated with drug absorption include ionization (pKa), and solubility; permeability can be determined by PAMPA and Caco-2. PAMPA is attractive as an early screen due to the low consumption of drug and the low cost compared to tests such as Caco-2, gastrointestinal tract (GIT) and Blood–brain barrier (BBB) with which there is a high correlation.

A range of parameters can be used to assess the quality of a compound, or a series of compounds, as proposed in the Lipinski's Rule of Five. Such parameters include calculated properties such as cLogP to estimate lipophilicity, molecular weight, polar surface area and measured properties, such as potency, in-vitro measurement of enzymatic clearance etc. Some descriptors such as ligand efficiency (LE) and lipophilic efficiency (LiPE) combine such parameters to assess druglikeness.

While HTS is a commonly used method for novel drug discovery, it is not the only method. It is often possible to start from a molecule which already has some of the desired properties. Such a molecule might be extracted from a natural product or even be a drug on the market which could be improved upon (so-called "me too" drugs). Other methods, such as virtual high throughput screening, where screening is done using computer-generated models and attempting to "dock" virtual libraries to a target, are also often used.

Another important method for drug discovery is *de novo* drug design, in which a prediction is made of the sorts of chemicals that might (e.g.) fit into an active site of the target enzyme. For example, virtual screening and computer-aided drug design are often used to identify new chemical moieties that may interact with a target protein. Molecular modelling and molecular dynamics simulations can be used as a guide to improve the potency and properties of new drug leads.

There is also a paradigm shift in the drug discovery community to shift away from HTS, which is expensive and may only cover limited chemical space, to the screening of smaller libraries (maximum a few thousand compounds). These include fragment-based lead discovery (FBDD) and protein-directed dynamic combinatorial chemistry. The ligands in these approaches are usually much smaller, and they bind to the target protein with weaker binding affinity than those hits that are identified from HTS. Further modified through organic synthesis into lead compounds are often required. Such modifications are often guided by protein X-ray crystallography of the protein-fragment complex. The advantages of these approaches are that they allow more efficient screening and the compound library, although small, typically covers a large chemical space when compared to HTS.

Once a lead compound series has been established with sufficient target potency and selectivity and favourable drug-like properties, one or two compounds will then be proposed for drug development. The best of these is generally called the lead compound, while the other will be designated as the "backup".

Nature as Source of Drugs

Traditionally many drugs and other chemicals with biological activity have been discovered by studying allelopathy - chemicals that organisms create that affect the activity of other organisms in the fight for survival.

Despite the rise of combinatorial chemistry as an integral part of lead discovery process, natural products still play a major role as starting material for drug discovery. A 2007 report found that of the 974 small molecule new chemical entities developed between 1981 and 2006, 63% were natural derived or semisynthetic derivatives of natural products. For certain therapy areas, such as antimicrobials, antineoplastics, antihypertensive and anti-inflammatory drugs, the numbers were higher. In many cases, these products have been used traditionally for many years.

Natural products may be useful as a source of novel chemical structures for modern techniques of development of antibacterial therapies.

Despite the implied potential, only a fraction of Earth's living species has been tested for bioactivity.

Plant-derived

Prior to Paracelsus, the vast majority of traditionally used crude drugs in Western medicine were plant-derived extracts. This has resulted in a pool of information about the potential of plant species as an important source of starting material for drug discovery. A different set of metabolites is sometimes produced in the different anatomical parts of the plant (e.g. root, leaves and flower), and botanical knowledge is crucial also for the correct identification of bioactive plant materials.

Microbial Metabolites

Microbes compete for living space and nutrients. To survive in these conditions, many microbes have developed abilities to prevent competing species from proliferating. Microbes are the main source of antimicrobial drugs. *Streptomyces* species have been a valuable source of antibiotics. The classical example of an antibiotic discovered as a defense mechanism against another microbe is the discovery of penicillin in bacterial cultures contaminated by *Penicillium* fungi in 1928.

Marine Invertebrates

Marine environments are potential sources for new bioactive agents. Arabinose nucleosides discovered from marine invertebrates in 1950s, demonstrating for the first time that sugar moieties other than ribose and deoxyribose can yield bioactive nucleoside structures. However, it was 2004 when the first marine-derived drug was approved. The cone snail toxin ziconotide, also known as Prialt, was approved by the Food and Drug Administration to treat severe neuropathic pain. Several other marine-derived agents are now in clinical trials for indications such as cancer, anti-inflammatory use and pain. One class of these agents are bryostatin-like compounds, under investigation as anti-cancer therapy.

Chemical Diversity of Natural Products

As above mentioned, combinatorial chemistry was a key technology enabling the efficient generation of large screening libraries for the needs of high-throughput screening. However, now, after two decades of combinatorial chemistry, it has been pointed out that despite the increased efficiency in chemical synthesis, no increase in lead or drug candidates has been reached. This has led to analysis of chemical characteristics of combinatorial chemistry products, compared to existing drugs or natural products. The chemoinformatics concept chemical diversity, depicted as distribution of compounds in the chemical space based on their physicochemical characteristics, is often used to describe the difference between the combinatorial chemistry libraries and natural products. The synthetic, combinatorial library compounds seem to cover only a limited and quite uniform chemical space, whereas existing drugs and particularly natural products, exhibit much greater chemical diversity, distributing more evenly to the chemical space. The most prominent differences between natural products and compounds in combinatorial chemistry libraries is the number of chiral centers (much higher in natural compounds), structure rigidity (higher in natural compounds) and number of aromatic moieties (higher in combinatorial chemistry libraries). Other chemical differences between these two groups include the nature of heteroatoms (O and N enriched in natural products, and S and halogen atoms more often present in synthetic compounds), as well as level of non-aromatic unsaturation (higher in natural products). As both structure rigidity and chirality are both well-established factors in medicinal chemistry known to enhance compounds specificity and efficacy as a drug, it has been suggested that natural products compare favourable to today's combinatorial chemistry libraries as potential lead molecules.

Natural Product Drug Discovery

Screening

Two main approaches exist for the finding of new bioactive chemical entities from natural sources.

The first is sometimes referred to as random collection and screening of material, but in fact the collection is often far from random in that biological (often botanical) knowledge is used about which families show promise, based on a number of factors, including past screening. This approach is based on the fact that only a small part of earth's biodiversity has ever been tested for pharmaceutical activity. It is also based on the fact that organisms living in a species-rich environment need to evolve defensive and competitive mechanisms to survive, mechanisms which might usefully be exploited in the development of drugs that can cure diseases affecting humans. A collection of plant, animal and microbial samples from rich ecosystems can potentially give rise to novel biological activities worth exploiting in the drug development process. One example of a successful use of this strategy is the screening for antitumour agents by the National Cancer Institute, started in the 1960s. Paclitaxel was identified from Pacific yew tree Taxus brevifolia. Paclitaxel showed anti-tumour activity by a previously undescribed mechanism (stabilization of microtubules) and is now approved for clinical use for the treatment of lung, breast and ovarian cancer, as well as for Kaposi's sarcoma. Early in the 21st century, Cabazitaxel (made by Sanofi, a French firm), another relative of taxol has been shown effective against prostate cancer, also because it works by preventing the formation of microtubules, which pull the chromosomes apart in dividing cells (such as cancer cells). Still another examples are: 1. Camptotheca (Camptothecin · Topotecan · Irinotecan · Rubitecan · Belotecan); 2. Podophyllum (Etoposide · Teniposide); 3a. Anthracyclines (Aclarubicin · Daunorubicin · Doxorubicin · Epirubicin · Idarubicin · Amrubicin · Pirarubicin · Valrubicin · Zorubicin); 3b. Anthracenediones (Mitoxantrone · Pixantrone).

Nor do all drugs developed in this manner come from plants. Professor Louise Rollins-Smith of Vanderbilt University's Medical Center, for example, has developed from the skin of frogs a compound which blocks AIDS. Professor Rollins-Smith is aware of declining amphibian populations and has said: "We need to protect these species long enough for us to understand their medicinal cabinet."

The second main approach involves Ethnobotany, the study of the general use of plants in society, and ethnopharmacology, an area inside ethnobotany, which is focused specifically on medicinal uses.

Both of these two main approaches can be used in selecting starting materials for future drugs. Artemisinin, an antimalarial agent from sweet wormtree Artemisia annua, used in Chinese medicine since 200BC is one drug used as part of combination therapy for multiresistant Plasmodium falciparum.

Structural Elucidation

The elucidation of the chemical structure is critical to avoid the re-discovery of a chemical agent that is already known for its structure and chemical activity. Mass spectrometry is a method in which individual compounds are identified based on their mass/charge ratio, after ionization. Chemical compounds exist in nature as mixtures, so the combination of liquid chromatography and mass spectrometry (LC-MS) is often used to separate the individual chemicals. Databases of mass spectras for known compounds are available, and can be used to assign a structure to an unknown mass spectrum. Nuclear magnetic resonance spectroscopy is the primary technique for determining chemical structures of natural products. NMR yields information about individual hydrogen and carbon atoms in the structure, allowing detailed reconstruction of the molecule's architecture.

Drug Development

Drug development is the process of bringing a new pharmaceutical drug to the market once a lead compound has been identified through the process of drug discovery. It includes pre-clinical research on microorganisms and animals, filing for regulatory status, such as via the United States Food and Drug Administration for an investigational new drug to initiate clinical trials on humans, and may include the step of obtaining regulatory approval with a new drug application to market the drug.

New Chemical Entity Development

Broadly, the process of drug development can be divided into pre-clinical and clinical work.

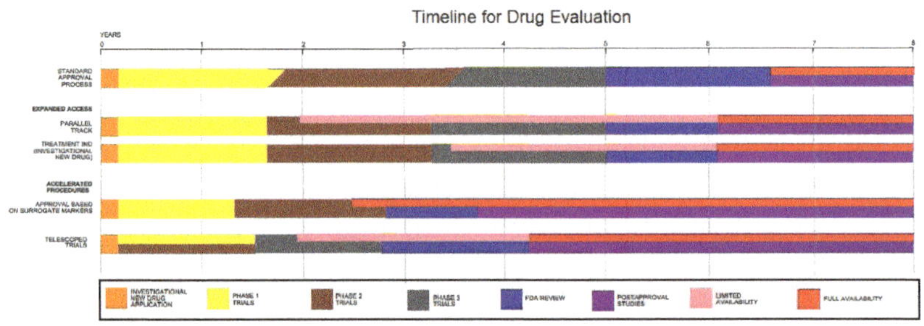

Timeline showing the various drug approval tracks and research phases

Pre-clinical

New chemical entities (NCEs, also known as new molecular entities or NMEs) are compounds that emerge from the process of drug discovery. These have promising activity against a particular biological target that is important in disease. However, little is known about the safety, toxicity, pharmacokinetics, and metabolism of this NCE in humans. It is the function of drug development to assess all of these parameters prior to human clinical trials. A further major objective of drug development is to recommend the dose and schedule for the first use in a human clinical trial ("first-in-man" [FIM] or First Human Dose [FHD]).

In addition, drug development must establish the physicochemical properties of the NCE: its chemical makeup, stability, and solubility. Manufacturers must optimize the process they use to make the chemical so they can scale up from a medicinal chemist producing milligrams, to manufacturing on the kilogram and ton scale. They further examine the product for suitability to package as capsules, tablets, aerosol, intramuscular injectable, subcutaneous injectable, or intravenous formulations. Together, these processes are known in preclinical development as *chemistry, manufacturing, and control* (CMC).

Many aspects of drug development focus on satisfying the regulatory requirements of drug licensing authorities. These generally constitute a number of tests designed to determine the major toxicities of a novel compound prior to first use in humans. It is a legal requirement that an assessment of major organ toxicity be performed (effects on the heart and lungs, brain, kidney, liver and digestive system), as well as effects on other parts of the body that might be affected by the drug (e.g., the skin if the new drug is to be delivered through the skin). Increasingly, these tests are made using *in vitro* methods (e.g., with isolated cells), but many tests can only be made by using experimental animals to demonstrate the complex interplay of metabolism and drug exposure on toxicity.

The information is gathered from this pre-clinical testing, as well as information on CMC, and submitted to regulatory authorities (in the US, to the FDA), as an Investigational New Drug application or IND. If the IND is approved, development moves to the clinical phase.

Clinical Phase

Clinical trials involve three or four steps:

- Phase I trials, usually in healthy volunteers, determine safety and dosing.
- Phase II trials are used to get an initial reading of efficacy and further explore safety in small numbers of patients having the disease targeted by the NCE.
- Phase III trials are large, pivotal trials to determine safety and efficacy in sufficiently large numbers of patients with the targeted disease. If safety and efficacy are adequately proved, clinical testing may stop at this step and the NCE advances to the new drug application (NDA) stage.
- Phase IV trials are post-approval trials that are sometimes a condition attached by the FDA, also called post-market surveillance studies.

The process of defining characteristics of the drug does not stop once an NCE begins human clinical trials. In addition to the tests required to move a novel drug into the clinic for the first time, manufacturers must ensure that any long-term or chronic toxicities are well-defined, including effects on systems not previously monitored (fertility, reproduction, immune system, among others). They must also test the compound for its potential to cause cancer (carcinogenicity testing).

If a compound emerges from these tests with an acceptable toxicity and safety profile, and the company can further show it has the desired effect in clinical trials, then the NCE portfolio of evidence can be submitted for marketing approval in the various countries where the manufacturer plans to sell it. In the United States, this process is called a "new drug application" or NDA.

Most NCEs fail during drug development, either because they have unacceptable toxicity or because they simply do not have the intended effect on the targeted disease as shown in clinical trials.

Cost

The full cost of bringing a new drug (i.e., new chemical entity) to market – from discovery through clinical trials to approval – is complex and controversial. Typically, companies spend tens to hundreds of millions of U.S. dollars. One element of the complexity is that the much-publicized final numbers often not only include the out-of-pocket expenses for conducting a series of Phase I-III clinical trials, but also the *capital costs* of the long period (10 or more years) during which the company must cover out-of-pocket costs for preclinical drug discovery. Additionally, companies often do not report whether a given figure includes the capitalized cost or comprises only out-of-pocket expenses, or both.

Another element of complexity is that all estimates are based on confidential information controlled by drug companies, released by them voluntarily, leading to inability to verify costs. The numbers are controversial, as drug companies use them to justify the prices of their drugs and various advocates for lower drug prices have challenged them. The controversy is not only between "high" and "low", but also the high numbers may vary considerably for the manifold factors in drug development.

One study assessed both capitalized and out-of-pocket costs as about US$1.8 billion and $870 million, respectively.

In an analysis of the drug development costs for 98 companies over a decade, the average cost per drug developed and approved by a single-drug company was $350 million. But for companies that approved between eight and 13 drugs over 10 years, the cost per drug went as high as $5.5 billion, due mainly to geographic expansion for marketing and ongoing costs for Phase IV trials and continuous monitoring for safety.

Alternatives to conventional drug development have the objective for universities, governments and pharmaceutical industry to collaborate and optimize resources.

Valuation

The nature of a drug development project is characterised by high attrition rates, large capital expenditures, and long timelines. This makes the valuation of such projects and companies a challenging task. Not all valuation methods can cope with these particularities. The most commonly used valuation methods are risk-adjusted net present value (rNPV), decision trees, real options, or comparables.

The most important value drivers are the cost of capital or discount rate that is used, phase attributes such as duration, success rates, and costs, and the forecasted sales, including cost of goods and marketing and sales expenses. Less objective aspects like quality of the management or novelty of the technology should be reflected in the cash flows estimation.

Success Rate

Candidates for a new drug to treat a disease might, theoretically, include from 5,000 to 10,000 chemical compounds. On average about 250 of these show sufficient promise for further evaluation using laboratory tests, mice and other test animals. Typically, about ten of these qualify for tests on humans. A study conducted by the Tufts Center for the Study of Drug Development covering the 1980s and 1990s found that only 21.5 percent of drugs that started Phase I trials were eventually approved for marketing. In the time period of 2006 to 2015, the success rate was 9.6%. The high failure rates associated with pharmaceutical development are referred to as the "attrition rate" problem. Careful decision making during drug development is essential to avoid costly failures. In many cases, intelligent programme and clinical trial design can prevent false negative results. Well designed dose-finding studies and comparisons against both a placebo and a gold-standard treatment arm play a major role in achieving reliable data.

Novel Initiatives to Boost Drug Development

Novel initiatives include partnering between governmental organizations and industry. The world's largest such initiative is the Innovative Medicines Initiative (IMI), and examples of major national initiatives are Top Institute Pharma in the Netherlands and Biopeople in Denmark. In 2004, the FDA created the "Critical Path Initiative" to guide the new drug development process.

Stages of Drug Development

Pre-clinical Development

In drug development, preclinical development, also named preclinical studies and nonclinical studies, is a stage of research that begins before clinical trials (testing in humans) can begin, and during which important feasibility, iterative testing and drug safety data are collected.

The main goals of pre-clinical studies are to determine the safe dose for first-in-man study and assess a product's safety profile. Products may include new medical devices, drugs, gene therapy solutions and diagnostic tools.

On average, only one in every 5,000 compounds that enters drug discovery to the stage of preclinical development becomes an approved drug.

Types of Preclinical Research

Each class of product may undergo different types of preclinical research. For instance, drugs may undergo pharmacodynamics (what the drug does to the body) (PD), pharmacokinetics (what the body does to the drug) (PK), ADME, and toxicology testing. This data allows researchers to allometrically estimate a safe starting dose of the drug for clinical trials in humans. Medical devices that do not have drug attached will not undergo these additional tests and may go directly to Good Laboratory Practices (GLP) testing for safety of the device and its components. Some medical devices will also undergo biocompatibility testing which helps to show whether a component of the device or all components are sustainable in a living model. Most preclinical studies must adhere to GLPs in ICH Guidelines to be acceptable for submission to regulatory agencies such as the Food & Drug Administration in the United States.

Typically, both *in vitro* and *in vivo* tests will be performed. Studies of a drug's toxicity include which organs are targeted by that drug, as well as if there are any long-term carcinogenic effects or toxic effects on mammalian reproduction.

Animal Testing

The information collected from these studies is vital so that safe human testing can begin. Typically, in drug development studies animal testing involves two species. The most commonly used models are murine and canine, although primate and porcine are also used.

Choice of Species

The choice of species is based on which will give the best correlation to human trials. Differences in the gut, enzyme activity, circulatory system, or other considerations make certain models more appropriate based on the dosage form, site of activity, or noxious metabolites. For example, canines may not be good models for solid oral dosage forms because the characteristic carnivore intestine is underdeveloped compared to the omnivore's, and gastric emptying rates are increased. Also, rodents can not act as models for antibiotic drugs because the resulting alteration to their intestinal flora causes significant adverse effects. Depending on a drug's functional groups, it may be metabolized in similar or different ways between species, which will affect both efficacy and toxicology.

Medical device studies also use this basic premise. Most studies are performed in larger species such as dogs, pigs and sheep which allow for testing in a similar sized model as that of a human. In addition, some species are used for similarity in specific organs or organ system physiology (swine for dermatological and coronary stent studies; goats for mammary implant studies; dogs for gastric and cancer studies; etc.).

Importantly, the regulatory guidelines of FDA, EMA, and other similar international and regional authorities usually require safety testing in at least two mammalian species, including one non-rodent species, prior to human trials authorization.

Ethical Issues

Animal testing in the research-based pharmaceutical industry has been reduced in recent years both for ethical and cost reasons. However, most research will still involve animal based testing for the need of similarity in anatomy and physiology that is required for diverse product development.

No Observable Effect Levels

Based on preclinical trials, No Observable Adverse Effect Levels (NOAEL) on drugs are established, which are used to determine initial phase 1 clinical trial dosage levels on a mass API per mass patient basis. Generally a 1/100 uncertainty factor or "safety margin" is included to account for interspecies (1/10) and inter-individual (1/10) differences.

Clinical Trial

Clinical trials are experiments done in clinical research. Such prospective biomedical or behavioral research studies on human participants are designed to answer specific questions about biomedical or behavioral interventions, including new treatments (such as novel vaccines, drugs, dietary choices, dietary supplements, and medical devices) and known interventions that warrant further study and comparison. Clinical trials generate data on safety and efficacy. They are conducted only after they have received health authority/ethics committee approval in the country where approval of the therapy is sought. These authorities are responsible for vetting the risk/benefit ratio of the trial - their approval does not mean that the therapy is 'safe' or effective, only that the trial may be conducted.

Depending on product type and development stage, investigators initially enroll volunteers and/or patients into small pilot studies, and subsequently conduct progressively larger scale comparative studies. Clinical trials can vary in size and cost, and they can involve a single research center or multiple centers, in one country or in multiple countries. Clinical study design aims to ensure the scientific validity and reproducibility of the results.

Trials can be quite costly, depending on a number of factors. The sponsor may be a governmental organization or a pharmaceutical, biotechnology or medical device company. Certain functions necessary to the trial, such as monitoring and lab work, may be managed by an outsourced partner, such as a contract research organization or a central laboratory.

Only 10% of all drugs started in human clinical trials become an approved drug.

Overview

Trials of Drugs

Some clinical trials involve healthy subjects with no pre-existing medical conditions. Other clinical trials pertain to patients with specific health conditions who are willing to try an experimental treatment.

When participants are healthy volunteers who receive financial incentives, the goals are different than when the participants are sick. During dosing periods, study subjects typically remain under supervision for one to 40 nights.

Usually pilot experiments are conducted to gain insights for design of the clinical trial to follow.

There are two goals to testing medical treatments: to learn whether they work well enough, called "efficacy" or "effectiveness"; and to learn whether they are safe enough, called "safety". Neither is an absolute criterion; both safety and efficacy are evaluated relative to how the treatment is intended to be used, what other treatments are available, and the severity of the disease or condition. The benefits must outweigh the risks. For example, many drugs to treat cancer have severe side effects that would not be acceptable for an over-the-counter pain medication, yet the cancer drugs have been approved since they are used under a physician's care, and are used for a life-threatening condition.

In the US, the elderly constitute only 14% of the population, while they consume over one-third of drugs. People over 55 (or a similar cutoff age) are often excluded from trials because their greater health issues and drug use complicate data interpretation, and because they have different physiological capacity than younger people. Women, children and people with unrelated medical conditions are also frequently excluded. For women, a major reason for exclusion is the possibility of pregnancy and the unknown risks to the fetus.

The sponsor designs the trial in coordination with a panel of expert clinical investigators, including what alternative/existing treatments to compare to the new drug and what type(s) of patients might benefit. If the sponsor cannot obtain enough subjects at one location, investigators at other locations are recruited to join the study.

During the trial, investigators recruit patients with the predetermined characteristics, administer the treatment(s) and collect data on the patients' health for a defined time period.

Subjects are volunteers who are not paid for participating. (However, the investigators are paid.)

Data include measurements such as vital signs, concentration of the study drug in the blood and/or tissues, changes to symptoms and whether health outcomes. The researchers send the data to the trial sponsor, who then analyzes the pooled data using statistical tests.

Examples of clinical trial goals include assessing the safety and (relative) effectiveness of a medication or device:

- On a specific kind of patient (e.g., patients who have been diagnosed with Alzheimer's disease)
- At a different dose (e.g., 10-mg dose instead of 5-mg dose)

- For a new indication
- Is more effective for the patient's condition than the standard therapy
- Relative to two or more already approved/common interventions for that disease (e.g., device A vs. device B, therapy A vs. therapy B)

While most clinical trials test one alternative to the novel intervention, some expand to three or four.

Except for small, single-location trials, the design and objectives are specified in a document called a clinical trial protocol. The protocol is the trial's 'operating manual' and ensures that all researchers perform the trial in the same way on similar patients and that the data is comparable across all patients.

Because a trial is designed to test hypotheses and rigorously monitor and assess outcomes, it can be seen as an application of the scientific method, specifically the experimental step.

The most common clinical trials evaluate new drugs, medical devices (such as a new catheter), biologics, psychological therapies, or other interventions. Clinical trials may be required before a national regulatory authority approves marketing of the innovation.

Trials of Devices

Similarly to drugs, medical devices are sometimes subjected to clinical trials. Device trials may compare a new device to an established therapy, or may compare similar devices to each other. An example of the former in the field of vascular surgery is the Open versus Endovascular Repair (OVER trial) for the treatment of abdominal aortic aneurysm, which compared the older open aortic repair technique to the newer endovascular aneurysm repair device. An example of the latter is the LEOPARD trial, which compares EVAR devices.

Trials of Procedures

Similarly to drugs, medical or surgical procedures may be subjected to clinical trials, such as case-controlled studies for surgical interventions.

History

The concepts behind clinical trials are ancient. The Book of Daniel chapter 1, verses 12 through 15, for instance, describes a planned experiment with both baseline and follow-up observations of two groups who either partook of, or did not partake of, "the King's meat" over a trial period of ten days. Persian physician Avicenna, in *The Canon of Medicine* (1025) gave similar advice for determining the efficacy of medical drugs and substances.

Development

Although early medical experimentation was often performed, the use of a control group to provide an accurate comparison for the demonstration of the intervention's efficacy, was generally lacking. For instance, Lady Mary Wortley Montagu, who campaigned for the introduction of inoculation (then called variolation) to prevent smallpox, arranged for seven prisoners who had been sentenced to death to undergo variolation in exchange for their life. Although they survived and did not contract smallpox, there was no control group to assess whether this result was due to the inoculation or some other factor. Similar experiments performed by Edward Jenner over his smallpox vaccine were equally conceptually flawed.

Edward Jenner vaccinating James Phipps, a boy of eight, on 14 May 1796. Jenner failed to use a control group.

The first proper clinical trial was conducted by the physician James Lind. The disease scurvy, now known to be caused by a Vitamin C deficiency, would often have terrible effects on the welfare of the crew of long distance voyages. In 1740, the catastrophic result of Anson's circumnavigation attracted much attention in Europe; out of 1900 men, 1400 had died, most of them allegedly from having contracted scurvy. John Woodall, an English military surgeon of the British East India Company, had recommended the consumption of citrus fruit (it has an antiscorbutic effect) from the 17th century, but their use did not become widespread.

Lind conducted the first systematic clinical trial in 1747. He included a dietary supplement of an acidic quality in the experiment after two months at sea, when the ship was already afflicted with scurvy. He divided twelve scorbutic sailors into six groups of two. They all received the same diet but, in addition, group one was given a quart of cider daily, group two twenty-five drops of elixir of vitriol (sulfuric acid), group three six spoonfuls of vinegar, group four half a pint of seawater, group five received two oranges and one lemon, and the last group a spicy paste plus a drink of barley water. The treatment of group five stopped after six days when they ran out of fruit, but by that time one sailor was fit for duty while the other had almost recovered. Apart from that, only group one also showed some effect of its treatment.

After 1750 the discipline began to take its modern shape. John Haygarth demonstrated the importance of a control group for the correct identification of the placebo effect in his celebrated study of the ineffective remedy called *Perkin's tractors*. Further work in that direction was carried out by the eminent physician Sir William Gull, 1st Baronet in the 1860s.

Frederick Akbar Mahomed (d. 1884), who worked at Guy's Hospital in London, made substantial contributions to the process of clinical trials, where "he separated chronic nephritis with secondary hypertension from what we now term essential hypertension. He also founded the Collective

Investigation Record for the British Medical Association; this organization collected data from physicians practicing outside the hospital setting and was the precursor of modern collaborative clinical trials."

Modern Trials

Austin Bradford Hill was a pivotal figure in the modern development of clinical trials.

Sir Ronald A. Fisher, while working for the Rothamsted experimental station in the field of agriculture, developed his *Principles of experimental design* in the 1920s as an accurate methodology for the proper design of experiments. Among his major ideas, was the importance of randomization - the random assignment of individuals to different groups for the experiment; replication - to reduce uncertainty, measurements should be repeated and experiments replicated to identify sources of variation; blocking - to arrange experimental units into groups of units that are similar to each other, and thus reducing irrelevant sources of variation; use of factorial experiments - efficient at evaluating the effects and possible interactions of several independent factors.

The British Medical Research Council officially recognized the importance of clinical trials from the 1930s. The Council established the *Therapeutic Trials Committee* to advise and assist in the arrangement of properly controlled clinical trials on new products that seem likely on experimental grounds to have value in the treatment of disease.

The first randomised curative trial was carried out at the MRC Tuberculosis Research Unit by Sir Geoffrey Marshall (1887–1982). The trial, carried out between 1946-1947, aimed to test the efficacy of the chemical streptomycin for curing pulmonary tuberculosis. The trial was both double-blind and placebo-controlled.

The methodology of clinical trials was further developed by Sir Austin Bradford Hill, who had been involved in the streptomcyin trials. From the 1920s, Hill applied statistics to medicine, attending the lectures of renowned mathematician Karl Pearson, amongst others. He became famous for a landmark study carried out in collaboration with Richard Doll on the correlation between smoking and lung cancer. They carried out a case-control study in 1950, which compared lung cancer patients with matched control and also began a sustained long-term prospective study into the broader issue of smoking and health, which involved studying the smoking habits and health of over 30,000 doctors over a period of several years. His certificate for election to the Royal Society called him "...the leader in the development in medicine of the precise experimental methods now used nationally and internationally in the evaluation of new therapeutic and prophylactic agents."

International clinical trials day is celebrated on 20 May.

Types

One way of classifying clinical trials is by the way the researchers behave.

- In a clinical observational study, the investigators observe the subjects and measure their outcomes. The researchers do not actively manage the study.
- In an interventional study, the investigators give the research subjects a particular medicine or other intervention to compare the treated subjects with those receiving no treatment or the standard treatment. Then the researchers measure how the subjects' health changes.

Another way of classifying trials is by their purpose. The U.S. National Institutes of Health (NIH) organizes trials into five different types:

- Prevention trials look for better ways to prevent disease in people who have never had the disease or to prevent a disease from returning. These approaches may include medicines, vitamins, vaccines, minerals, or lifestyle changes.
- Screening trials test the best way to detect certain diseases or health conditions.
- Diagnostic trials are conducted to find better tests or procedures for diagnosing a particular disease or condition.
- Treatment trials test experimental treatments, new combinations of drugs, or new approaches to surgery or radiation therapy.
- Quality of life trials (supportive care trials) explore ways to improve comfort and the quality of life for individuals with a chronic illness.
- Compassionate use trials or expanded access trials provide partially tested, unapproved therapeutics to a small number of patients who have no other realistic options. Usually, this involves a disease for which no effective therapy has been approved, or a patient who has already failed all standard treatments and whose health is too compromised to qualify for participation in randomized clinical trials. Usually, case-by-case approval must be granted by both the United States Food and Drug Administration and the pharmaceutical company for such exceptions.

A third classification is whether the trial design allows changes based on data accumulated during the trial.

- Fixed trials consider existing data only during the trial's design, do not modify the trial after it begins and do not assess the results until the study is complete.

- Adaptive clinical trials use existing data to design the trial, and then use interim results to modify the trial as it proceeds. Modifications include dosage, sample size, drug undergoing trial, patient selection criteria and "cocktail" mix. Adaptive trials often employ a Bayesian experimental design to assess the trial's progress. In some cases, trials have become an ongoing process that regularly adds and drops therapies and patient groups as more information is gained. The aim is to more quickly identify drugs that have a therapeutic effect and to zero in on patient populations for whom the drug is appropriate.

Finally, a common way of distinguishing trials is by phase, which in simple terms, relates to how close the drug is to being clinically proven both effective for its stated purpose and accepted by the regulatory authorities for use for that purpose.

Phases

Clinical trials involving new drugs are commonly classified into five phases. Each phase of the drug approval process is treated as a separate clinical trial. The drug-development process will normally proceed through all four phases over many years. If the drug successfully passes through phases 0, 1, 2, and 3, it will usually be approved by the national regulatory authority for use in the general population. Before pharmaceutical companies start clinical trials on a drug, they will also have conducted extensive preclinical studies. Each phase has a different purpose and helps scientists answer a different question.

Phase	Aim	Notes
Phase 0	Pharmacodynamics and pharmacokinetics in humans	Phase 0 trials are the first-in-human trials. Single subtherapeutic doses of the study drug or treatment are given to a small number of subjects (10 to 15) to gather preliminary data on the agent's pharmacodynamics (what the drug does to the body) and pharmacokinetics (what the body does to the drugs). For a test drug, the trial documents the absorption, distribution, metabolization, and removal (excretion) of the drug, and the drug's interactions within the body, to confirm that these appear to be as expected.
Phase 1	Screening for safety.	Testing within a small group of people (20–80) to evaluate safety, determine safe dosage ranges, and begin to identify side effects. A drug's side effects could be subtle or long term, or may only happen with a few people, so phase 1 trials are not expected to identify all side effects.
Phase 2	Establishing the efficacy of the drug, usually against a placebo.	Testing with a larger group of people (100–300) to determine efficacy and to further evaluate its safety. The gradual increase in test group size allows for the evocation of less-common side effects.
Phase 3	Final confirmation of safety and efficacy.	Testing with large groups of people (1,000–3,000) to confirm its efficacy, evaluate its effectiveness, monitor side effects, compare it to commonly used treatments, and collect information that will allow it to be used safely.
Phase 4	Safety studies during sales.	Postmarketing studies delineate additional information, including the treatment's risks, benefits, and optimal use. As such, they are ongoing during the drug's lifetime of active medical use. (Particularly relevant after approval under FDA Accelerated Approval Program)

Trial Design

A fundamental distinction in evidence-based practice is between observational studies and randomized controlled trials. Types of observational studies in epidemiology, such as the cohort study and the case-control study, provide less compelling evidence than the randomized controlled trial. In observational studies, the investigators only observe associations (correlations) between the treatments experienced by participants and their health status. However, under certain conditions, causal effects can be inferred from observational studies.

A randomized controlled trial can provide compelling evidence that the study treatment causes an effect on human health.

Currently, some phase 2 and most phase 3 drug trials are designed as randomized, double-blind, and placebo-controlled.

- Randomized: Each study subject is randomly assigned to receive either the study treatment or a placebo.
- Blind: The subjects involved in the study do not know which study treatment they receive. If the study is double-blind, the researchers also do not know which treatment a subject receives. This intent is to prevent researchers from treating the two groups differently. A form of double-blind study called a "double-dummy" design allows additional insurance against bias. In this kind of study, all patients are given both placebo and active doses in alternating periods.
- Placebo-controlled: The use of a placebo (fake treatment) allows the researchers to isolate the effect of the study treatment from the placebo effect.

Although the term "clinical trials" is most commonly associated with the large, randomized studies typical of phase 3, many clinical trials are small. They may be "sponsored" by single researchers or a small group of researchers, and are designed to test simple questions. In the field of rare diseases, sometimes the number of patients is the limiting factor for the size of a clinical trial.

Active Comparator Studies

Of note, during the last 10 years or so, it has become a common practice to conduct "active comparator" studies (also known as "active control" trials). In other words, when a treatment is clearly better than doing nothing for the subject (*i.e.* giving them the placebo), the alternate treatment would be a standard-of-care therapy. The study would compare the 'test' treatment to standard-of-care therapy.

A growing trend in the pharmacology field involves the use of third-party contractors to obtain the required comparator compounds. Such third parties provide expertise in the logistics of obtaining, storing, and shipping the comparators. As an advantage to the manufacturer of the comparator compounds, a well-established comparator sourcing agency can alleviate the problem of parallel importing (importing a patented compound for sale in a country outside the patenting agency's sphere of influence).

Master Protocol

In such studies, multiple experimental treatments are tested in a single trial. Genetic testing enables researchers to group patients according to their genetic profile, deliver drugs based on that profile to that group and compare the results. Multiple companies can participate, each bringing a different drug. The first such approach targets squamous cell cancer, which includes varying genetic disruptions from patient to patient. Amgen, AstraZeneca and Pfizer are involved, the first time they have worked together in a late-stage trial. Patients whose genomic profiles do not match any of the trial drugs receive a drug designed to stimulate the immune system to attack cancer.

Clinical Trial Protocol

A clinical trial protocol is a document used to define and manage the trial. It is prepared by a panel of experts. All study investigators are expected to strictly observe the protocol.

The protocol describes the scientific rationale, objective(s), design, methodology, statistical considerations and organization of the planned trial. Details of the trial are provided in documents referenced in the protocol, such as an investigator's brochure.

The protocol contains a precise study plan to assure safety and health of the trial subjects and to provide an exact template for trial conduct by investigators. This allows data to be combined across all investigators/sites. The protocol also informs the study administrators (often a contract research organization).

The format and content of clinical trial protocols sponsored by pharmaceutical, biotechnology or medical device companies in the United States, European Union, or Japan have been standardized to follow Good Clinical Practice guidance issued by the International Conference on Harmonization of Technical Requirements for Registration of Pharmaceuticals for Human Use (ICH). Reg-ulatory authorities in Canada and Australia also follow ICH guidelines.

Design Features

Informed Consent

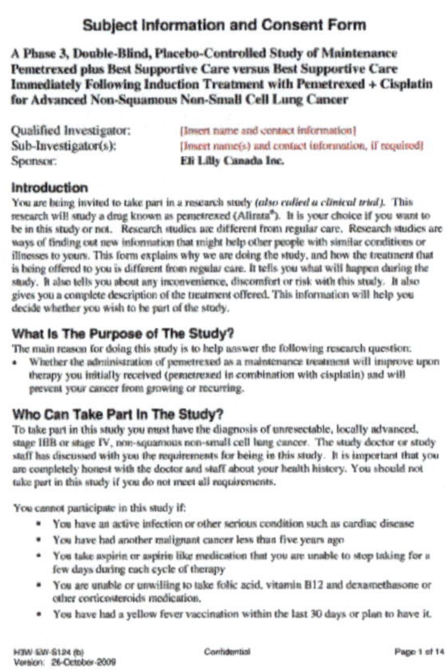

Subject Information and Consent Form

A Phase 3, Double-Blind, Placebo-Controlled Study of Maintenance Pemetrexed plus Best Supportive Care versus Best Supportive Care Immediately Following Induction Treatment with Pemetrexed + Cisplatin for Advanced Non-Squamous Non-Small Cell Lung Cancer

Qualified Investigator: [Insert name and contact information]
Sub-Investigator(s): [Insert name(s) and contact information, if required]
Sponsor: **Eli Lilly Canada Inc.**

Introduction

You are being invited to take part in a research study *(also called a clinical trial)*. This research will study a drug known as pemetrexed (Alimta®). It is your choice if you want to be in this study or not. Research studies are different from regular care. Research studies are ways of finding out new information that might help other people with similar conditions or illnesses to yours. This form explains why we are doing the study, and how the treatment that is being offered to you is different from regular care. It tells you what will happen during the study. It also tells you about any inconvenience, discomfort or risk with this study. It also gives you a complete description of the treatment offered. This information will help you decide whether you wish to be part of the study.

What Is The Purpose of The Study?

The main reason for doing this study is to help answer the following research question:

- Whether the administration of pemetrexed as a maintenance treatment will improve upon therapy you initially received (pemetrexed in combination with cisplatin) and will prevent your cancer from growing or recurring.

Who Can Take Part In The Study?

To take part in this study you must have the diagnosis of unresectable, locally advanced, stage IIIB or stage IV, non-squamous non-small cell lung cancer. The study doctor or study staff has discussed with you the requirements for being in this study. It is important that you are completely honest with the doctor and staff about your health history. You should not take part in this study if you do not meet all requirements.

You cannot participate in this study if:

- You have an active infection or other serious condition such as cardiac disease
- You have had another malignant cancer less than five years ago
- You take aspirin or aspirin like medication that you are unable to stop taking for a few days during each cycle of therapy
- You are unable or unwilling to take folic acid, vitamin B12 and dexamethasone or other corticosteroids medication.
- You have had a yellow fever vaccination within the last 30 days or plan to have it.

H3W-SW-S124 (b) Confidential Page 1 of 14
Version: 26-October-2009

Example of informed consent document from the PARAMOUNT trial

Clinical trials recruit study subjects to sign a document representing their "informed consent". The document includes details such as its purpose, duration, required procedures, risks, potential benefits, key contacts and institutional requirements. The participant then decides whether to sign the document. The document is not a contract, as the participant can withdraw at any time without penalty.

Informed consent is a legal process in which a recruit is instructed about key facts before deciding whether to participate. Researchers explain the details of the study in terms the subject can understand. The information is presented in the subject's native language. Generally, children cannot autonomously provide informed consent, but depending on their age and other factors, may be required to provide informed assent.

Statistical Power

The number of subjects has a large impact on the ability to reliably detect and measure effects of the intervention. This is described as its "power". The larger the number of participants, the greater the statistical power and the greater the cost.

The statistical power estimates the ability of a trial to detect a difference of a particular size (or larger) between the treatment and control groups. For example, a trial of a lipid-lowering drug versus placebo with 100 patients in each group might have a power of 0.90 to detect a difference between placebo and trial groups receiving dosage of 10 mg/dL or more, but only 0.70 to detect a difference of 6 mg/dL.

Placebo Groups

Merely giving a treatment can have nonspecific effects. These are controlled for by the inclusion of patients who receive only a placebo. Subjects are assigned randomly without informing them to which group they belonged. Many trials are doubled-blinded so that researchers do not know to which group a subject is assigned.

Assigning a subject to a placebo group can pose an ethical problem if it violates his or her right to receive the best available treatment. The Declaration of Helsinki provides guidelines on this issue.

Duration

Clinical trials are only a small part of the research that goes into developing a new treatment. Potential drugs, for example, first have to be discovered, purified, characterized, and tested in labs (in cell and animal studies) before ever undergoing clinical trials. In all, about 1,000 potential drugs are tested before just one reaches the point of being tested in a clinical trial. For example, a new cancer drug has, on average, six years of research behind it before it even makes it to clinical trials. But the major holdup in making new cancer drugs available is the time it takes to complete clinical trials themselves. On average, about eight years pass from the time a cancer drug enters clinical trials until it receives approval from regulatory agencies for sale to the public. Drugs for other diseases have similar timelines.

Some reasons a clinical trial might last several years:

- For chronic conditions such as cancer, it takes months, if not years, to see if a cancer treatment has an effect on a patient.

- For drugs that are not expected to have a strong effect (meaning a large number of patients must be recruited to observe 'any' effect), recruiting enough patients to test the drug's effectiveness (i.e., getting statistical power) can take several years.
- Only certain people who have the target disease condition are eligible to take part in each clinical trial. Researchers who treat these particular patients must participate in the trial. Then they must identify the desirable patients and obtain consent from them or their families to take part in the trial.

The biggest barrier to completing studies is the shortage of people who take part. All drug and many device trials target a subset of the population, meaning not everyone can participate. Some drug trials require patients to have unusual combinations of disease characteristics. It is a challenge to find the appropriate patients and obtain their consent, especially when they may receive no direct benefit (because they are not paid, the study drug is not yet proven to work, or the patient may receive a placebo). In the case of cancer patients, fewer than 5% of adults with cancer will participate in drug trials. According to the Pharmaceutical Research and Manufacturers of America (PhRMA), about 400 cancer medicines were being tested in clinical trials in 2005. Not all of these will prove to be useful, but those that are may be delayed in getting approved because the number of participants is so low.

For clinical trials involving a seasonal indication (such as airborne allergies, seasonal affective disorder, influenza, and others), the study can only be done during a limited part of the year (such as spring for pollen allergies), when the drug can be tested. This can be an additional complication on the length of the study, yet proper planning and the use of trial sites in the Southern, as well as the Northern Hemisphere allows for year-round trials, which can reduce the length of the studies.

Clinical trials that do not involve a new drug usually have a much shorter duration. (Exceptions are epidemiological studies, such as the Nurses' Health Study).

Administration

Clinical trials designed by a local investigator, and (in the US) federally funded clinical trials, are almost always administered by the researcher who designed the study and applied for the grant. Small-scale device studies may be administered by the sponsoring company. Clinical trials of new drugs are usually administered by a contract research organization (CRO) hired by the sponsoring company. The sponsor provides the drug and medical oversight. A CRO is contracted to perform all the administrative work on a clinical trial. For phases 2, 3 and 4, the CRO recruits participating researchers, trains them, provides them with supplies, coordinates study administration and data collection, sets up meetings, monitors the sites for compliance with the clinical protocol, and ensures the sponsor receives data from every site. Specialist site management organizations can also be hired to coordinate with the CRO to ensure rapid IRB/IEC approval and faster site initiation and patient recruitment. Phase 1 clinical trials of new medicines are often conducted in a specialist clinical trial clinic, with dedicated pharmacologists, where the subjects can be observed by full-time staff. These clinics are often run by a CRO which specialises in these studies.

At a participating site, one or more research assistants (often nurses) do most of the work in conducting the clinical trial. The research assistant's job can include some or all of the following:

providing the local institutional review board (IRB) with the documentation necessary to obtain its permission to conduct the study, assisting with study start-up, identifying eligible patients, obtaining consent from them or their families, administering study treatment(s), collecting and statistically analyzing data, maintaining and updating data files during followup, and communicating with the IRB, as well as the sponsor and CRO.

Marketing

Janet Yang uses the Interactional Justice Model to test the effects of willingness to talk with a doctor and clinical trial enrollment. Results found that potential clinical trial candidates were less likely to enroll in clinical trials if the patient is more willing to talk with their doctor. The reasoning behind this discovery may be patients are happy with their current care. Another reason for the negative relationship between perceived fairness and clinical trial enrollment is the lack of independence from the care provider. Results found that there is a positive relationship between a lack of willingness to talk with their doctor and clinical trial enrollment. Lack of willingness to talk about clinical trials with current care providers may be due to patients' independence from the doctor. Patients who are less likely to talk about clinical trials are more willing to use other sources of information to gain a better insight of alternative treatments. Clinical trial enrollment should be motivated to utilize websites and television advertising to inform the public about clinical trial enrollment.

Information Technology

The last decade has seen a proliferation of information technology use in the planning and conduct of clinical trials. Clinical trial management systems are often used by research sponsors or CROs to help plan and manage the operational aspects of a clinical trial, particularly with respect to investigational sites. Advanced analytics for identifying researchers and research sites with expertise in a given area utilize public and private information about ongoing research. Web-based electronic data capture (EDC) and clinical data management systems are used in a majority of clinical trials to collect case report data from sites, manage its quality and prepare it for analysis. Interactive voice response systems are used by sites to register the enrollment of patients using a phone and to allocate patients to a particular treatment arm (although phones are being increasingly replaced with web-based (IWRS) tools which are sometimes part of the EDC system). While patient-reported outcome were often paper based in the past, measurements are increasingly being collected using web portals or hand-held ePRO (or eDiary) devices, sometimes wireless. Statistical software is used to analyze the collected data and prepare them for regulatory submission. Access to many of these applications are increasingly aggregated in web-based clinical trial portals. In 2011, the FDA approved a phase 1 trial that used telemonitoring, also known as remote patient monitoring, to collect biometric data in patients' homes and transmit it electronically to the trial database. This technology provides many more data points and is far more convenient for patients, because they have fewer visits to trial sites.

Ethical Aspects

Clinical trials are closely supervised by appropriate regulatory authorities. All studies involving a medical or therapeutic intervention on patients must be approved by a supervising ethics committee before permission is granted to run the trial. The local ethics committee has discretion on how it will supervise noninterventional studies (observational studies or those using already collected data). In the US, this body is called the Institutional Review Board (IRB); in the EU, they are called Ethics committees. Most IRBs are located at the local investigator's hospital or institution, but some sponsors allow the use of a central (independent/for profit) IRB for investigators who work at smaller institutions.

To be ethical, researchers must obtain the full and informed consent of participating human subjects. (One of the IRB's main functions is to ensure potential patients are adequately informed about the clinical trial.) If the patient is unable to consent for him/herself, researchers can seek consent from the patient's legally authorized representative. In California, the state has prioritized the individuals who can serve as the legally authorized representative.

In some US locations, the local IRB must certify researchers and their staff before they can conduct clinical trials. They must understand the federal patient privacy (HIPAA) law and good clinical practice. The International Conference of Harmonisation Guidelines for Good Clinical Practice is a set of standards used internationally for the conduct of clinical trials. The guidelines aim to ensure the "rights, safety and well being of trial subjects are protected".

The notion of informed consent of participating human subjects exists in many countries all over the world, but its precise definition may still vary.

Informed consent is clearly a 'necessary' condition for ethical conduct but does not 'ensure' ethical conduct. In compassionate use trials the latter becomes a particularly difficult problem. The final objective is to serve the community of patients or future patients in a best-possible and most responsible way. However, it may be hard to turn this objective into a well-defined, quantified, objective function. In some cases this can be done, however, for instance, for questions of when to stop sequential treatments, and then quantified methods may play an important role.

Additional ethical concerns are present when conducting clinical trials on children (pediatrics), and in emergency or epidemic situations.

Conflicts of Interest and Unfavorable Studies

In response to specific cases in which unfavorable data from pharmaceutical company-sponsored research were not published, the Pharmaceutical Research and Manufacturers of America published new guidelines urging companies to report all findings and limit the financial involvement in drug companies by researchers. The US Congress signed into law a bill which requires phase II and phase III clinical trials to be registered by the sponsor on the the clinical trials website compiled by the National Institutes of Health.

Drug researchers not directly employed by pharmaceutical companies often seek grants from manufacturers, and manufacturers often look to academic researchers to conduct studies within networks of universities and their hospitals, e.g., for translational cancer research. Similarly, competition for tenured academic positions, government grants and prestige create conflicts of interest among academic scientists. According to one study, approximately 75% of articles retracted for misconduct-related reasons have no declared industry financial support. Seeding trials are particularly controversial.

In the United States, all clinical trials submitted to the FDA as part of a drug approval process are independently assessed by clinical experts within the Food and Drug Administration, including inspections of primary data collection at selected clinical trial sites.

In 2001, the editors of 12 major journals issued a joint editorial, published in each journal, on the control over clinical trials exerted by sponsors, particularly targeting the use of contracts which allow sponsors to review the studies prior to publication and withhold publication. They strengthened editorial restrictions to counter the effect. The editorial noted that contract research organizations had, by 2000, received 60% of the grants from pharmaceutical companies in the US. Researchers may be restricted from contributing to the trial design, accessing the raw data, and interpreting the results.

Safety

Responsibility for the safety of the subjects in a clinical trial is shared between the sponsor, the local site investigators (if different from the sponsor), the various IRBs that supervise the study, and (in some cases, if the study involves a marketable drug or device), the regulatory agency for the country where the drug or device will be sold.

For safety reasons, many clinical trials of drugs are designed to exclude women of childbearing age, pregnant women, and/or women who become pregnant during the study. In some cases, the male partners of these women are also excluded or required to take birth control measures.

Sponsor

Throughout the clinical trial, the sponsor is responsible for accurately informing the local site investigators of the true historical safety record of the drug, device or other medical treatments to be tested, and of any potential interactions of the study treatment(s) with already approved treatments. This allows the local investigators to make an informed judgment on whether to participate in the study or not. The sponsor is also responsible for monitoring the results of the study as they come in from the various sites, as the trial proceeds. In larger clinical trials, a sponsor will use the services of a data monitoring committee (DMC, known in the US as a data safety monitoring board). This independent group of clinicians and statisticians meets periodically to review the unblinded data the sponsor has received so far. The DMC has the power to recommend termination of the study based on their review, for example if the study treatment is causing more deaths than the standard treatment, or seems to be causing unexpected and study-related serious adverse events.The sponsor is responsible for collecting adverse event reports from all site investigators in the study, and for informing all the investigators of the sponsor's judgment as to whether these adverse events were related or not related to the study treatment. This is an area where sponsors can slant their judgment to favor the study treatment.

The sponsor and the local site investigators are jointly responsible for writing a site-specific informed consent that accurately informs the potential subjects of the true risks and potential benefits of participating in the study, while at the same time presenting the material as briefly as possible and in ordinary language. FDA regulations and ICH guidelines both require "the information that is given to the subject or the representative shall be in language understandable to the subject or the representative." If the participant's native language is not English, the sponsor must translate the informed consent into the language of the participant.

Local Site Investigators

The ethical principle of *primum non nocere* guides the trial, and if an investigator believes the study treatment may be harming subjects in the study, the investigator can stop participating at any time. On the other hand, investigators often have a financial interest in recruiting subjects, and can act unethically to obtain and maintain their participation.

The local investigators are responsible for conducting the study according to the study protocol, and supervising the study staff throughout the duration of the study. The local investigator or his/her study staff are also responsible for ensuring the potential subjects in the study understand the risks and potential benefits of participating in the study; in other words, they (or their legally authorized representatives) must give truly informed consent. They are responsible for reviewing all adverse event reports sent by the sponsor. (These adverse event reports contain the opinion of both the investigator at the site where the adverse event occurred, and the sponsor, regarding the relationship of the adverse event to the study treatments). They also are responsible for making an independent judgment of these reports, and promptly informing the local IRB of all serious and study treatment-related adverse events.

When a local investigator is the sponsor, there may not be formal adverse event reports, but study staff at all locations are responsible for informing the coordinating investigator of anything unexpected. The local investigator is responsible for being truthful to the local IRB in all communications relating to the study.

Institutional Review Boards (IRBs)

Approval by an Institutional Review Board (IRB), or ethics board, is necessary before all but the most informal research can begin. In commercial clinical trials, the study protocol is not approved by an IRB before the sponsor recruits sites to conduct the trial. However, the study protocol and procedures have been tailored to fit generic IRB submission requirements. In this case, and where there is no independent sponsor, each local site investigator submits the study protocol, the consent(s), the data collection forms, and supporting documentation to the local IRB. Universities and most hospitals have in-house IRBs. Other researchers (such as in walk-in clinics) use independent IRBs.

The IRB scrutinizes the study for both medical safety and protection of the patients involved in the study, before it allows the researcher to begin the study. It may require changes in study procedures or in the explanations given to the patient. A required yearly "continuing review" report from the investigator updates the IRB on the progress of the study and any new safety information related to the study.

Regulatory Agencies

In the US, the FDA can audit the files of local site investigators after they have finished participating in a study, to see if they were correctly following study procedures. This audit may be random, or for cause (because the investigator is suspected of fraudulent data). Avoiding an audit is an incentive for investigators to follow study procedures.

Alternatively, many American pharmaceutical companies have moved some clinical trials overseas. Benefits of conducting trials abroad include lower costs (in some countries) and the ability to run larger trials in shorter timeframes. Critics have argued that clinical trials performed outside the U.S. allow companies to avoid many of the FDA's regulations, since the FDA audits these trials less frequently than U.S. studies. For drug applications approved by the FDA in 2008, 0.7 percent of foreign clinical study sites were audited by the FDA compared to 1.9 percent domestically. Other criticisms of foreign clinical studies, especially in developing countries, relate to the rights and welfare of study participants, integrity of study data, and relevance of data to the U.S. population.

Different countries have different regulatory requirements and enforcement abilities. An estimated 40% of all clinical trials now take place in Asia, Eastern Europe, and Central and South America. "There is no compulsory registration system for clinical trials in these countries and many do not follow European directives in their operations", says Jacob Sijtsma of the Netherlands-based WEMOS, an advocacy health organisation tracking clinical trials in developing countries.

Beginning in the 1980s, harmonization of clinical trial protocols was shown as feasible across countries of the European Union. At the same time, coordination between Europe, Japan and the United States led to a joint regulatory-industry initiative on international harmonization named after 1990 as the International Conference on Harmonisation of Technical Requirements for Registration of Pharmaceuticals for Human Use (ICH) Currently, most clinical trial programs follow ICH guidelines, aimed at "ensuring that good quality, safe and effective medicines are developed and registered in the most efficient and cost-effective manner. These activities are pursued in the interest of the consumer and public health, to prevent unnecessary duplication of clinical trials in humans and to minimize the use of animal testing without compromising the regulatory obligations of safety and effectiveness."

Economics

Sponsor

The cost of a study depends on many factors, especially the number of sites conducting the study, the number of patients required, and whether the study treatment is already approved for medical use. Clinical trials follow a standardized process.

The costs to a pharmaceutical company of administering a phase 3 or 4 clinical trial may include, among others:

- manufacturing the drug(s)/device(s) tested
- staff salaries for the designers and administrators of the trial
- payments to the contract research organization, the site management organization (if used) and any outside consultants
- payments to local researchers (and their staffs) for their time and effort in recruiting patients and collecting data for the sponsor
- study materials and shipping
- communication with the local researchers, including on-site monitoring by the CRO before and (in some cases) multiple times during the study

- one or more investigator training meetings
- costs incurred by the local researchers, such as pharmacy fees, IRB fees and postage
- any payments to patients enrolled in the trial (all payments are strictly overseen by the IRBs to ensure the patients do not feel coerced to take part in the trial by overly attractive payments)

These costs are incurred over several years.

In the US, sponsors may receive a 50% tax credit for clinical trials of drugs being developed for the treatment of rare (orphan) diseases.

National health agencies, such as the US National Institutes of Health, offer grants to investigators who design clinical trials that attempt to answer research questions of interest to the agency. In these cases, the investigator who writes the grant and administers the study acts as the sponsor, and coordinates data collection from any other sites. These other sites may or may not be paid for participating in the study, depending on the amount of the grant and the amount of effort expected from them.

Clinical trials are traditionally expensive and difficult to undertake. Using internet resources can, in some cases, reduce the economic burden. New technologies enable sponsors and CRO's to reduce trial costs by executing online feasibility assessments and better collaborate with research centers such as ViS Research Institute.

Investigators

Many clinical trials do not involve any money. However, when the sponsor is a private company or a national health agency, investigators are almost always paid to participate. These amounts can be small, just covering a partial salary for research assistants and the cost of any supplies (usually the case with national health agency studies), or be substantial and include 'overhead' that allows the investigator to pay the research staff during times between clinical trials.

Subjects

Participants in phase 1 drug trials do not gain any direct benefit from taking part. They are generally paid an inconvenience allowance because they give up their time (sometimes away from their homes); the amounts paid are regulated and are not related to the level of risk involved. In most other trials, subjects are not paid to ensure their motivation for participating is the hope of getting better or contributing to medical knowledge, without their judgment being skewed by financial considerations. However, they are often given small payments for study-related expenses such as travel or as compensation for their time in providing follow-up information about their health after they are discharged from medical care.

Participation as Labour

It has been suggested that clinical trial participants be considered to be performing 'experimental' or 'clinical labour'. Re-classifying clinical trials as labour is supported by the fact that information gained from clinical trials contributes to biomedical knowledge, and thus increases the profits of pharmaceutical companies. The labour performed by those participants in clinical trials includes the provision of tissue samples and information, the performance of other tasks, such as adhering to a special diet, or (in the case of phase I trials particularly) exposing themselves to risk. The participants in exchange are offered potential access to medical treatment. For some, this may be a treatment with the potential to succeed where other treatments have failed. For other individuals, particularly those situated in countries such as China or India, they may be given access to healthcare which they otherwise would be unable to afford, for the duration of the trial. Thus, the exchange which exists may serve to classify clinical trials as a form of labour.

Participant Recruitment and Participation

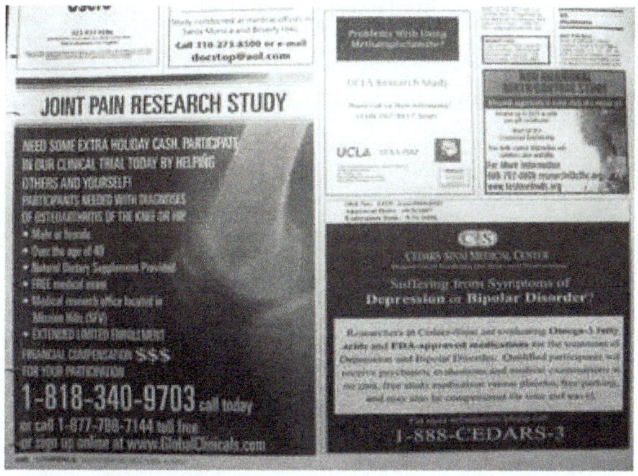

Newspaper advertisements seeking patients and healthy volunteers to participate in clinical trials

Phase 0 and phase 1 drug trials seek healthy volunteers. Most other clinical trials seek patients who have a specific disease or medical condition. The diversity observed in society, by consensus, should be reflected in clinical trials through the appropriate inclusion of ethnic minority populations. Patient recruitment or participant recruitment (as some participants in clinical trials are considered 'healthy' and not patients) plays a significant role in the activities and responsibilities of sites conducting clinical trials.

Locating Trials

Depending on the kind of participants required, sponsors of clinical trials, or contract research organizations working on their behalf, try to find sites with qualified personnel as well as access to patients who could participate in the trial. Working with those sites, they may use various recruitment strategies, including patient databases, newspaper and radio advertisements, flyers, posters in places the patients might go (such as doctor's offices), and personal recruitment of patients by investigators.

Volunteers with specific conditions or diseases have additional online resources to help them locate clinical trials. For example, the Fox Trial Finder connects Parkinson's disease trials around the world to volunteers who have a specific set of criteria such as location, age, and symptoms. Other disease-specific services exist for volunteers to find trials related to their condition. Volunteers may search directly on ClinicalTrials.gov to locate trials using a registry run by the U.S. National Institutes of Health and National Library of Medicine.

However, many clinical trials will not accept participants who contact them directly to volunteer, as it is believed this may bias the characteristics of the population being studied. Such trials typically recruit via networks of medical professionals who ask their individual patients to consider enrollment.

Steps for Volunteers

Before participating in a clinical trial, interested volunteers should speak with their doctors, family members, and others who have participated in trials in the past. After locating a trial, volunteers will often have the opportunity to speak or e-mail the clinical trial coordinator for more information and to answer any questions. After receiving consent from their doctors, volunteers then arrange an appointment for a screening visit with the trial coordinator.

All volunteers being considered for a trial are required to undertake a medical screening. Requirements differ for different trials, but typically volunteers will have the following tests in a medical laboratory:

- Measurement of the electrical activity of the heart (ECG)
- Measurement of blood pressure, heart rate and temperature
- Blood sampling
- Urine sampling
- Weight and height measurement
- Drugs abuse testing
- Pregnancy testing (females only)

Research

In 2012, Z. Janet Yang, Katherine A. McComas, Geri K. Gay, John P. Leonard, Andrew J. Dannenberg, and Hildy Dillon conducted research on the attitudes towards clinical trial treatment and the decision making of signing up for such trials by cancer patients and the general population. They used the risk information seeking and processing (RISP) model to analyze the social implications that affect attitudes and decision making pertaining to clinical trials. People who hold a higher stake or interest in clinical trial treatment showed a greater likelihood of seeking information about clinical trials. Those with networks that stress the importance of learning about clinical trials are also more likely to seek and process information more deeply. People with more knowledge about clinical trials tend to have to a greater likelihood of signing up. In the study, cancer patients reported more optimistic attitudes towards clinical trials than the general population. Having a more optimistic outlook on clinical trials also leads to greater likelihood of enrolling.

Clinical Trials Around the World

- USA Clinical Trials:
- European Clinical Trials:

References

- World Health Organization (2003). Introduction to drug utilization research (PDF). Geneva: World Health Organization. p. 33. ISBN 924156234X.
- Finkel, Richard; Cubeddu, Luigi; Clark, Michelle (2009). Lippencott's Illustrated Reviews: Pharmacology 4th Edition. Lippencott Williams & Wilkins. pp. 1–4. ISBN 978-0-7817-7155-9.
- H. F. J. Horstmanshoff, Marten Stol, Cornelis Tilburg (2004), Magic and Rationality in Ancient Near Eastern and Graeco-Roman Medicine, p. 99, Brill Publishers, ISBN 90-04-13666-5.
- Kenneth G. Zysk, Asceticism and Healing in Ancient India: Medicine in the Buddhist Monastery, Oxford University Press, rev. ed. (1998) ISBN 0-19-505956-5.
- Miller, AA; Miller, PF (editor) (2011). Emerging Trends in Antibacterial Discovery: Answering the Call to Arms. Caister Academic Press. ISBN 978-1-904455-89-9.
- Walsh G (2003). Biopharmaceuticals: Biochemistry and Biotechnology, Second Edition. John Wiley & Sons Ltd. ISBN 978-0-470-84326-0.
- Madsen U, Krogsgaard-Larsen P, Liljefors T (2002). Textbook of Drug Design and Discovery. Washington, DC: Taylor & Francis. ISBN 0-415-28288-8.
- Reynolds CH, Merz KM, Ringe D, eds. (2010). Drug Design: Structure- and Ligand-Based Approaches (1 ed.). Cambridge, UK: Cambridge University Press. ISBN 978-0521887236.
- Wu-Pong S, Rojanasakul Y (2008). Biopharmaceutical drug design and development (2nd ed.). Totowa, NJ Humana Press: Humana Press. ISBN 978-1-59745-532-9.
- Hopkins AL (2011). "Chapter 25: Pharmacological space". In Wermuth CG. The Practice of Medicinal Chemistry (3 ed.). Academic Press. pp. 521–527. ISBN 978-0-12-374194-3.
- Guner OF (2000). Pharmacophore Perception, Development, and use in Drug Design. La Jolla, Calif: International University Line. ISBN 0-9636817-6-1.
- Tropsha A (2010). "QSAR in Drug Discovery". In Reynolds CH, Merz KM, Ringe D. Drug Design: Structure- and Ligand-Based Approaches (1 ed.). Cambridge, UK: Cambridge University Press. pp. 151–164. ISBN 978-0521887236.
- Creswell, J.W. (2008). Educational research: Planning, conducting, and evaluating quantitative and qualitative research (3rd). Upper Saddle River, NJ: Prentice Hall. 2008, p. 300. ISBN 0-13-613550-1
- Peter W. Huber (12 November 2013). The Cure in the Code: How 20th Century Law Is Undermining 21st Century Medicine. Basic Books. ISBN 978-0-465-06981-1.
- Maloney, Dennis M (1984). Protection of Human Research Subjects: A Practical Guide to Federal Laws and Regulations. Boston, MA: Springer US. p. 151. ISBN 9781461327035.
- Gramatica P (2011). "Chapter 17: Modeling Chemicals in the Environment". In Gramatica P, Livingstone DJ, Davis AM. Drug Design Strategies: Quantitative Approaches. Royal Society of Chemistry. p. 466. doi:10.1039/9781849733410-00458. ISBN 978-1849731669.

5

Allied Fields of Medicinal Chemistry

This chapter provides a plethora of allied fields of medicinal chemistry. They allow for more complex developments in the field of medicinal chemistry and provide a better comprehension of medicinal chemistry. Chemistry, organic chemistry, biochemistry, chemical biology and organometallic chemistry are the associated fields that are listed in this chapter.

Chemistry

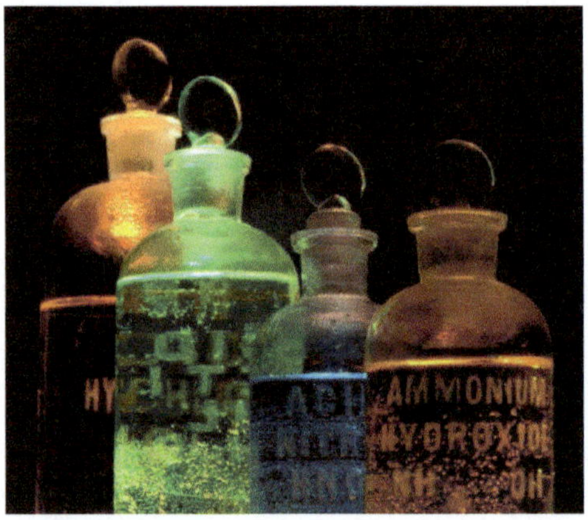

Solutions of substances in reagent bottles, including ammonium hydroxide and nitric acid, illuminated in different colors

Chemistry is a branch of physical science that studies the composition, structure, properties and change of matter. Chemistry includes topics such as the properties of individual atoms, how atoms form chemical bonds to create chemical compounds, the interactions of substances through intermolecular forces that give matter its general properties, and the interactions between substances through chemical reactions to form different substances.

Chemistry is sometimes called the central science because it bridges other natural sciences, including physics, geology and biology. For the differences between chemistry and physics see comparison of chemistry and physics.

Scholars disagree about the etymology of the word *chemistry*. The history of chemistry can be traced to alchemy, which had been practiced for several millennia in various parts of the world.

Etymology

The word *chemistry* comes from the word *alchemy* which was an earlier set of practices that encompassed elements of chemistry, metallurgy, philosophy, astrology, astronomy, mysticism and medicine. It is often seen as linked to the quest to turn lead or another common starting material into gold. Alchemy, which was practiced around 330, is the study of the composition of waters, movement, growth, embodying, disembodying, drawing the spirits from bodies and bonding the spirits within bodies (Zosimos). An alchemist was called a 'chemist' in popular speech, and later the suffix "-ry" was added to this to describe the art of the chemist as "chemistry".

The word *alchemy* in turn is derived from the Arabic word *al-kīmīā* (ءايميكلا). In origin, the term is borrowed from the Greek. This may have Egyptian origins since *al-kīmīā* is derived from the Greek, which is in turn derived from the word Chemi or Kimi, which is the ancient name of Egypt in Egyptian. Alternately,*al-kīmīā* may derive from χημεία, meaning "cast together".

Definition

In retrospect, the definition of chemistry has changed over time, as new discoveries and theories add to the functionality of the science. The term "chymistry", in the view of noted scientist Robert Boyle in 1661, meant the subject of the material principles of mixed bodies. In 1663 the chemist Christopher Glaser described "chymistry" as a scientific art, by which one learns to dissolve bodies, and draw from them the different substances on their composition, and how to unite them again, and exalt them to a higher perfection.

The 1730 definition of the word "chemistry", as used by Georg Ernst Stahl, meant the art of resolving mixed, compound, or aggregate bodies into their principles; and of composing such bodies from those principles. In 1837, Jean-Baptiste Dumas considered the word "chemistry" to refer to the science concerned with the laws and effects of molecular forces. This definition further evolved until, in 1947, it came to mean the science of substances: their structure, their properties, and the reactions that change them into other substances - a characterization accepted by Linus Pauling. More recently, in 1998, Professor Raymond Chang broadened the definition of "chemistry" to mean the study of matter and the changes it undergoes.

History

Democritus' atomist philosophy was later adopted by Epicurus (341–270 BCE).

Early civilizations, such as the Egyptians Babylonians, Indians amassed practical knowledge concerning the arts of metallurgy, pottery and dyes, but didn't develop a systematic theory.

A basic chemical hypothesis first emerged in Classical Greece with the theory of four elements as propounded definitively by Aristotle stating that that fire, air, earth and water were the fundamental elements from which everything is formed as a combination. Greek atomism dates back to 440 BC, arising in works by philosophers such as Democritus and Epicurus. In 50 BC, the Roman philosopher Lucretius expanded upon the theory in his book *De rerum natura* (On The Nature of Things). Unlike modern concepts of science, Greek atomism was purely philosophical in nature, with little concern for empirical observations and no concern for chemical experiments.

In the Hellenistic world the art of alchemy first proliferated, mingling magic and occultism into the study of natural substances with the ultimate goal of transmuting elements into gold and discovering the elixir of eternal life. Alchemy was discovered and practised widely throughout the Arab world after the Muslim conquests, and from there, diffused into medieval and Renaissance Europe through Latin translations.

Chemistry as Science

Under the influence of the new empirical methods propounded by Sir Francis Bacon and others, a group of chemists at Oxford, Robert Boyle, Robert Hooke and John Mayow began to reshape the old alchemical traditions into a scientific discipline. Boyle in particular is regarded as the founding father of chemistry due to his most important work, the classic chemistry text *The Sceptical Chymist* where the differentiation is made between the claims of alchemy and the empirical scientific discoveries of the new chemistry. He formulated Boyle's law, rejected the classical "four elements" and proposed a mechanistic alternative of atoms and chemical reactions that could be subject to rigorous experiment.

Antoine-Laurent de Lavoisier is considered the "Father of Modern Chemistry".

The theory of phlogiston (a substance at the root of all combustion) was propounded by the German Georg Ernst Stahl in the early 18th century and was only overturned by the end of the century by the French chemist Antoine Lavoisier, the chemical analogue of Newton in physics; who did more than any other to establish the new science on proper theoretical footing, by elucidating the principle of conservation of mass and developing a new system of chemical nomenclature used to this day.

Prior to his work, though, many important discoveries had been made, specifically relating to the nature of 'air' which was discovered to be composed of many different gases. The Scottish chemist Joseph Black (the first experimental chemist) and the Dutchman J. B. van Helmont discovered carbon dioxide, or what Black called 'fixed air' in 1754; Henry Cavendish discovered hydrogen and elucidated its properties and Joseph Priestley and, independently, Carl Wilhelm Scheele isolated pure oxygen.

In his periodic table, Dmitri Mendeleev predicted the existence of 7 new elements, and placed all 60 elements known at the time in their correct places.

English scientist John Dalton proposed the modern theory of atoms; that all substances are composed of indivisible 'atoms' of matter and that different atoms have varying atomic weights.

The development of the electrochemical theory of chemical combinations occurred in the early 19th century as the result of the work of two scientists in particular, J. J. Berzelius and Humphry Davy, made possible by the prior invention of the voltaic pile by Alessandro Volta. Davy discovered nine new elements including the alkali metals by extracting them from their oxides with electric current.

British William Prout first proposed ordering all the elements by their atomic weight as all atoms had a weight that was an exact multiple of the atomic weight of hydrogen. J. A. R. Newlands devised an early table of elements, which was then developed into the modern periodic table of elements in the 1860s by Dmitri Mendeleev and independently by several other scientists including Julius Lothar Meyer. The inert gases, later called the noble gases were discovered by William Ramsay in collaboration with Lord Rayleigh at the end of the century, thereby filling in the basic structure of the table.

Organic chemistry was developed by Justus von Liebig and others, following Friedrich Wöhler's synthesis of urea which proved that living organisms were, in theory, reducible to chemistry. Other crucial 19th century advances were; an understanding of valence bonding (Edward Frankland in 1852) and the application of thermodynamics to chemistry (J. W. Gibbs and Svante Arrhenius in the 1870s).

Chemical Structure

At the turn of the twentieth century the theoretical underpinnings of chemistry were finally understood due to a series of remarkable discoveries that succeeded in probing and discovering the very nature of the internal structure of atoms. In 1897, J. J. Thomson of Cambridge University discovered the electron and soon after the French scientist Becquerel as well as the couple Pierre and Marie Curie investigated the phenomenon of radioactivity. In a series of pioneering scattering experiments Ernest Rutherford at the University of Manchester discovered the internal structure of the atom and the existence of the proton, classified and explained the different types of radioactivity and successfully transmuted the first element by bombarding nitrogen with alpha particles.

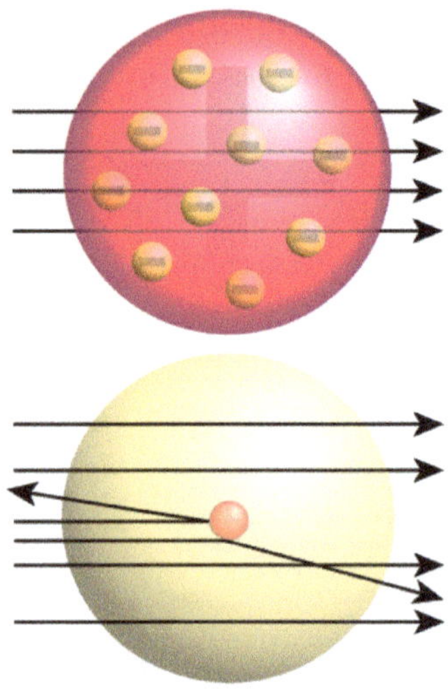

Top: Expected results: alpha particles passing through the plum pudding model of the atom undisturbed. *Bottom:* Observed results: a small portion of the particles were deflected, indicating a small, concentrated charge.

His work on atomic structure was improved on by his students, the Danish physicist Niels Bohr and Henry Moseley. The electronic theory of chemical bonds and molecular orbitals was developed by the American scientists Linus Pauling and Gilbert N. Lewis.

The year 2011 was declared by the United Nations as the International Year of Chemistry. It was an initiative of the International Union of Pure and Applied Chemistry, and of the United Nations Educational, Scientific, and Cultural Organization and involves chemical societies, academics, and institutions worldwide and relied on individual initiatives to organize local and regional activities.

Principles of Modern Chemistry

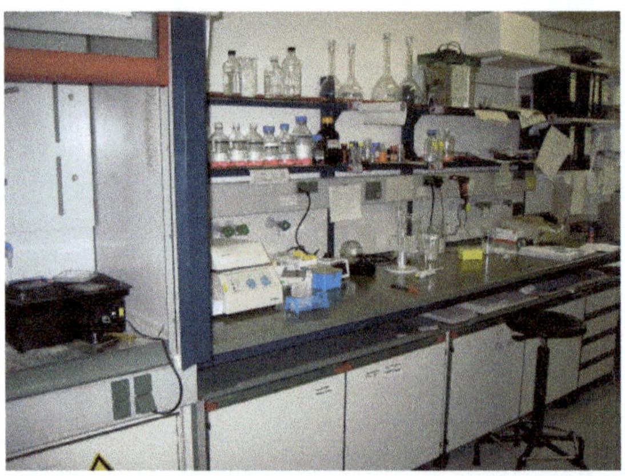

Laboratory, Institute of Biochemistry, University of Cologne.

The current model of atomic structure is the quantum mechanical model. Traditional chemistry starts with the study of elementary particles, atoms, molecules, substances, metals, crystals and other aggregates of matter. This matter can be studied in solid, liquid, or gas states, in isolation or in combination. The interactions, reactions and transformations that are studied in chemistry are usually the result of interactions between atoms, leading to rearrangements of the chemical bonds which hold atoms together. Such behaviors are studied in a chemistry laboratory.

The chemistry laboratory stereotypically uses various forms of laboratory glassware. However glassware is not central to chemistry, and a great deal of experimental (as well as applied/industrial) chemistry is done without it.

A chemical reaction is a transformation of some substances into one or more different substances. The basis of such a chemical transformation is the rearrangement of electrons in the chemical bonds between atoms. It can be symbolically depicted through a chemical equation, which usually involves atoms as subjects. The number of atoms on the left and the right in the equation for a chemical transformation is equal. (When the number of atoms on either side is unequal, the transformation is referred to as a nuclear reaction or radioactive decay.) The type of chemical reactions a substance may undergo and the energy changes that may accompany it are constrained by certain basic rules, known as chemical laws.

Energy and entropy considerations are invariably important in almost all chemical studies. Chemical substances are classified in terms of their structure, phase, as well as their chemical compositions. They can be analyzed using the tools of chemical analysis, e.g. spectroscopy and chromatography. Scientists engaged in chemical research are known as chemists. Most chemists specialize in one or more sub-disciplines. Several concepts are essential for the study of chemistry; some of them are:

Matter

In chemistry, matter is defined as anything that has rest mass and volume (it takes up space) and is made up of particles. The particles that make up matter have rest mass as well - not all particles have rest mass, such as the photon. Matter can be a pure chemical substance or a mixture of substances.

Atom

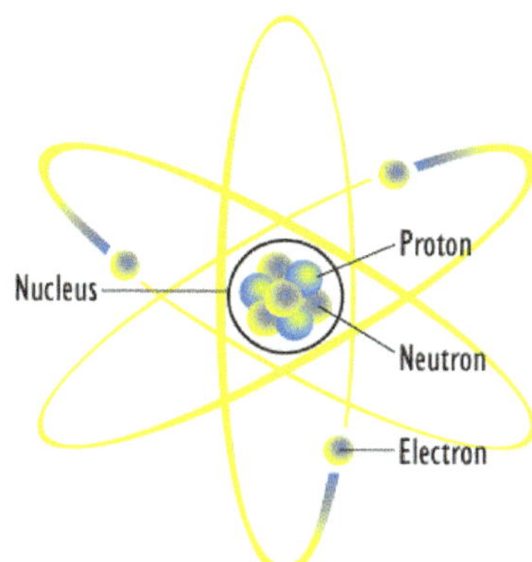

A diagram of an atom based on the Rutherford model

The atom is the basic unit of chemistry. It consists of a dense core called the atomic nucleus surrounded by a space called the electron cloud. The nucleus is made up of positively charged protons and uncharged neutrons (together called nucleons), while the electron cloud consists of negatively charged electrons which orbit the nucleus. In a neutral atom, the negatively charged electrons balance out the positive charge of the protons. The nucleus is dense; the mass of a nucleon is 1,836 times that of an electron, yet the radius of an atom is about 10,000 times that of its nucleus.

The atom is also the smallest entity that can be envisaged to retain the chemical properties of the element, such as electronegativity, ionization potential, preferred oxidation state(s), coordination number, and preferred types of bonds to form (e.g., metallic, ionic, covalent).

Element

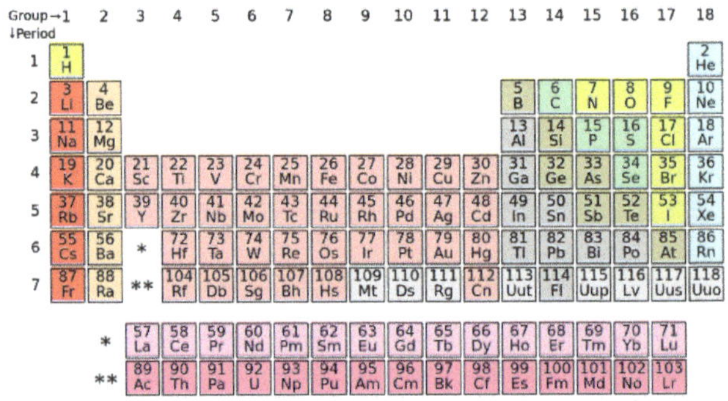

Standard form of the periodic table of chemical elements. The colors represent different categories of elements

A chemical element is a pure substance which is composed of a single type of atom, characterized by its particular number of protons in the nuclei of its atoms, known as the atomic number and represented by the symbol *Z*. The mass number is the sum of the number of protons and neutrons in a nucleus. Although all the nuclei of all atoms belonging to one element will have the same atomic number, they may not necessarily have the same mass number; atoms of an element which have different mass numbers are known as isotopes. For example, all atoms with 6 protons in their nuclei are atoms of the chemical element carbon, but atoms of carbon may have mass numbers of 12 or 13.

The standard presentation of the chemical elements is in the periodic table, which orders elements by atomic number. The periodic table is arranged in groups, or columns, and periods, or rows. The periodic table is useful in identifying periodic trends.

Compound

Carbon dioxide (CO_2), an example of a chemical compound

A *compound* is a pure chemical substance composed of more than one element. The properties of a compound bear little similarity to those of its elements. The standard nomenclature of compounds is set by the International Union of Pure and Applied Chemistry (IUPAC). Organic compounds are named according to the organic nomenclature system. Inorganic compounds are named according to the inorganic nomenclature system. In addition the Chemical Abstracts Service has devised a method to index chemical substances. In this scheme each chemical substance is identifiable by a number known as its CAS registry number.

Molecule

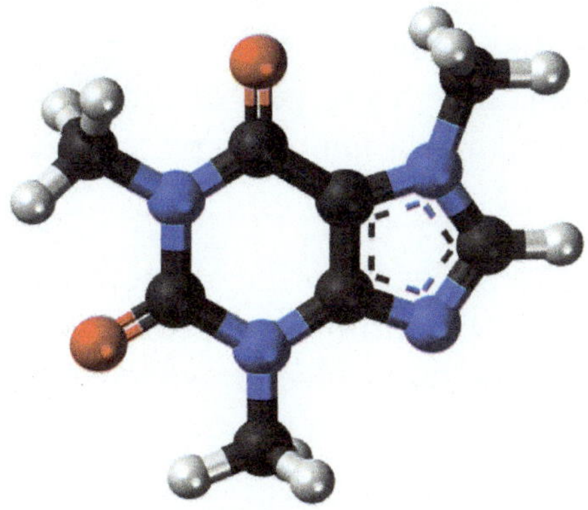

A ball-and-stick representation of the caffeine molecule ($C_8H_{10}N_4O_2$).

A *molecule* is the smallest indivisible portion of a pure chemical substance that has its unique set of chemical properties, that is, its potential to undergo a certain set of chemical reactions with other substances. However, this definition only works well for substances that are composed of molecules, which is not true of many substances. Molecules are typically a set of atoms bound together by covalent bonds, such that the structure is electrically neutral and all valence electrons are paired with other electrons either in bonds or in lone pairs.

Thus, molecules exist as electrically neutral units, unlike ions. When this rule is broken, giving the "molecule" a charge, the result is sometimes named a molecular ion or a polyatomic ion. However, the discrete and separate nature of the molecular concept usually requires that molecular ions be present only in well-separated form, such as a directed beam in a vacuum in a mass spectrometer. Charged polyatomic collections residing in solids (for example, common sulfate or nitrate ions) are generally not considered "molecules" in chemistry.

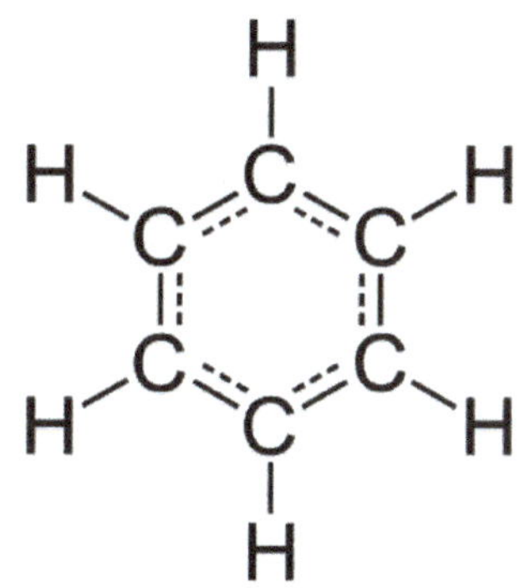

A 2-D skeletal model of a benzene molecule (C_6H_6)

The "inert" or noble gas elements (helium, neon, argon, krypton, xenon and radon) are composed of lone atoms as their smallest discrete unit, but the other isolated chemical elements consist of either molecules or networks of atoms bonded to each other in some way. Identifiable molecules compose familiar substances such as water, air, and many organic compounds like alcohol, sugar, gasoline, and the various pharmaceuticals.

However, not all substances or chemical compounds consist of discrete molecules, and indeed most of the solid substances that make up the solid crust, mantle, and core of the Earth are chemical compounds without molecules. These other types of substances, such as ionic compounds and network solids, are organized in such a way as to lack the existence of identifiable molecules *per se*. Instead, these substances are discussed in terms of formula units or unit cells as the smallest repeating structure within the substance. Examples of such substances are mineral salts (such as table salt), solids like carbon and diamond, metals, and familiar silica and silicate minerals such as quartz and granite.

One of the main characteristics of a molecule is its geometry often called its structure. While the structure of diatomic, triatomic or tetra atomic molecules may be trivial, (linear, angular pyramidal etc.) the structure of polyatomic molecules, that are constituted of more than six atoms (of several elements) can be crucial for its chemical nature.

Substance and Mixture

A chemical substance is a kind of matter with a definite composition and set of properties. A collection of substances is called a mixture. Examples of mixtures are air and alloys.

Mole and Amount of Substance

The mole is a unit of measurement that denotes an amount of substance (also called chemical amount). The mole is defined as the number of atoms found in exactly 0.012 kilogram (or 12 grams) of carbon-12, where the carbon-12 atoms are unbound, at rest and in their ground state. The number of entities per mole is known as the Avogadro constant, and is determined empirically to be approximately 6.022×10^{23} mol^{-1}. Molar concentration is the amount of a particular substance per volume of solution, and is commonly reported in $moldm^{-3}$.

Phase

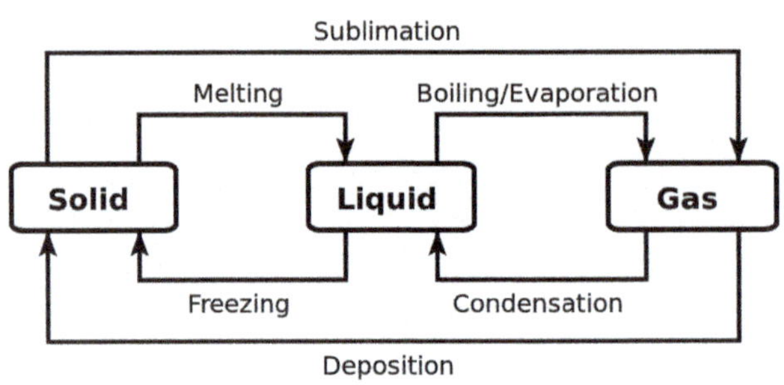

Example of phase changes

In addition to the specific chemical properties that distinguish different chemical classifications, chemicals can exist in several phases. For the most part, the chemical classifications are independent of these bulk phase classifications; however, some more exotic phases are incompatible with certain chemical properties. A *phase* is a set of states of a chemical system that have similar bulk structural properties, over a range of conditions, such as pressure or temperature.

Physical properties, such as density and refractive index tend to fall within values characteristic of the phase. The phase of matter is defined by the *phase transition*, which is when energy put into or taken out of the system goes into rearranging the structure of the system, instead of changing the bulk conditions.

Sometimes the distinction between phases can be continuous instead of having a discrete boundary, in this case the matter is considered to be in a supercritical state. When three states meet based on the conditions, it is known as a triple point and since this is invariant, it is a convenient way to define a set of conditions.

The most familiar examples of phases are solids, liquids, and gases. Many substances exhibit multiple solid phases. For example, there are three phases of solid iron (alpha, gamma, and delta) that vary based on temperature and pressure. A principal difference between solid phases is the crystal structure, or arrangement, of the atoms. Another phase commonly encountered in the study of chemistry is the *aqueous* phase, which is the state of substances dissolved in aqueous solution (that is, in water).

Less familiar phases include plasmas, Bose–Einstein condensates and fermionic condensates and the paramagnetic and ferromagnetic phases of magnetic materials. While most familiar phases deal with three-dimensional systems, it is also possible to define analogs in two-dimensional systems, which has received attention for its relevance to systems in biology.

Bonding

Atoms sticking together in molecules or crystals are said to be bonded with one another. A chemical bond may be visualized as the multipole balance between the positive charges in the nuclei and the negative charges oscillating about them. More than simple attraction and repulsion, the energies and distributions characterize the availability of an electron to bond to another atom.

A chemical bond can be a covalent bond, an ionic bond, a hydrogen bond or just because of Van der Waals force. Each of these kinds of bonds is ascribed to some potential. These potentials create the interactions which hold atoms together in molecules or crystals. In many simple compounds, valence bond theory, the Valence Shell Electron Pair Repulsion model (VSEPR), and the concept of oxidation number can be used to explain molecular structure and composition.

An ionic bond is formed when a metal loses one or more of its electrons, becoming a positively charged cation, and the electrons are then gained by the non-metal atom, becoming a negatively charged anion. The two oppositely charged ions attract one another, and the ionic bond is the electrostatic force of attraction between them. For example, sodium (Na), a metal, loses one electron to become an Na^+ cation while chlorine (Cl), a non-metal, gains this electron to become Cl^-. The ions are held together due to electrostatic attraction, and that compound sodium chloride (NaCl), or common table salt, is formed.

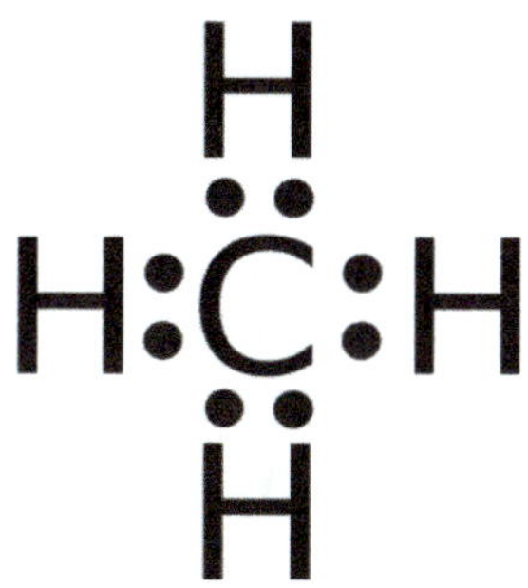

In the methane molecule (CH_4), the carbon atom shares a pair of valence electrons with each of the four hydrogen atoms. Thus, the octet rule is satisfied for C-atom (it has eight electrons in its valence shell) and the duet rule is satisfied for the H-atoms (they have two electrons in their valence shells).

In a covalent bond, one or more pairs of valence electrons are shared by two atoms: the resulting electrically neutral group of bonded atoms is termed a molecule. Atoms will share valence electrons in such a way as to create a noble gas electron configuration (eight electrons in their outermost shell) for each atom. Atoms that tend to combine in such a way that they each have eight electrons in their valence shell are said to follow the octet rule. However, some elements like hydrogen and lithium need only two electrons in their outermost shell to attain this stable configuration; these atoms are said to follow the *duet rule*, and in this way they are reaching the electron configuration of the noble gas helium, which has two electrons in its outer shell.

Similarly, theories from classical physics can be used to predict many ionic structures. With more complicated compounds, such as metal complexes, valence bond theory is less applicable and alternative approaches, such as the molecular orbital theory, are generally used. See diagram on electronic orbitals.

Energy

In the context of chemistry, energy is an attribute of a substance as a consequence of its atomic, molecular or aggregate structure. Since a chemical transformation is accompanied by a change in one or more of these kinds of structures, it is invariably accompanied by an increase or decrease of energy of the substances involved. Some energy is transferred between the surroundings and the reactants of the reaction in the form of heat or light; thus the products of a reaction may have more or less energy than the reactants.

A reaction is said to be exergonic if the final state is lower on the energy scale than the initial state; in the case of endergonic reactions the situation is the reverse. A reaction is said to be exothermic if the reaction releases heat to the surroundings; in the case of endothermic reactions, the reaction absorbs heat from the surroundings.

Chemical reactions are invariably not possible unless the reactants surmount an energy barrier known as the activation energy. The *speed* of a chemical reaction (at given temperature T) is related to the activation energy E, by the Boltzmann's population factor $e^{-E/kT}$ - that is the probability of a molecule to have energy greater than or equal to E at the given temperature T. This exponential dependence of a reaction rate on temperature is known as the Arrhenius equation. The activation energy necessary for a chemical reaction to occur can be in the form of heat, light, electricity or mechanical force in the form of ultrasound.

A related concept free energy, which also incorporates entropy considerations, is a very useful means for predicting the feasibility of a reaction and determining the state of equilibrium of a chemical reaction, in chemical thermodynamics. A reaction is feasible only if the total change in the Gibbs free energy is negative, $\Delta G \leq 0$; if it is equal to zero the chemical reaction is said to be at equilibrium.

There exist only limited possible states of energy for electrons, atoms and molecules. These are determined by the rules of quantum mechanics, which require quantization of energy of a bound system. The atoms/molecules in a higher energy state are said to be excited. The molecules/atoms of substance in an excited energy state are often much more reactive; that is, more amenable to chemical reactions.

The phase of a substance is invariably determined by its energy and the energy of its surroundings. When the intermolecular forces of a substance are such that the energy of the surroundings is not sufficient to overcome them, it occurs in a more ordered phase like liquid or solid as is the case with water (H_2O); a liquid at room temperature because its molecules are bound by hydrogen bonds. Whereas hydrogen sulfide (H_2S) is a gas at room temperature and standard pressure, as its molecules are bound by weaker dipole-dipole interactions.

The transfer of energy from one chemical substance to another depends on the *size* of energy quanta emitted from one substance. However, heat energy is often transferred more easily from almost any substance to another because the phonons responsible for vibrational and rotational energy levels in a substance have much less energy than photons invoked for the electronic energy transfer. Thus, because vibrational and rotational energy levels are more closely spaced than electronic energy levels, heat is more easily transferred between substances relative to light or other forms of electronic energy. For example, ultraviolet electromagnetic radiation is not transferred with as much efficacy from one substance to another as thermal or electrical energy.

The existence of characteristic energy levels for different chemical substances is useful for their identification by the analysis of spectral lines. Different kinds of spectra are often used in chemical spectroscopy, e.g. IR, microwave, NMR, ESR, etc. Spectroscopy is also used to identify the composition of remote objects - like stars and distant galaxies - by analyzing their radiation spectra.

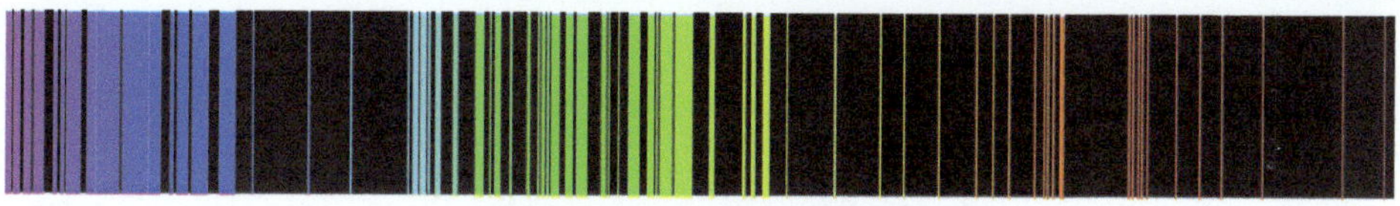

Emission spectrum of iron

The term chemical energy is often used to indicate the potential of a chemical substance to undergo a transformation through a chemical reaction or to transform other chemical substances.

Reaction

During chemical reactions, bonds between atoms break and form, resulting in different substances with different properties. In a blast furnace, iron oxide, a compound, reacts with carbon monoxide to form iron, one of the chemical elements, and carbon dioxide.

When a chemical substance is transformed as a result of its interaction with another substance or with energy, a chemical reaction is said to have occurred. A *chemical reaction* is therefore a concept related to the "reaction" of a substance when it comes in close contact with another, whether as a mixture or a solution; exposure to some form of energy, or both. It results in some energy exchange between the constituents of the reaction as well as with the system environment, which may be designed vessels—often laboratory glassware.

Chemical reactions can result in the formation or dissociation of molecules, that is, molecules breaking apart to form two or more smaller molecules, or rearrangement of atoms within or across molecules. Chemical reactions usually involve the making or breaking of chemical bonds. Oxidation, reduction, dissociation, acid-base neutralization and molecular rearrangement are some of the commonly used kinds of chemical reactions.

A chemical reaction can be symbolically depicted through a chemical equation. While in a non-nuclear chemical reaction the number and kind of atoms on both sides of the equation are equal, for a nuclear reaction this holds true only for the nuclear particles viz. protons and neutrons.

The sequence of steps in which the reorganization of chemical bonds may be taking place in the course of a chemical reaction is called its mechanism. A chemical reaction can be envisioned to take place in a number of steps, each of which may have a different speed. Many reaction intermediates with variable stability can thus be envisaged during the course of a reaction. Reaction mechanisms are proposed to explain the kinetics and the relative product mix of a reaction. Many physical chemists specialize in exploring and proposing the mechanisms of various chemical reactions. Several empirical rules, like the Woodward–Hoffmann rules often come in handy while proposing a mechanism for a chemical reaction.

According to the IUPAC gold book, a chemical reaction is "a process that results in the interconversion of chemical species." Accordingly, a chemical reaction may be an elementary reaction or a stepwise reaction. An additional caveat is made, in that this definition includes cases where the interconversion of conformers is experimentally observable. Such detectable chemical reactions normally involve sets of molecular entities as indicated by this definition, but it is often conceptually convenient to use the term also for changes involving single molecular entities (i.e. 'microscopic chemical events').

Ions and Salts

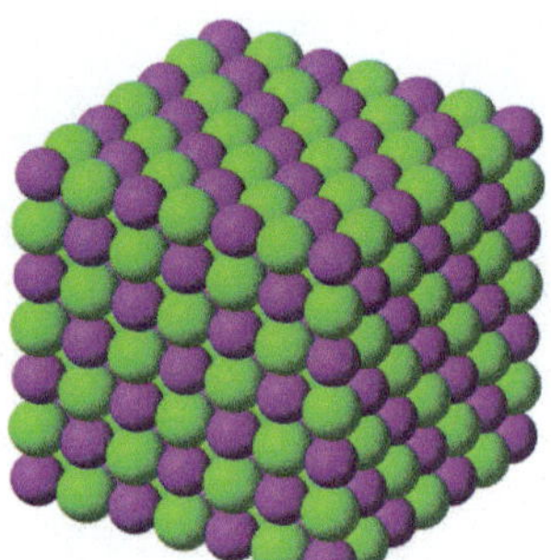

The crystal lattice structure of potassium chloride (KCl), a salt which is formed due to the attraction of K^+ cations and Cl^- anions. Note how the overall charge of the ionic compound is zero.

An *ion* is a charged species, an atom or a molecule, that has lost or gained one or more electrons. When an atom loses an electron and thus has more protons than electrons, the atom is a positively charged ion or cation. When an atom gains an electron and thus has more electrons than protons, the atom is a negatively charged ion or anion. Cations and anions can form a crystalline lattice of neutral salts, such as the Na^+ and Cl^- ions forming sodium chloride, or NaCl. Examples of polyatomic ions that do not split up during acid-base reactions are hydroxide (OH^-) and phosphate (PO_4^{3-}).

Plasma is composed of gaseous matter that has been completely ionized, usually through high temperature.

Acidity and Basicity

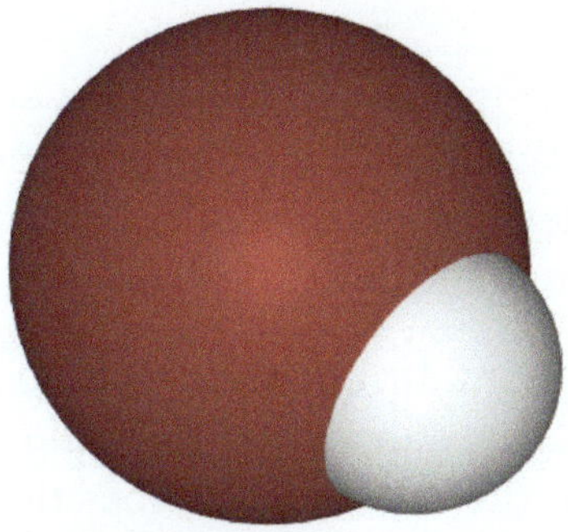

When hydrogen bromide (HBr), pictured, is dissolved in water, it forms the strong acid hydrobromic acid

A substance can often be classified as an acid or a base. There are several different theories which explain acid-base behavior. The simplest is Arrhenius theory, which states than an acid is a substance that produces hydronium ions when it is dissolved in water, and a base is one that produces hydroxide ions when dissolved in water. According to Brønsted–Lowry acid–base theory, acids are substances that donate a positive hydrogen ion to another substance in a chemical reaction; by extension, a base is the substance which receives that hydrogen ion.

A third common theory is Lewis acid-base theory, which is based on the formation of new chemical bonds. Lewis theory explains that an acid is a substance which is capable of accepting a pair of electrons from another substance during the process of bond formation, while a base is a substance which can provide a pair of electrons to form a new bond. According to this theory, the crucial things being exchanged are charges. There are several other ways in which a substance may be classified as an acid or a base, as is evident in the history of this concept.

Acid strength is commonly measured by two methods. One measurement, based on the Arrhenius definition of acidity, is pH, which is a measurement of the hydronium ion concentration in a solution, as expressed on a negative logarithmic scale. Thus, solutions that have a low pH have a high hydronium ion concentration, and can be said to be more acidic. The other measurement, based on the Brønsted–Lowry definition, is the acid dissociation constant (K_a), which measures the relative ability of a substance to act as an acid under the Brønsted–Lowry definition of an acid. That is, substances with a higher K_a are more likely to donate hydrogen ions in chemical reactions than those with lower K_a values.

Redox

Redox (*red*uction-*ox*idation) reactions include all chemical reactions in which atoms have their oxidation state changed by either gaining electrons (reduction) or losing electrons (oxidation). Substances that have the ability to oxidize other substances are said to be oxidative and are known as oxidizing agents, oxidants or oxidizers. An oxidant removes electrons from another substance. Similarly, substances that have the ability to reduce other substances are said to be reductive and are known as reducing agents, reductants, or reducers.

A reductant transfers electrons to another substance, and is thus oxidized itself. And because it "donates" electrons it is also called an electron donor. Oxidation and reduction properly refer to a change in oxidation number—the actual transfer of electrons may never occur. Thus, oxidation is better defined as an increase in oxidation number, and reduction as a decrease in oxidation number.

Equilibrium

Although the concept of equilibrium is widely used across sciences, in the context of chemistry, it arises whenever a number of different states of the chemical composition are possible, as for example, in a mixture of several chemical compounds that can react with one another, or when a substance can be present in more than one kind of phase.

A system of chemical substances at equilibrium, even though having an unchanging composition, is most often not static; molecules of the substances continue to react with one another thus giving rise to a dynamic equilibrium. Thus the concept describes the state in which the parameters such as chemical composition remain unchanged over time.

Chemical Laws

Chemical reactions are governed by certain laws, which have become fundamental concepts in chemistry. Some of them are:

- Avogadro's law
- Beer–Lambert law
- Boyle's law (1662, relating pressure and volume)
- Charles's law (1787, relating volume and temperature)
- Fick's laws of diffusion
- Gay-Lussac's law (1809, relating pressure and temperature)
- Le Chatelier's principle
- Henry's law
- Hess's law
- Law of conservation of energy leads to the important concepts of equilibrium, thermodynamics, and kinetics.
- Law of conservation of mass continues to be conserved in isolated systems, even in modern physics. However, special relativity shows that due to mass–energy equivalence, whenever non-material "energy" (heat, light, kinetic energy) is removed from a non-isolated system, some mass will be lost with it. High energy losses result in loss of weighable amounts of mass, an important topic in nuclear chemistry.
- Law of definite composition, although in many systems (notably biomacromolecules and minerals) the ratios tend to require large numbers, and are frequently represented as a fraction.

- Law of multiple proportions
- Raoult's law

Practice

Subdisciplines

Chemistry is typically divided into several major sub-disciplines. There are also several main cross-disciplinary and more specialized fields of chemistry.

- Analytical chemistry is the analysis of material samples to gain an understanding of their chemical composition and structure. Analytical chemistry incorporates standardized experimental methods in chemistry. These methods may be used in all subdisciplines of chemistry, excluding purely theoretical chemistry.
- Biochemistry is the study of the chemicals, chemical reactions and chemical interactions that take place in living organisms. Biochemistry and organic chemistry are closely related, as in medicinal chemistry or neurochemistry. Biochemistry is also associated with molecular biology and genetics.
- Inorganic chemistry is the study of the properties and reactions of inorganic compounds. The distinction between organic and inorganic disciplines is not absolute and there is much overlap, most importantly in the sub-discipline of organometallic chemistry.
- Materials chemistry is the preparation, characterization, and understanding of substances with a useful function. The field is a new breadth of study in graduate programs, and it integrates elements from all classical areas of chemistry with a focus on fundamental issues that are unique to materials. Primary systems of study include the chemistry of condensed phases (solids, liquids, polymers) and interfaces between different phases.
- Neurochemistry is the study of neurochemicals; including transmitters, peptides, proteins, lipids, sugars, and nucleic acids; their interactions, and the roles they play in forming, maintaining, and modifying the nervous system.
- Nuclear chemistry is the study of how subatomic particles come together and make nuclei. Modern Transmutation is a large component of nuclear chemistry, and the table of nuclides is an important result and tool for this field.
- Organic chemistry is the study of the structure, properties, composition, mechanisms, and reactions of organic compounds. An organic compound is defined as any compound based on a carbon skeleton.
- Physical chemistry is the study of the physical and fundamental basis of chemical systems and processes. In particular, the energetics and dynamics of such systems and processes are of interest to physical chemists. Important areas of study include chemical thermodynamics, chemical kinetics, electrochemistry, statistical mechanics, spectroscopy, and more recently, astrochemistry. Physical chemistry has large overlap with molecular physics. Physical chemistry involves the use of infinitesimal calculus in deriving equations. It is

usually associated with quantum chemistry and theoretical chemistry. Physical chemistry is a distinct discipline from chemical physics, but again, there is very strong overlap.

- Theoretical chemistry is the study of chemistry via fundamental theoretical reasoning (usually within mathematics or physics). In particular the application of quantum mechanics to chemistry is called quantum chemistry. Since the end of the Second World War, the development of computers has allowed a systematic development of computational chemistry, which is the art of developing and applying computer programs for solving chemical problems. Theoretical chemistry has large overlap with (theoretical and experimental) condensed matter physics and molecular physics.

Other disciplines within chemistry are traditionally grouped by the type of matter being studied or the kind of study. These include inorganic chemistry, the study of inorganic matter; organic chemistry, the study of organic (carbon-based) matter; biochemistry, the study of substances found in biological organisms; physical chemistry, the study of chemical processes using physical concepts such as thermodynamics and quantum mechanics; and analytical chemistry, the analysis of material samples to gain an understanding of their chemical composition and structure. Many more specialized disciplines have emerged in recent years, e.g. neurochemistry the chemical study of the nervous system.

Other fields include agrochemistry, astrochemistry (and cosmochemistry), atmospheric chemistry, chemical engineering, chemical biology, chemo-informatics, electrochemistry, environmental chemistry, femtochemistry, flavor chemistry, flow chemistry, geochemistry, green chemistry, histochemistry, history of chemistry, hydrogenation chemistry, immunochemistry, marine chemistry, materials science, mathematical chemistry, mechanochemistry, medicinal chemistry, molecular biology, molecular mechanics, nanotechnology, natural product chemistry, oenology, organometallic chemistry, petrochemistry, pharmacology, photochemistry, physical organic chemistry, phytochemistry, polymer chemistry, radiochemistry, solid-state chemistry, sonochemistry, supramolecular chemistry, surface chemistry, synthetic chemistry, thermochemistry, and many others.

Chemical Industry

The chemical industry represents an important economic activity worldwide. The global top 50 chemical producers in 2013 had sales of US$980.5 billion with a profit margin of 10.3%.

Professional Societies

- American Chemical Society
- American Society for Neurochemistry
- Chemical Institute of Canada
- Chemical Society of Peru
- International Union of Pure and Applied Chemistry
- Royal Australian Chemical Institute

- Royal Netherlands Chemical Society
- Royal Society of Chemistry
- Society of Chemical Industry
- World Association of Theoretical and Computational Chemists
- List of chemistry societies

Organic Chemistry

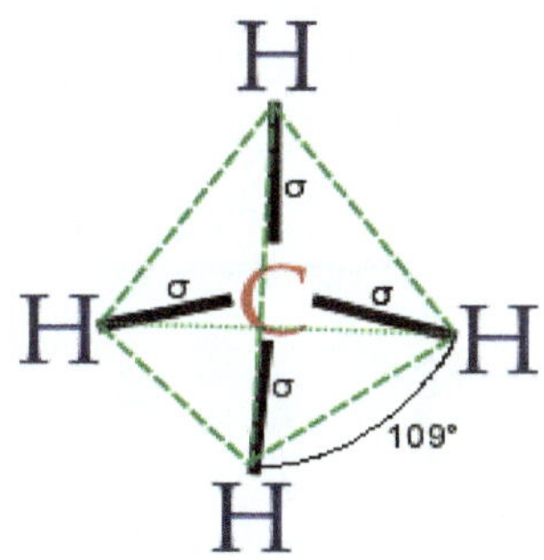

Methane, CH_4, in line-angle representation showing four carbon-hydrogen single (σ) bonds in black, and the 3D shape of such tetrahedral molecules, with ~109° interior bond angles, in green. Methane is the simplest organic chemical and simplest hydrocarbon, and molecules can be built up conceptually from it by exchanging up to all 4 hydrogens with carbon or other atoms.

Organic chemistry is a chemistry subdiscipline involving the scientific study of the structure, properties, and reactions of organic compounds and organic materials, i.e., matter in its various forms that contain carbon atoms. Study of structure includes many physical and chemical methods to determine the chemical composition and the chemical constitution of organic compounds and materials. Study of properties includes both physical properties and chemical properties, and uses similar methods as well as methods to evaluate chemical reactivity, with the aim to understand the behavior of the organic matter in its pure form (when possible), but also in solutions, mixtures, and fabricated forms. The study of organic reactions includes probing their scope through use in preparation of target compounds (e.g., natural products, drugs, polymers, etc.) by chemical synthesis, as well as the focused study of the reactivities of individual organic molecules, both in the laboratory and via theoretical (in silico) study.

The range of chemicals studied in organic chemistry include hydrocarbons (compounds containing only carbon and hydrogen), as well as myriad compositions based always on carbon, but also containing other elements, especially oxygen, nitrogen, sulfur, phosphorus (these, included in many organic chemicals in biology) and the radiostable elements of the halogens.

In the modern era, the range extends further into the periodic table, with main group elements, including:

- Group 1 and 2 organometallic compounds, i.e., involving alkali (e.g., lithium, sodium, and potassium) or alkaline earth metals (e.g., magnesium)
- Metalloids (e.g., boron and silicon) or other metals (e.g., aluminium and tin)

In addition, much modern research focuses on organic chemistry involving further organometallics, including the lanthanides, but especially the transition metals; (e.g., zinc, copper, palladium, nickel, cobalt, titanium and chromium)

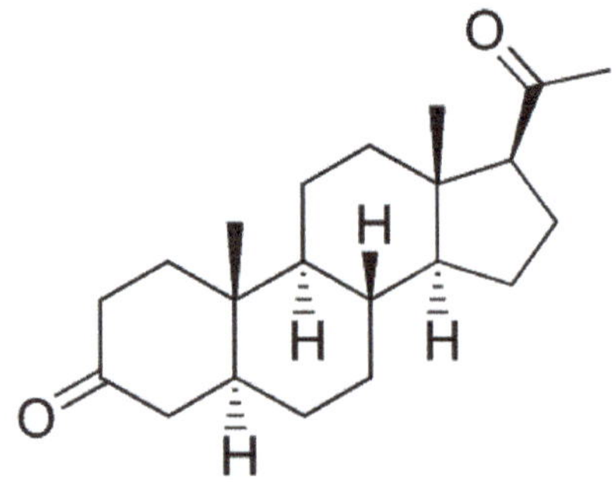

Line-angle representation

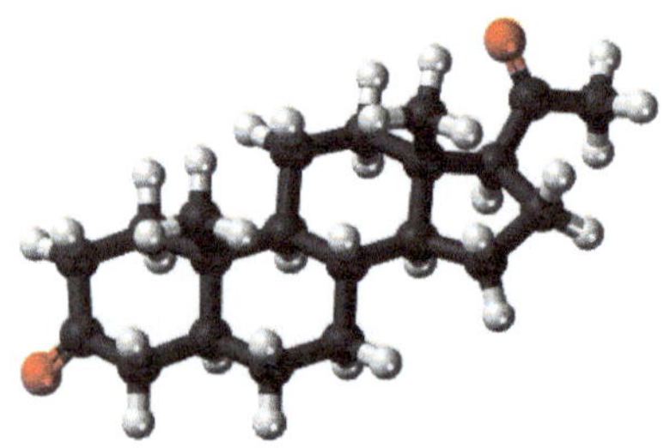

Ball-and-stick representation

Space-filling representation

Three representations of an organic compound, 5α-Dihydroprogesterone (5α-DHP), a steroid hormone. For molecules showing color, the carbon atoms are in black, hydrogens in gray, and oxygens in red. In the line angle representation, carbon atoms are implied at every terminus of a line and vertex of multiple lines, and hydrogen atoms are implied to fill the remaining needed valences (up to 4).

Finally, organic compounds form the basis of all earthly life and constitute a significant part of human endeavors in chemistry. The bonding patterns open to carbon, with its valence of four—formal single, double, and triple bonds, as well as various structures with delocalized electrons—make the array of organic compounds structurally diverse, and their range of applications enormous. They either form the basis of, or are important constituents of, many commercial products including pharmaceuticals; petrochemicals and products made from them (including lubricants, solvents, etc.); plastics; fuels and explosives; etc. As indicated, the study of organic chemistry overlaps with organometallic chemistry and biochemistry, but also with medicinal chemistry, polymer chemistry, as well as many aspects of materials science.

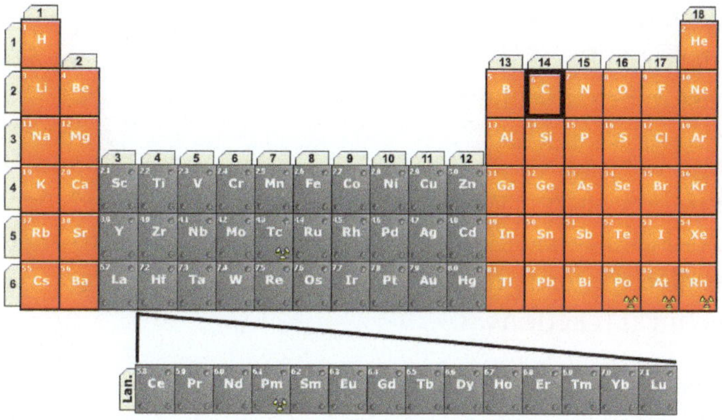

Periodic table of elements of interest in organic chemistry. The table illustrates all elements of current interest in modern organic and organometallic chemistry, indicating main group elements in orange, and transition metals and lanthanides (Lan) in grey.

History

Friedrich Wöhler

Before the nineteenth century, chemists generally believed that compounds obtained from living organisms were endowed with a vital force that distinguished them from inorganic compounds. According to the concept of vitalism (vital force theory), organic matter was endowed with a "vital force". During the first half of the nineteenth century, some of the first systematic studies of organic compounds were reported. Around 1816 Michel Chevreul started a study of soaps made from various fats and alkalis. He separated the different acids that, in combination with the alkali, produced the soap. Since these were all individual compounds, he demonstrated that it was possible to make a chemical change in various fats (which traditionally come from organic sources), producing new compounds, without "vital force". In 1828 Friedrich Wöhler produced the organic chemical urea (carbamide), a constituent of urine, from the inorganic ammonium cyanate NH_4CNO, in what is now called the Wöhler synthesis. Although Wöhler was always cautious about claiming that he had disproved the theory of vital force, this event has often been thought of as a turning point.

In 1856 William Henry Perkin, while trying to manufacture quinine, accidentally manufactured the organic dye now known as Perkin's mauve. Through its great financial success, this discovery greatly increased interest in organic chemistry.

The crucial breakthrough for organic chemistry was the concept of chemical structure, developed independently and simultaneously by Friedrich August Kekulé and Archibald Scott Couper in 1858. Both men suggested that tetravalent carbon atoms could link to each other to form a carbon lattice, and that the detailed patterns of atomic bonding could be discerned by skillful interpretations of appropriate chemical reactions.

The pharmaceutical industry began in the last decade of the 19th century when the manufacturing of acetylsalicylic acid—more commonly referred to as aspirin—in Germany was started by Bayer. The first time a drug was systematically improved was with arsphenamine (Salvarsan). Though numerous derivatives of the dangerous toxic atoxyl were examined by Paul Ehrlich and his group, the compound with best effectiveness and toxicity characteristics was selected for production.

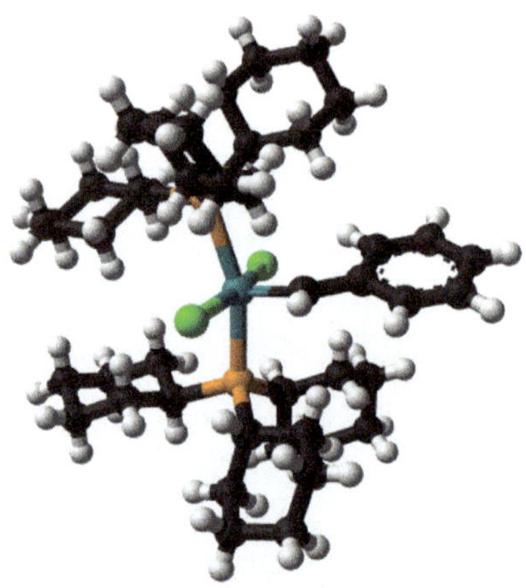

An example of an organometallic molecule, a catalyst called Grubbs' catalyst, as a ball-and-stick model based on an X-ray crystal structure. The formula of the catalyst is often given as $RuCl_2(P\text{-}Cy_3)_2(=CHPh)$, where the ruthenium metal atom, Ru, is at very center in turquoise, carbons are in black, hydrogens in gray-white, chlorine in green, and phosphorus in orange. The metal ligand at the bottom is a tricyclohexyl phosphine, abbreviated PCy, and another of these appears at the top of the image (where its rings are obscuring one another). The group projecting out to the right has a metal-carbon double bond, as is known as an alkylidene. Robert Grubbs shared the 2005 Nobel prize in chemistry with Richard R. Schrock and Yves Chauvin for their work on the reactions such catalysts mediate, called olefin metathesis.

Early examples of organic reactions and applications were often found because of a combination of luck and preparation for unexpected observations. The latter half of the 19th century however witnessed systematic studies of organic compounds. The development of synthetic indigo is illustrative. The production of indigo from plant sources dropped from 19,000 tons in 1897 to 1,000 tons by 1914 thanks to the synthetic methods developed by Adolf von Baeyer. In 2002, 17,000 tons of synthetic indigo were produced from petrochemicals.

In the early part of the 20th Century, polymers and enzymes were shown to be large organic molecules, and petroleum was shown to be of biological origin.

The multiple-step synthesis of complex organic compounds is called total synthesis. Total synthesis of complex natural compounds increased in complexity to glucose and terpineol. For example, cholesterol-related compounds have opened ways to synthesize complex human hormones and their modified derivatives. Since the start of the 20th century, complexity of total syntheses has been increased to include molecules of high complexity such as lysergic acid and vitamin B12.

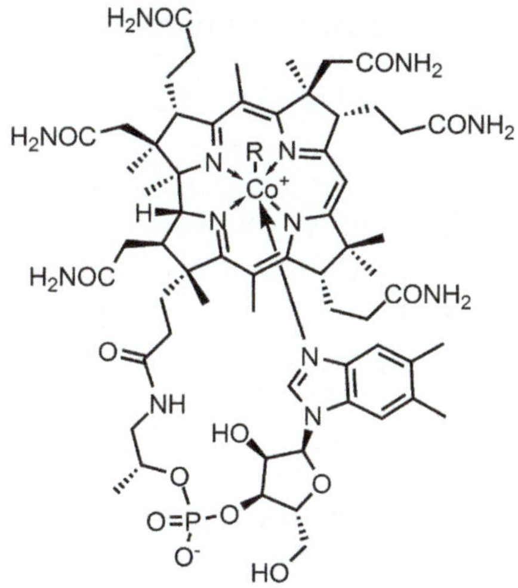

R = 5'-deoxyadenosyl, Me, OH, CN

The total synthesis of vitamin B12 marked a major achievement in organic chemistry.

The development of organic chemistry benefited from the discovery of petroleum and the development of the petrochemical industry. The conversion of individual compounds obtained from petroleum into different compound types by various chemical processes led to the birth of the petrochemical industry, which successfully manufactured artificial rubbers, various organic adhesives, property-modifying petroleum additives, and plastics.

The majority of chemical compounds occurring in biological organisms are in fact carbon compounds, so the association between organic chemistry and biochemistry is so close that biochemistry might be regarded as in essence a branch of organic chemistry. Although the history of biochemistry might be taken to span some four centuries, fundamental understanding of the field only began to develop in the late 19th century and the actual term *biochemistry* was coined around the start of 20th century. Research in the field increased throughout the twentieth century, without any indication of slackening in the rate of increase, as may be verified by inspection of abstraction and indexing services such as BIOSIS Previews and Biological Abstracts, which began in the 1920s as a single annual volume, but has grown so drastically that by the end of the 20th century it was only available to the everyday user as an online electronic database.

Characterization

Since organic compounds often exist as mixtures, a variety of techniques have also been developed to assess purity, especially important being chromatography techniques such as HPLC and gas chromatography. Traditional methods of separation include distillation, crystallization, and solvent extraction.

Organic compounds were traditionally characterized by a variety of chemical tests, called "wet methods", but such tests have been largely displaced by spectroscopic or other computer-intensive methods of analysis. Listed in approximate order of utility, the chief analytical methods are:

- Nuclear magnetic resonance (NMR) spectroscopy is the most commonly used technique, often permitting complete assignment of atom connectivity and even stereochemistry using correlation spectroscopy. The principal constituent atoms of organic chemistry – hydrogen and carbon – exist naturally with NMR-responsive isotopes, respectively ^{1}H and ^{13}C.
- Elemental analysis: A destructive method used to determine the elemental composition of a molecule.
- Mass spectrometry indicates the molecular weight of a compound and, from the fragmentation patterns, its structure. High resolution mass spectrometry can usually identify the exact formula of a compound and is used in lieu of elemental analysis. In former times, mass spectrometry was restricted to neutral molecules exhibiting some volatility, but advanced ionization techniques allow one to obtain the "mass spec" of virtually any organic compound.
- Crystallography is an unambiguous method for determining molecular geometry, the proviso being that single crystals of the material must be available and the crystal must be representative of the sample. Highly automated software allows a structure to be determined within hours of obtaining a suitable crystal.

Traditional spectroscopic methods such as infrared spectroscopy, optical rotation, UV/VIS spectroscopy provide relatively nonspecific structural information but remain in use for specific classes of compounds. Traditionally refractive index and density were also important for substance identification.

Properties

Physical properties of organic compounds typically of interest include both quantitative and qualitative features. Quantitative information includes melting point, boiling point, and index of refraction. Qualitative properties include odor, consistency, solubility, and color.

Melting and Boiling Properties

Organic compounds typically melt and many boil. In contrast, while inorganic materials generally can be melted, many do not boil, tending instead to degrade. In earlier times, the melting point (m.p.) and boiling point (b.p.) provided crucial information on the purity and identity of organic compounds. The melting and boiling points correlate with the polarity of the molecules and their molecular weight. Some organic compounds, especially symmetrical ones, sublime, that is they evaporate without melting. A well-known example of a sublimable organic compound is para-dichlorobenzene, the odiferous constituent of modern mothballs. Organic compounds are usually not very stable at temperatures above 300 °C, although some exceptions exist.

Solubility

Neutral organic compounds tend to be hydrophobic; that is, they are less soluble in water than in organic solvents. Exceptions include organic compounds that contain ionizable (which can be converted in ions) groups as well as low molecular weight alcohols, amines, and carboxylic acids where hydrogen bonding occurs. Organic compounds tend to dissolve in organic solvents. Solvents can be either pure substances like ether or ethyl alcohol, or mixtures, such as the paraffinic solvents such as the various petroleum ethers and white spirits, or the range of pure or mixed aromatic solvents obtained from petroleum or tar fractions by physical separation or by chemical conversion. Solubility in the different solvents depends upon the solvent type and on the functional groups if present.

Solid State Properties

Various specialized properties of molecular crystals and organic polymers with conjugated systems are of interest depending on applications, e.g. thermo-mechanical and electro-mechanical such as piezoelectricity, electrical conductivity (conductive polymers and organic semiconduc-tors), and electro-optical (e.g. non-linear optics) properties. For historical reasons, such properties are mainly the subjects of the areas of polymer science and materials science.

Nomenclature

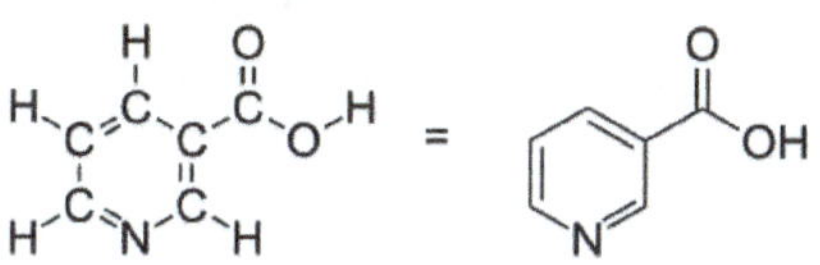

= $C_5H_4N\text{-}3\text{-}CO_2H$ = pyridine-3-carboxylic acid = niacin = vitamin B_3

Various names and depictions for one organic compound.

The names of organic compounds are either systematic, following logically from a set of rules, or nonsystematic, following various traditions. Systematic nomenclature is stipulated by specifications from IUPAC. Systematic nomenclature starts with the name for a parent structure within the molecule of interest. This parent name is then modified by prefixes, suffixes, and numbers to unambiguously convey the structure. Given that millions of organic compounds are known, rigorous use of systematic names can be cumbersome. Thus, IUPAC recommendations are more closely followed for simple compounds, but not complex molecules. To use the systematic naming, one must know the structures and names of the parent structures. Parent structures include unsubstituted hydrocarbons, heterocycles, and monofunctionalized derivatives thereof.

Nonsystematic nomenclature is simpler and unambiguous, at least to organic chemists. Nonsystematic names do not indicate the structure of the compound. They are common for complex molecules, which includes most natural products. Thus, the informally named lysergic acid diethylamide is systematically named (6a*R*,9*R*)-*N*,*N*-diethyl-7-methyl-4,6,6a,7,8,9-hexahydroindolo-[4,3-*fg*] quinoline-9-carboxamide.

With the increased use of computing, other naming methods have evolved that are intended to be interpreted by machines. Two popular formats are SMILES and InChI.

Structural Drawings

Organic molecules are described more commonly by drawings or structural formulas, combinations of drawings and chemical symbols. The line-angle formula is simple and unambiguous. In this system, the endpoints and intersections of each line represent one carbon, and hydrogen atoms can either be notated explicitly or assumed to be present as implied by tetravalent carbon. The depiction of organic compounds with drawings is greatly simplified by the fact that carbon in almost all organic compounds has four bonds, nitrogen three, oxygen two, and hydrogen one.

Classification of Organic Compounds

Functional Groups

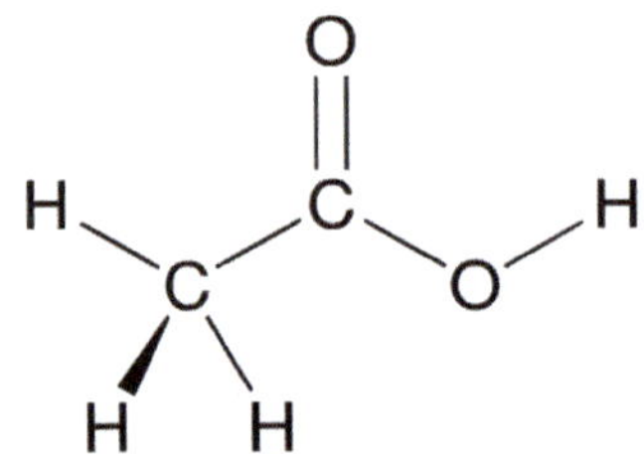

The family of carboxylic acids contains a carboxyl (-COOH) functional group. Acetic acid, shown here, is an example.

The concept of functional groups is central in organic chemistry, both as a means to classify structures and for predicting properties. A functional group is a molecular module, and the reactivity of that functional group is assumed, within limits, to be the same in a variety of molecules. Functional groups can have decisive influence on the chemical and physical properties of organic compounds. Molecules are classified on the basis of their functional groups. Alcohols, for example, all have the subunit C-O-H. All alcohols tend to be somewhat hydrophilic, usually form esters, and usually can be converted to the corresponding halides. Most functional groups feature heteroatoms (atoms other than C and H). Organic compounds are classified according to functional groups, alcohols, carboxylic acids, amines, etc.

Aliphatic Compounds

The aliphatic hydrocarbons are subdivided into three groups of homologous series according to their state of saturation:

- paraffins, which are alkanes without any double or triple bonds,
- olefins or alkenes which contain one or more double bonds, i.e. di-olefins (dienes) or poly-olefins.
- alkynes, which have one or more triple bonds.

The rest of the group is classed according to the functional groups present. Such compounds can be "straight-chain", branched-chain or cyclic. The degree of branching affects characteristics, such as the octane number or cetane number in petroleum chemistry.

Both saturated (alicyclic) compounds and unsaturated compounds exist as cyclic derivatives. The most stable rings contain five or six carbon atoms, but large rings (macrocycles) and smaller rings are common. The smallest cycloalkane family is the three-membered cyclopropane ($(CH_2)_3$). Saturated cyclic compounds contain single bonds only, whereas aromatic rings have an alternating (or conjugated) double bond. Cycloalkanes do not contain multiple bonds, whereas the cycloalkenes and the cycloalkynes do.

Aromatic Compounds

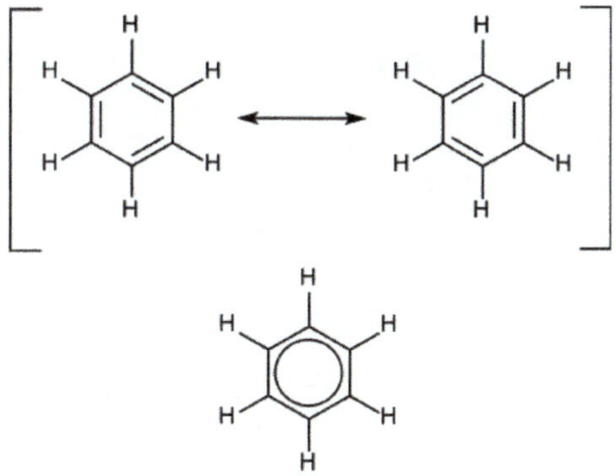

Benzene is one of the best-known aromatic compounds as it is one of the simplest and most stable aromatics.

Aromatic hydrocarbons contain conjugated double bonds. This means that every carbon atom in the ring is sp2 hybridized, allowing for added stability. The most important example is benzene, the structure of which was formulated by Kekulé who first proposed the delocalization or resonance principle for explaining its structure. For "conventional" cyclic compounds, aromaticity is conferred by the presence of $4n + 2$ delocalized pi electrons, where n is an integer. Particular instability (antiaromaticity) is conferred by the presence of $4n$ conjugated pi electrons.

Heterocyclic Compounds

The characteristics of the cyclic hydrocarbons are again altered if heteroatoms are present, which can exist as either substituents attached externally to the ring (exocyclic) or as a member of the ring itself (endocyclic). In the case of the latter, the ring is termed a heterocycle. Pyridine and furan are examples of aromatic heterocycles while piperidine and tetrahydrofuran are the corresponding alicyclic heterocycles. The heteroatom of heterocyclic molecules is generally oxygen, sulfur, or nitrogen, with the latter being particularly common in biochemical systems.

Heterocycles are commonly found in a wide range of products including aniline dyes and medicines. Additionally, they are prevalent in a wide range of biochemical compounds such as alkaloids, vitamins, steroids, and nucleic acids (e.g. DNA, RNA).

Rings can fuse with other rings on an edge to give polycyclic compounds. The purine nucleoside bases are notable polycyclic aromatic heterocycles. Rings can also fuse on a "corner" such that one atom (almost always carbon) has two bonds going to one ring and two to another. Such compounds are termed spiro and are important in a number of natural products.

Polymers

One important property of carbon is that it readily forms chains, or networks, that are linked by carbon-carbon (carbon-to-carbon) bonds. The linking process is called polymerization, while the chains, or networks, are called polymers. The source compound is called a monomer.

Two main groups of polymers exist: synthetic polymers and biopolymers. Synthetic polymers are artificially manufactured, and are commonly referred to as industrial polymers. Biopolymers occur within a respectfully natural environment, or without human intervention.

This swimming board is made of polystyrene, an example of a polymer.

Since the invention of the first synthetic polymer product, bakelite, synthetic polymer products have frequently been invented.

Common synthetic organic polymers are polyethylene (polythene), polypropylene, nylon, teflon (PTFE), polystyrene, polyesters, polymethylmethacrylate (called perspex and plexiglas), and polyvinylchloride (PVC).

Both synthetic and natural rubber are polymers.

Varieties of each synthetic polymer product may exist, for purposes of a specific use. Changing the conditions of polymerization alters the chemical composition of the product and its properties. These alterations include the chain length, or branching, or the tacticity.

With a single monomer as a start, the product is a homopolymer.

Secondary component(s) may be added to create a heteropolymer (co-polymer) and the degree of clustering of the different components can also be controlled.

Physical characteristics, such as hardness, density, mechanical or tensile strength, abrasion resistance, heat resistance, transparency, colour, etc. will depend on the final composition.

Biomolecules

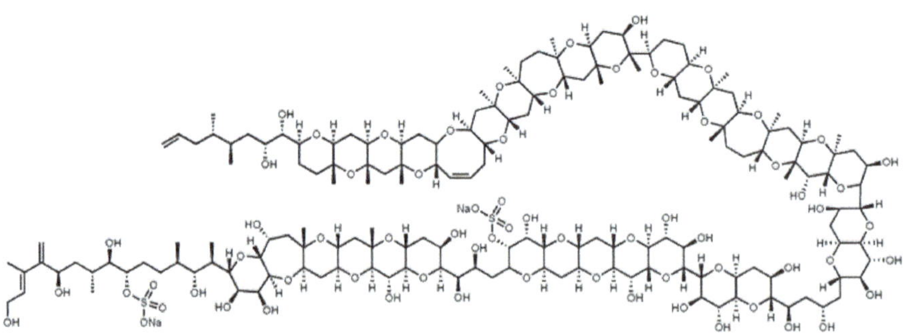

Maitotoxin, a complex organic biological toxin.

Biomolecular chemistry is a major category within organic chemistry which is frequently studied by biochemists. Many complex multi-functional group molecules are important in living organisms. Some are long-chain biopolymers, and these include peptides, DNA, RNA and the polysaccharides such as starches in animals and celluloses in plants. The other main classes are amino acids (monomer building blocks of peptides and proteins), carbohydrates (which includes the polysaccharides), the nucleic acids (which include DNA and RNA as polymers), and the lipids. In addition, animal biochemistry contains many small molecule intermediates which assist in energy production through the Krebs cycle, and produces isoprene, the most common hydrocarbon in animals. Isoprenes in animals form the important steroid structural (cholesterol) and steroid hormone compounds; and in plants form terpenes, terpenoids, some alkaloids, and a class of hydrocarbons called biopolymer polyisoprenoids present in the latex of various species of plants, which is the basis for making rubber.

Small Molecules

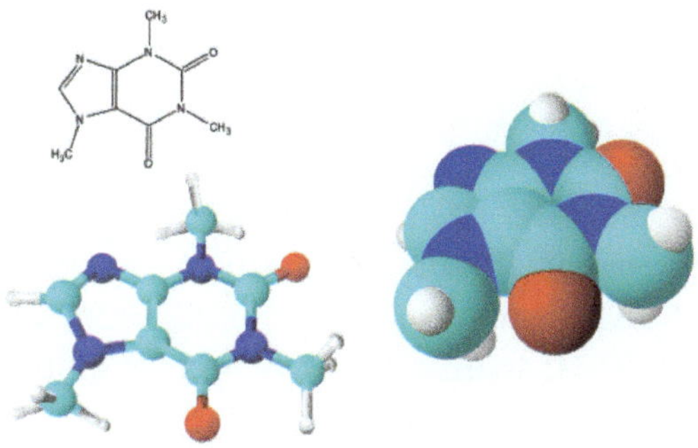

Molecular models of caffeine.

In pharmacology, an important group of organic compounds is small molecules, also referred to as 'small organic compounds'. In this context, a small molecule is a small organic compound that is biologically active, but is not a polymer. In practice, small molecules have a molar mass less than approximately 1000 g/mol.

Fullerenes

Fullerenes and carbon nanotubes, carbon compounds with spheroidal and tubular structures, have stimulated much research into the related field of materials science. The first fullerene was discovered in 1985 by Sir Harold W. Kroto (one of the authors of this article) of the United Kingdom and by Richard E. Smalley and Robert F. Curl, Jr., of the United States. Using a laser to vaporize graphite rods in an atmosphere of helium gas, these chemists and their assistants obtained cagelike molecules composed of 60 carbon atoms (C60) joined together by single and double bonds to form a hollow sphere with 12 pentagonal and 20 hexagonal faces—a design that resembles a football, or soccer ball. In 1996 the trio was awarded the Nobel Prize for their pioneering efforts. The C60 molecule was named buckminsterfullerene (or, more simply, the buckyball) after the American architect R. Buckminster Fuller, whose geodesic dome is constructed on the same structural principles.

Others

Organic compounds containing bonds of carbon to nitrogen, oxygen and the halogens are not normally grouped separately. Others are sometimes put into major groups within organic chemistry and discussed under titles such as organosulfur chemistry, organometallic chemistry, organophosphorus chemistry and organosilicon chemistry.

Organic Reactions

Organic reactions are chemical reactions involving organic compounds. Many of these reactions are associated with functional groups. The general theory of these reactions involves careful analysis of such properties as the electron affinity of key atoms, bond strengths and steric hindrance. These factors can determine the relative stability of short-lived reactive intermediates, which usually directly determine the path of the reaction.

The basic reaction types are: addition reactions, elimination reactions, substitution reactions, pericyclic reactions, rearrangement reactions and redox reactions. An example of a common reaction is a substitution reaction written as:

$$Nu^- + C\text{-}X \rightarrow C\text{-}Nu + X^-$$

where X is some functional group and Nu is a nucleophile.

The number of possible organic reactions is basically infinite. However, certain general patterns are observed that can be used to describe many common or useful reactions. Each reaction has a stepwise reaction mechanism that explains how it happens in sequence—although the detailed description of steps is not always clear from a list of reactants alone.

The stepwise course of any given reaction mechanism can be represented using arrow pushing techniques in which curved arrows are used to track the movement of electrons as starting materials transition through intermediates to final products.

Organic Synthesis

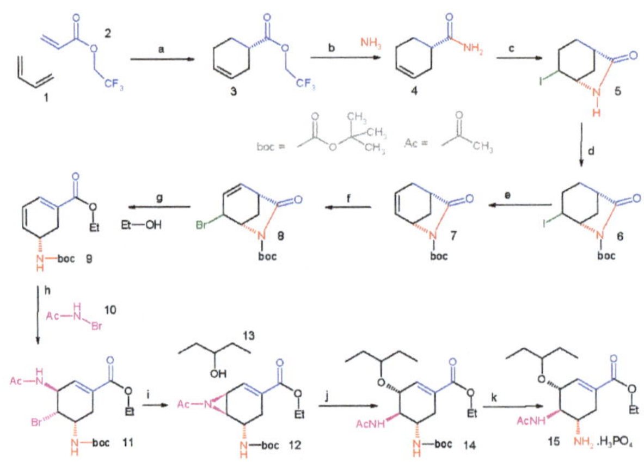

A synthesis designed by E.J. Corey for oseltamivir (Tamiflu). This synthesis has 11 distinct reactions.

Synthetic organic chemistry is an applied science as it borders engineering, the "design, analysis, and/or construction of works for practical purposes". Organic synthesis of a novel compound is a problem solving task, where a synthesis is designed for a target molecule by selecting optimal reactions from optimal starting materials. Complex compounds can have tens of reaction steps that sequentially build the desired molecule. The synthesis proceeds by utilizing the reactivity of the functional groups in the molecule. For example, a carbonyl compound can be used as a nucleophile by converting it into an enolate, or as an electrophile; the combination of the two is called the aldol reaction. Designing practically useful syntheses always requires conducting the actual synthesis in the laboratory. The scientific practice of creating novel synthetic routes for complex molecules is called total synthesis.

Strategies to design a synthesis include retrosynthesis, popularized by E.J. Corey, starts with the target molecule and splices it to pieces according to known reactions. The pieces, or the proposed precursors, receive the same treatment, until available and ideally inexpensive starting materials are reached. Then, the retrosynthesis is written in the opposite direction to give the synthesis. A "synthetic tree" can be constructed, because each compound and also each precursor has multiple syntheses.

Biochemistry

Biochemistry, sometimes called biological chemistry, is the study of chemical processes within and relating to living organisms. By controlling information flow through biochemical signaling and the flow of chemical energy through metabolism, biochemical processes give rise to the complexity of life. Over the last decades of the 20th century, biochemistry has become so successful at explaining living processes that now almost all areas of the life sciences from botany to medicine to genetics are engaged in biochemical research. Today, the main focus of pure biochemistry is on understanding how biological molecules give rise to the processes that occur within living cells, which in turn relates greatly to the study and understanding of tissues, organs, and whole organisms—that is, all of biology.

Biochemistry is closely related to molecular biology, the study of the molecular mechanisms by which genetic information encoded in DNA is able to result in the processes of life. Depending on the exact definition of the terms used, molecular biology can be thought of as a branch of biochemistry, or biochemistry as a tool with which to investigate and study molecular biology.

Much of biochemistry deals with the structures, functions and interactions of biological macromolecules, such as proteins, nucleic acids, carbohydrates and lipids, which provide the structure of cells and perform many of the functions associated with life. The chemistry of the cell also depends on the reactions of smaller molecules and ions. These can be inorganic, for example water and metal ions, or organic, for example the amino acids, which are used to synthesize proteins. The mechanisms by which cells harness energy from their environment via chemical reactions are known as metabolism. The findings of biochemistry are applied primarily in medicine, nutrition, and agriculture. In medicine, biochemists investigate the causes and cures of diseases. In nutrition, they study how to maintain health and study the effects of nutritional deficiencies. In agriculture, biochemists investigate soil and fertilizers, and try to discover ways to improve crop cultivation, crop storage and pest control.

History

At its broadest definition, biochemistry can be seen as a study of the components and composition of living things and how they come together to become life, and the history of biochemistry may therefore go back as far as the ancient Greeks. However, biochemistry as a specific scientific discipline has its beginning some time in the 19th century, or a little earlier, depending on which aspect of biochemistry one is being focused on. Some argued that the beginning of biochemistry may have been the discovery of the first enzyme, diastase (today called amylase), in 1833 by Anselme Payen, while others considered Eduard Buchner's first demonstration of a complex biochemical process alcoholic fermentation in cell-free extracts in 1897 to be the birth of biochemistry. Some might also point as its beginning to the influential 1842 work by Justus von Liebig, *Animal chemistry, or, Organic chemistry in its applications to physiology and pathology*, which presented a chemical theory of metabolism, or even earlier to the 18th century studies on fermentation and respiration by Antoine Lavoisier. Many other pioneers in the field who helped to uncover the layers of complexity of biochemistry have been proclaimed founders of modern biochemistry, for example Emil Fischer for his work on the chemistry of proteins, and F. Gowland Hopkins on enzymes and the dynamic nature of biochemistry.

Gerty Cori and Carl Cori jointly won the Nobel Prize in 1947 for their discovery of the Cori cycle at RPMI.

The term "biochemistry" itself is derived from a combination of biology and chemistry. In 1877, Felix Hoppe-Seyler used the term (*biochemie* in German) as a synonym for physiological chemistry in the foreword to the first issue of *Zeitschrift für Physiologische Chemie* where he argued for the setting up of institutes dedicated to this field of study. The German chemist Carl Neuberg however is often cited to have been coined the word in 1903, while some credited it to Franz Hofmeister.

It was once generally believed that life and its materials had some essential property or substance (often referred to as the "vital principle") distinct from any found in non-living matter, and it was thought that only living beings could produce the molecules of life. Then, in 1828, Friedrich Wöhler published a paper on the synthesis of urea, proving that organic compounds can be created artificially. Since then, biochemistry has advanced, especially since the mid-20th century, with the development of new techniques such as chromatography, X-ray diffraction, dual polarisation interferometry, NMR spectroscopy, radioisotopic labeling, electron microscopy, and molecular dynamics simulations. These techniques allowed for the discovery and detailed analysis of many molecules and metabolic pathways of the cell, such as glycolysis and the Krebs cycle (citric acid cycle).

Another significant historic event in biochemistry is the discovery of the gene and its role in the transfer of information in the cell. This part of biochemistry is often called molecular biology. In the 1950s, James D. Watson, Francis Crick, Rosalind Franklin, and Maurice Wilkins were instrumental in solving DNA structure and suggesting its relationship with genetic transfer of information. In 1958, George Beadle and Edward Tatum received the Nobel Prize for work in fungi showing that one gene produces one enzyme. In 1988, Colin Pitchfork was the first person convicted of murder with DNA evidence, which led to growth of forensic science. More recently, Andrew Z. Fire and Craig C. Mello received the 2006 Nobel Prize for discovering the role of RNA interference (RNAi), in the silencing of gene expression.

Starting Materials: the Chemical Elements of Life

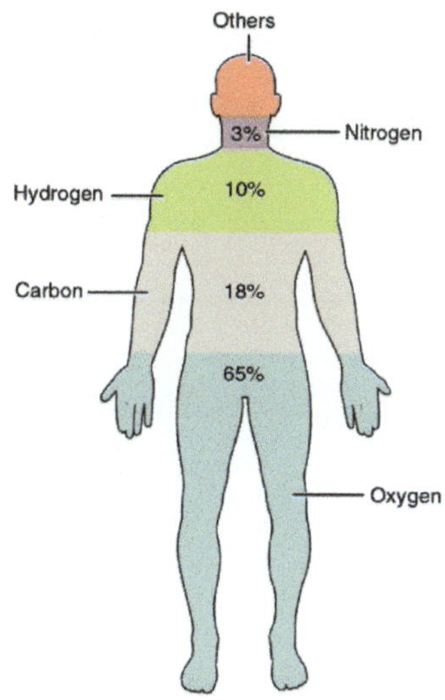

The main elements that compose the human body are shown from most abundant (by mass) to least abundant.

Around two dozen of the 92 naturally occurring chemical elements are essential to various kinds of biological life. Most rare elements on Earth are not needed by life (exceptions being selenium and iodine), while a few common ones (aluminum and titanium) are not used. Most organisms share element needs, but there are a few differences between plants and animals. For example, ocean algae use bromine, but land plants and animals seem to need none. All animals require sodium, but some plants do not. Plants need boron and silicon, but animals may not (or may need ultra-small amounts).

Just six elements—carbon, hydrogen, nitrogen, oxygen, calcium, and phosphorus—make up almost 99% of the mass of living cells, including those in the human body (composition of the human body for a complete list). In addition to the six major elements that compose most of the human body, humans require smaller amounts of possibly 18 more.

Biomolecules

The four main classes of molecules in biochemistry (often called biomolecules) are carbohydrates, lipids, proteins, and nucleic acids. Many biological molecules are polymers: in this terminology, *monomers* are relatively small micromolecules that are linked together to create large macromolecules known as *polymers*. When monomers are linked together to synthesize a biological polymer, they undergo a process called dehydration synthesis. Different macromolecules can assemble in larger complexes, often needed for biological activity.

Carbohydrates

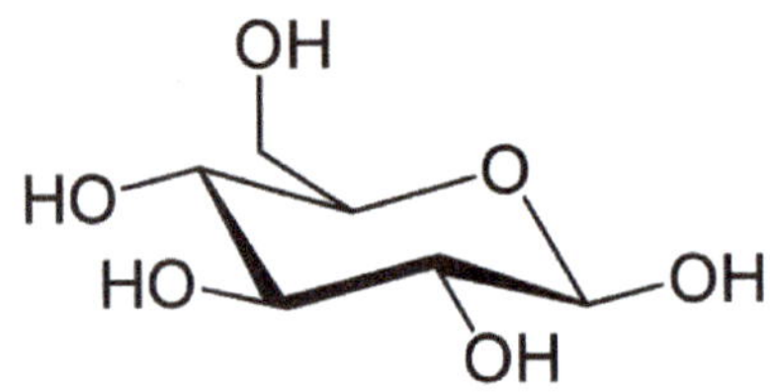

Glucose, a monosaccharide

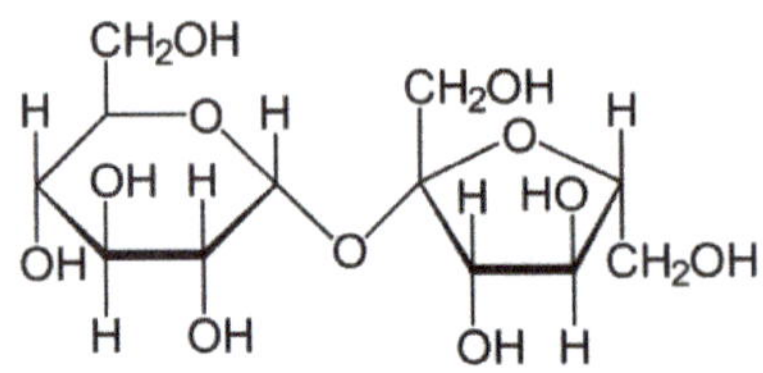

A molecule of sucrose (glucose + fructose), a disaccharide

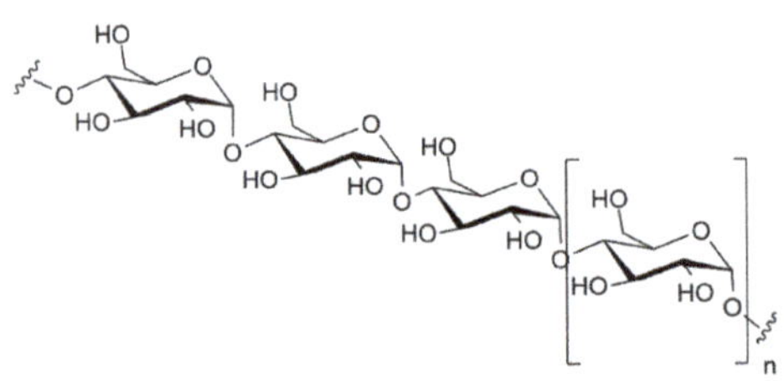

Amylose, a polysaccharide made up of several thousand glucose units

The function of carbohydrates includes energy storage and providing structure. Sugars are carbohydrates, but not all carbohydrates are sugars. There are more carbohydrates on Earth than any other known type of biomolecule; they are used to store energy and genetic information, as well as play important roles in cell to cell interactions and communications.

The simplest type of carbohydrate is a monosaccharide, which among other properties contains carbon, hydrogen, and oxygen, mostly in a ratio of 1:2:1 (generalized formula $C_nH_{2n}O_n$, where n is at least 3). Glucose ($C_6H_{12}O_6$) is one of the most important carbohydrates, others include fructose ($C_6H_{12}O_6$), the sugar commonly associated with the sweet taste of fruits,[a] and deoxyribose ($C_5H_{10}O_4$).

A monosaccharide can switch from the acyclic (open-chain) form to a cyclic form, through a nucleophilic addition reaction between the carbonyl group and one of the hydroxyls of the same molecule. The reaction creates a ring of carbon atoms closed by one bridging oxygen atom. The resulting molecule has an hemiacetal or hemiketal group, depending on whether the linear form was an aldose or a ketose. The reaction is easily reversed, yielding the original open-chain form.

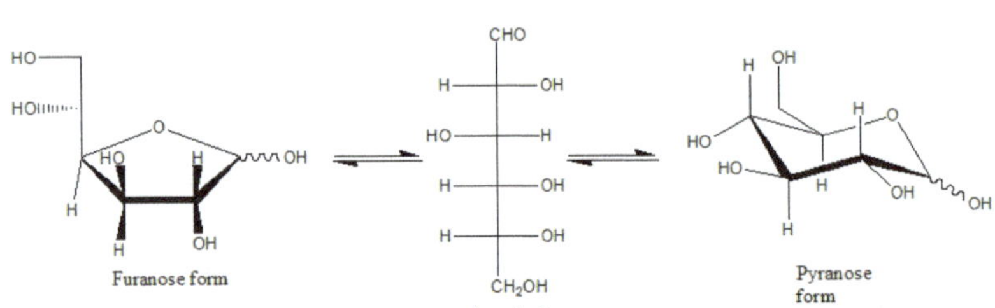

Conversion between the furanose, acyclic, and pyranose forms of D-glucose.

In these cyclic forms, the ring usually has 5 or 6 atoms. These forms are called furanoses and pyranoses, respectively — by analogy with furan and pyran, the simplest compounds with the same carbon-oxygen ring (although they lack the double bonds of these two molecules). For example, the aldohexose glucose may form a hemiacetal linkage between the hydroxyl on carbon 1 and the oxygen on carbon 4, yielding a molecule with a 5-membered ring, called glucofuranose. The same reaction can take place between carbons 1 and 5 to form a molecule with a 6-membered ring, called glucopyranose. Cyclic forms with a 7-atom ring (the same of oxepane), rarely encountered, are called heptoses.

When two monosaccharides undergo dehydration synthesis whereby a molecule of water is released, as two hydrogen atoms and one oxygen atom are lost from the two monosaccharides. The new molecule, consisting of two monosaccharides, is called a *disaccharide* and is conjoined together by a glycosidic or ether bond. The reverse reaction can also occur, using a molecule of water to split up a disaccharide and break the glycosidic bond; this is termed *hydrolysis*. The most well-known disaccharide is sucrose, ordinary sugar (in scientific contexts, called *table sugar* or *cane sugar* to differentiate it from other sugars). Sucrose consists of a glucose molecule and a fructose molecule joined together. Another important disaccharide is lactose, consisting of a glucose molecule and a galactose molecule. As most humans age, the production of lactase, the enzyme that hydrolyzes lactose back into glucose and galactose, typically decreases. This results in lactase deficiency, also called *lactose intolerance*.

When a few (around three to six) monosaccharides are joined, it is called an *oligosaccharide* (*oligo-* meaning "few"). These molecules tend to be used as markers and signals, as well as having some other uses. Many monosaccharides joined together make a polysaccharide. They can be joined together in one long linear chain, or they may be branched. Two of the most common polysaccharides are cellulose and glycogen, both consisting of repeating glucose monomers. Examples are *Cellulose* which is an important structural component of plant's cell walls, and *glycogen*, used as a form of energy storage in animals.

Sugar can be characterized by having reducing or non-reducing ends. A reducing end of a carbohydrate is a carbon atom that can be in equilibrium with the open-chain aldehyde (aldose) or keto form (ketose). If the joining of monomers takes place at such a carbon atom, the free hydroxy group of the pyranose or furanose form is exchanged with an OH-side-chain of another sugar, yielding a full acetal. This prevents opening of the chain to the aldehyde or keto form and renders the modified residue non-reducing. Lactose contains a reducing end at its glucose moiety, whereas the galactose moiety form a full acetal with the C4-OH group of glucose. Saccharose does not have a reducing end because of full acetal formation between the aldehyde carbon of glucose (C1) and the keto carbon of fructose (C2).

Lipids

Lipids comprises a diverse range of molecules and to some extent is a catchall for relatively water-insoluble or nonpolar compounds of biological origin, including waxes, fatty acids, fatty-acid derived phospholipids, sphingolipids, glycolipids, and terpenoids (e.g., retinoids and steroids). Some lipids are linear aliphatic molecules, while others have ring structures. Some are aromatic, while others are not. Some are flexible, while others are rigid.

Lipids are usually made from one molecule of glycerol combined with other molecules. In triglycerides, the main group of bulk lipids, there is one molecule of glycerol and three fatty acids.

Fatty acids are considered the monomer in that case, and may be saturated (no double bonds in the carbon chain) or unsaturated (one or more double bonds in the carbon chain).

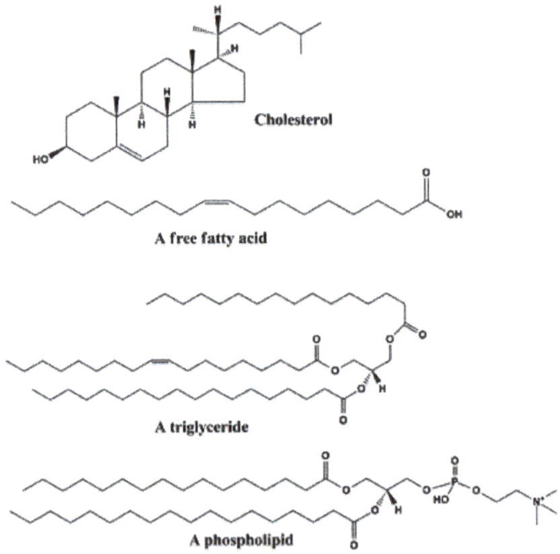

Structures of some common lipids. At the top are cholesterol and oleic acid. The middle structure is a triglyceride composed of oleoyl, stearoyl, and palmitoyl chains attached to a glycerol backbone. At the bottom is the common phospholipid, phosphatidylcholine.

Most lipids have some polar character in addition to being largely nonpolar. In general, the bulk of their structure is nonpolar or hydrophobic ("water-fearing"), meaning that it does not interact well with polar solvents like water. Another part of their structure is polar or hydrophilic ("water-loving") and will tend to associate with polar solvents like water. This makes them amphiphilic molecules (having both hydrophobic and hydrophilic portions). In the case of cholesterol, the polar group is a mere -OH (hydroxyl or alcohol). In the case of phospholipids, the polar groups are considerably larger and more polar, as described below.

Lipids are an integral part of our daily diet. Most oils and milk products that we use for cooking and eating like butter, cheese, ghee etc., are composed of fats. Vegetable oils are rich in various polyunsaturated fatty acids (PUFA). Lipid-containing foods undergo digestion within the body and are broken into fatty acids and glycerol, which are the final degradation products of fats and lipids. Lipids, especially phospholipids, are also used in various pharmaceutical products, either as co-solubilisers (e.g., in parenteral infusions) or else as drug carrier components (e.g., in a liposome or transfersome).

Proteins

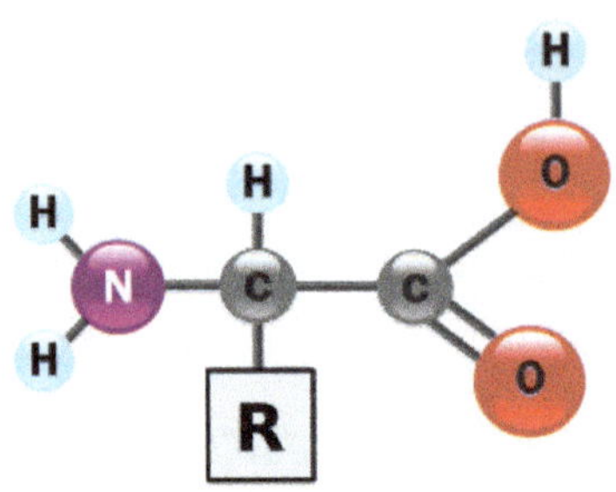

The general structure of an α-amino acid, with the amino group on the left and the carboxyl group on the right.

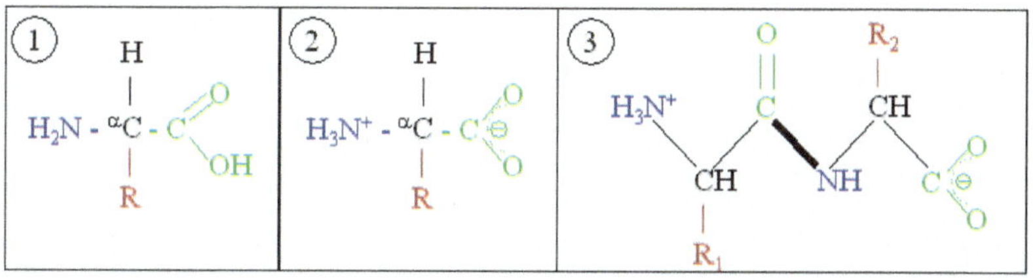

Generic amino acids (1) in neutral form, (2) as they exist physiologically, and (3) joined together as a dipeptide.

Proteins are very large molecules – macro-biopolymers – made from monomers called amino acids. An amino acid consists of a carbon atom bound to four groups. One is an amino group, $—NH_2$, and one is a carboxylic acid group, —COOH (although these exist as $—NH_3^+$ and $—COO^-$ under physiologic conditions). The third is a simple hydrogen atom. The fourth is commonly denoted "—R" and is different for each amino acid. There are 20 standard amino acids, each containing a carboxyl group, an amino group, and a side-chain (known as an "R" group). The "R" group is what makes each amino acid different, and the properties of the side-chains greatly influence the overall three-dimensional conformation of a protein. Some amino acids have functions by themselves or in a modified form; for instance, glutamate functions as an important neurotransmitter. Amino acids can be joined via a peptide bond. In this dehydration synthesis, a water molecule is removed and the peptide bond connects the nitrogen of one amino acid's amino group to the carbon of the other's carboxylic acid group. The resulting molecule is called a *dipeptide*, and short stretches of amino acids (usually, fewer than thirty) are called *peptides* or polypeptides. Longer stretches merit the title *proteins*. As an example, the important blood serum protein albumin contains 585 amino acid residues.

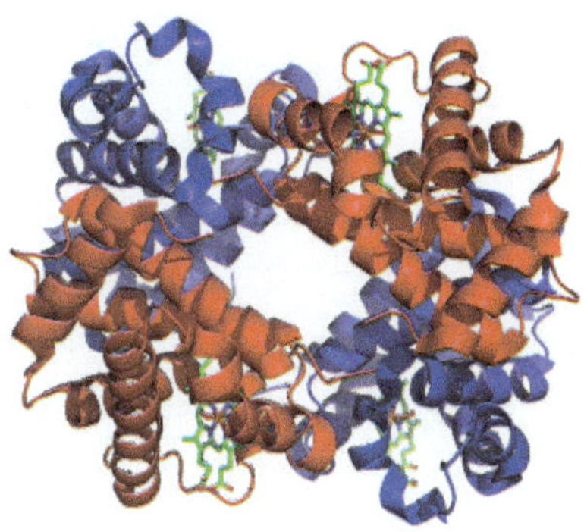

A schematic of hemoglobin. The red and blue ribbons represent the protein globin; the green structures are the heme groups.

Some proteins perform largely structural roles. For instance, movements of the proteins actin and myosin ultimately are responsible for the contraction of skeletal muscle. One property many proteins have is that they specifically bind to a certain molecule or class of molecules—they may be *extremely* selective in what they bind. Antibodies are an example of proteins that attach to one specific type of molecule. In fact, the enzyme-linked immunosorbent assay (ELISA), which uses antibodies, is one of the most sensitive tests modern medicine uses to detect various biomolecules. Probably the most important proteins, however, are the enzymes. Virtually every reaction in a living cell requires an enzyme to lower the activation energy of the reaction. These molecules recognize specific reactant molecules called *substrates*; they then catalyze the reaction between them. By lowering the activation energy, the enzyme speeds up that reaction by a rate of 10^{11} or more; a reaction that would normally take over 3,000 years to complete spontaneously might take less than a second with an enzyme. The enzyme itself is not used up in the process, and is free to catalyze the same reaction with a new set of substrates. Using various modifiers, the activity of the enzyme can be regulated, enabling control of the biochemistry of the cell as a whole.

The structure of proteins is traditionally described in a hierarchy of four levels. The primary structure of a protein simply consists of its linear sequence of amino acids; for instance, "alanine-glycine-tryptophan-serine-glutamate-asparagine-glycine-lysine-...". Secondary structure is concerned with local morphology (morphology being the study of structure). Some combinations of amino acids will tend to curl up in a coil called an α-helix or into a sheet called a β-sheet; some α-helixes can be seen in the hemoglobin schematic above. Tertiary structure is the entire three-dimensional shape of the protein. This shape is determined by the sequence of amino acids. In fact, a single change can change the entire structure. The alpha chain of hemoglobin contains 146 amino acid residues; substitution of the glutamate residue at position 6 with a valine residue changes the behavior of hemoglobin so much that it results in sickle-cell disease. Finally, quaternary structure is concerned with the structure of a protein with multiple peptide subunits, like hemoglobin with its four subunits. Not all proteins have more than one subunit.

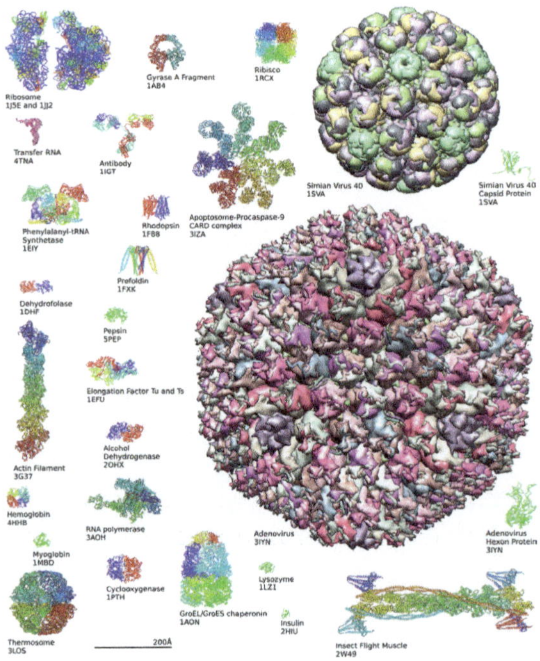

Examples of protein structures from the Protein Data Bank

Ingested proteins are usually broken up into single amino acids or dipeptides in the small intestine, and then absorbed. They can then be joined to make new proteins. Intermediate products of glycolysis, the citric acid cycle, and the pentose phosphate pathway can be used to make all twenty amino acids, and most bacteria and plants possess all the necessary enzymes to synthesize them. Humans and other mammals, however, can synthesize only half of them. They cannot synthesize isoleucine, leucine, lysine, methionine, phenylalanine, threonine, tryptophan, and valine. These are the essential amino acids, since it is essential to ingest them. Mammals do possess the enzymes to synthesize alanine, asparagine, aspartate, cysteine, glutamate, glutamine, glycine, proline, serine, and tyrosine, the nonessential amino acids. While they can synthesize arginine and histidine, they cannot produce it in sufficient amounts for young, growing animals, and so these are often considered essential amino acids.

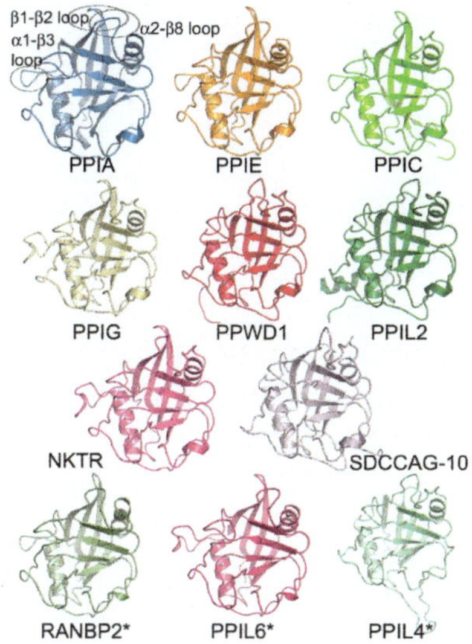

Members of a protein family, as represented by the structures of the isomerase domains.

If the amino group is removed from an amino acid, it leaves behind a carbon skeleton called an α-keto acid. Enzymes called transaminases can easily transfer the amino group from one amino acid (making it an α-keto acid) to another α-keto acid (making it an amino acid). This is important in the biosynthesis of amino acids, as for many of the pathways, intermediates from other biochemical pathways are converted to the α-keto acid skeleton, and then an amino group is added, often via transamination. The amino acids may then be linked together to make a protein.

A similar process is used to break down proteins. It is first hydrolyzed into its component amino acids. Free ammonia (NH_3), existing as the ammonium ion (NH_4^+) in blood, is toxic to life forms. A suitable method for excreting it must therefore exist. Different tactics have evolved in different animals, depending on the animals' needs. Unicellular organisms, of course, simply release the ammonia into the environment. Likewise, bony fish can release the ammonia into the water where it is quickly diluted. In general, mammals convert the ammonia into urea, via the urea cycle.

In order to determine whether two proteins are related, or in other words to decide whether they are homologous or not, scientists use sequence-comparison methods. Methods like sequence alignments and structural alignments are powerful tools that help scientists identify homologies between related molecules. The relevance of finding homologies among proteins goes beyond forming an evolutionary pattern of protein families. By finding how similar two protein sequences are, we acquire knowledge about their structure and therefore their function.

Nucleic Acids

Nucleic acids, so called because of its prevalence in cellular nuclei, is the generic name of the family of biopolymers. They are complex, high-molecular-weight biochemical macromolecules that can convey genetic information in all living cells and viruses. The monomers are called nucleotides, and each consists of three components: a nitrogenous heterocyclic base (either a purine or a pyrimidine), a pentose sugar, and a phosphate group.

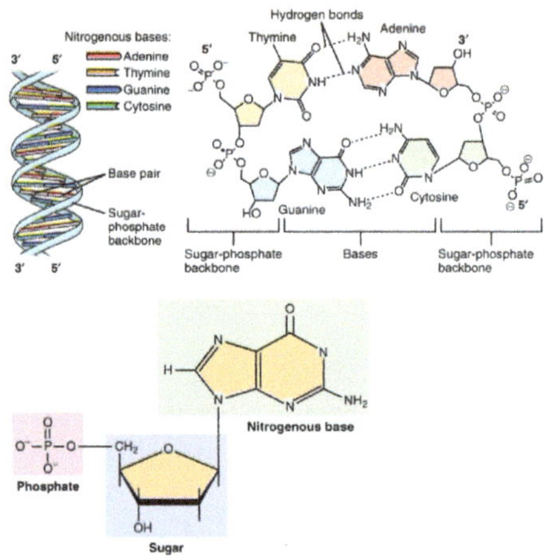

The structure of deoxyribonucleic acid (DNA), the picture shows the monomers being put together.

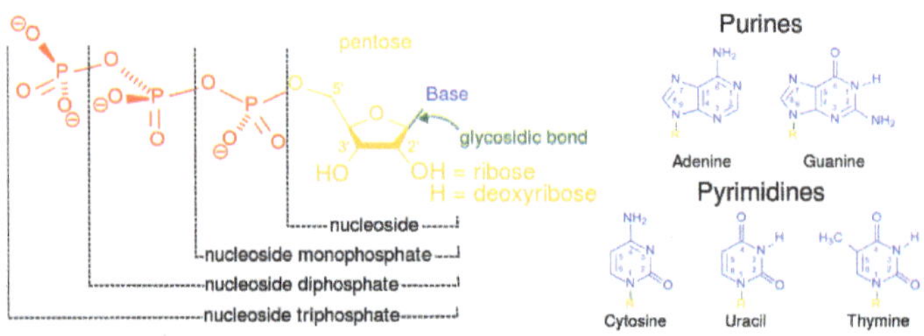

Structural elements of common nucleic acid constituents. Because they contain at least one phosphate group, the compounds marked *nucleoside monophosphate, nucleoside diphosphate* and *nucleoside triphosphate* are all nucleotides (not simply phosphate-lacking nucleosides).

The most common nucleic acids are deoxyribonucleic acid (DNA) and ribonucleic acid (RNA). The phosphate group and the sugar of each nucleotide bond with each other to form the backbone of the nucleic acid, while the sequence of nitrogenous bases stores the information. The most common nitrogenous bases are adenine, cytosine, guanine, thymine, and uracil. The nitrogenous bases of each strand of a nucleic acid will form hydrogen bonds with certain other nitrogenous bases in a complementary strand of nucleic acid (similar to a zipper). Adenine binds with thymine and uracil; Thymine binds only with adenine; and cytosine and guanine can bind only with one another.

Aside from the genetic material of the cell, nucleic acids often play a role as second messengers, as well as forming the base molecule for adenosine triphosphate (ATP), the primary energy-carrier molecule found in all living organisms. Also, the nitrogenous bases possible in the two nucleic acids are different: adenine, cytosine, and guanine occur in both RNA and DNA, while thymine occurs only in DNA and uracil occurs in RNA.

Metabolism

Carbohydrates as Energy Source

Glucose is the major energy source in most life forms. For instance, polysaccharides are broken down into their monomers (glycogen phosphorylase removes glucose residues from glycogen). Disaccharides like lactose or sucrose are cleaved into their two component monosaccharides.

Glycolysis (Anaerobic)

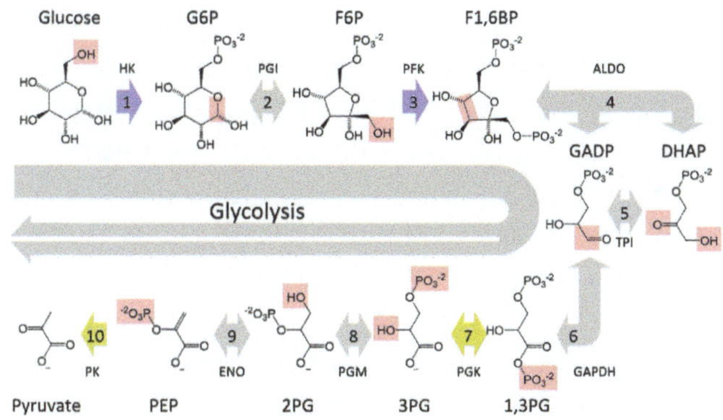

The metabolic pathway of glycolysis converts glucose to pyruvate by via a series of intermediate metabolites. Each chemical modification (red box) is performed by a different enzyme. Steps 1 and 3 consume ATP (blue) and steps 7 and 10 produce ATP (yellow). Since steps 6-10 occur twice per glucose molecule, this leads to a net production of ATP.

Glucose is mainly metabolized by a very important ten-step pathway called glycolysis, the net result of which is to break down one molecule of glucose into two molecules of pyruvate. This also produces a net two molecules of ATP, the energy currency of cells, along with two reducing equivalents of converting NAD^+ (nicotinamide adenine dinucleotide:oxidised form) to NADH (nicotinamide adenine dinucleotide:reduced form). This does not require oxygen; if no oxygen is available (or the cell cannot use oxygen), the NAD is restored by converting the pyruvate to lactate (lactic acid) (e.g., in humans) or to ethanol plus carbon dioxide (e.g., in yeast). Other monosaccharides like galactose and fructose can be converted into intermediates of the glycolytic pathway.

Aerobic

In aerobic cells with sufficient oxygen, as in most human cells, the pyruvate is further metabolized. It is irreversibly converted to acetyl-CoA, giving off one carbon atom as the waste product carbon dioxide, generating another reducing equivalent as NADH. The two molecules acetyl-CoA (from one molecule of glucose) then enter the citric acid cycle, producing two more molecules of ATP, six more NADH molecules and two reduced (ubi)quinones (via $FADH_2$ as enzyme-bound cofactor), and releasing the remaining carbon atoms as carbon dioxide. The produced NADH and quinol molecules then feed into the enzyme complexes of the respiratory chain, an electron transport system transferring the electrons ultimately to oxygen and conserving the released energy in the form of a proton gradient over a membrane (inner mitochondrial membrane in eukaryotes). Thus, oxygen is reduced to water and the original electron acceptors NAD^+ and quinone are regenerated. This is why humans breathe in oxygen and breathe out carbon dioxide. The energy released from transferring the electrons from high-energy states in NADH and quinol is conserved first as proton gradient and converted to ATP via ATP synthase. This generates an additional *28* molecules of ATP (24 from the 8 NADH + 4 from the 2 quinols), totaling to 32 molecules of ATP conserved per degraded glucose (two from glycolysis + two from the citrate cycle). It is clear that using oxygen to completely oxidize glucose provides an organism with far more energy than any oxygen-independent metabolic feature, and this is thought to be the reason why complex life appeared only after Earth's atmosphere accumulated large amounts of oxygen.

Gluconeogenesis

In vertebrates, vigorously contracting skeletal muscles (during weightlifting or sprinting, for example) do not receive enough oxygen to meet the energy demand, and so they shift to anaerobic metabolism, converting glucose to lactate. The liver regenerates the glucose, using a process called gluconeogenesis. This process is not quite the opposite of glycolysis, and actually requires three times the amount of energy gained from glycolysis (six molecules of ATP are used, compared to the two gained in glycolysis). Analogous to the above reactions, the glucose produced can then undergo glycolysis in tissues that need energy, be stored as glycogen (or starch in plants), or be converted to other monosaccharides or joined into di- or oligosaccharides. The combined pathways of glycolysis during exercise, lactate's crossing via the bloodstream to the liver, subsequent gluconeogenesis and release of glucose into the bloodstream is called the Cori cycle.

Relationship to Other "Molecular-scale" Biological Sciences

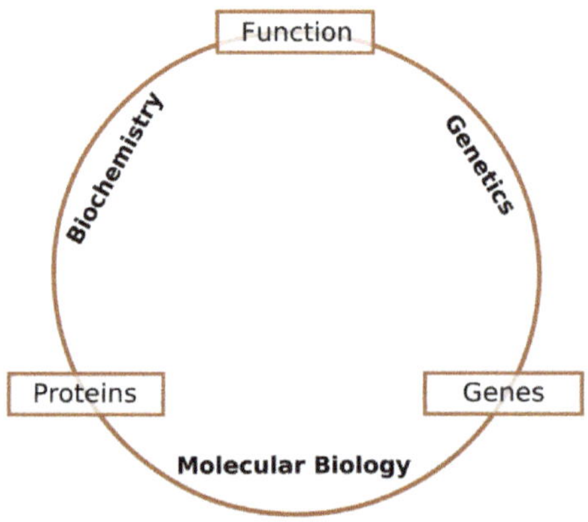

Schematic relationship between biochemistry, genetics, and molecular biology

Researchers in biochemistry use specific techniques native to biochemistry, but increasingly combine these with techniques and ideas developed in the fields of genetics, molecular biology and biophysics. There has never been a hard-line among these disciplines in terms of content and technique. Today, the terms *molecular biology* and *biochemistry* are nearly interchangeable. The following figure is a schematic that depicts one possible view of the relationship between the fields:

- *Biochemistry* is the study of the chemical substances and vital processes occurring in living organisms. Biochemists focus heavily on the role, function, and structure of biomolecules. The study of the chemistry behind biological processes and the synthesis of biologically active molecules are examples of biochemistry.

- *Genetics* is the study of the effect of genetic differences on organisms. Often this can be inferred by the absence of a normal component (e.g., one gene). The study of "mutants" – organisms with a changed gene that leads to the organism being different with respect to the so-called "wild type" or normal phenotype. Genetic interactions (epistasis) can often confound simple interpretations of such "knock-out" or "knock-in" studies.

- *Molecular biology* is the study of molecular underpinnings of the process of replication, transcription and translation of the genetic material. The central dogma of molecular biology where genetic material is transcribed into RNA and then translated into protein, despite being an oversimplified picture of molecular biology, still provides a good starting point for understanding the field. This picture, however, is undergoing revision in light of emerging novel roles for RNA.

- *Chemical biology* seeks to develop new tools based on small molecules that allow minimal perturbation of biological systems while providing detailed information about their function. Further, chemical biology employs biological systems to create non-natural hybrids between biomolecules and synthetic devices (for example emptied viral capsids that can deliver gene therapy or drug molecules).

Chemical Biology

Chemical biology is a scientific discipline spanning the fields of chemistry, biology, and physics. It involves the application of chemical techniques, tools, and analyses, and often compounds produced through synthetic chemistry, to the study and manipulation of biological systems. Chemical biologists attempt to use chemical principles to modulate systems to either investigate the underlying biology or create new function. Research done by chemical biologists is often closer related to that of cell biology than biochemistry. Biochemists study the chemistry of biomolecules and regulation of biochemical pathways within cells and tissues, e.g. cAMP or cGMP, while chemical biologists deal with novel chemical compounds *applied to* biology.

Introduction

Some forms of chemical biology attempt to answer biological questions by directly probing living systems at the chemical level. In contrast to research using biochemistry, genetics, or molecular biology, where mutagenesis can provide a new version of the organism or cell of interest, chemical biology studies probe systems *in vitro* and *in vivo* with small molecules that have been designed for a specific purpose or identified on the basis of biochemical or cell-based screening.

Chemical biology is one of many interfacial sciences that are characteristic of a general trend away from older, reductionist fields toward those whose goals are to achieve a description of scientific holism. In this sense, it is related to other fields such as proteomics. Chemical biology has scientific, historical and philosophical roots in medicinal chemistry, supramolecular chemistry (particularly host-guest chemistry), bioorganic chemistry, pharmacology, genetics, biochemistry, and metabolic engineering.

Systems of Interest

Proteomics

Proteomics investigates the proteome, the set of expressed proteins at a given time under defined conditions. As a discipline, proteomics has moved past rapid protein identification and has developed into a biological assay for quantitative analysis of complex protein samples by comparing protein changes in differently perturbed systems. Current goals in proteomics include determining protein sequences, abundance and any post-translational modifications. Also of interest are protein–protein interactions, cellular distribution of proteins and understanding protein activity. Another important aspect of proteomics is the advancement of technology to achieve these goals.

Protein levels, modifications, locations, and interactions are complex and dynamic properties. With this complexity in mind, experiments need to be carefully designed to answer specific questions especially in the face of the massive amounts of data that are generated by these analyses. The most valuable information comes from proteins that are expressed differently in a system being studied. These proteins can be compared relative to each other using quantitative proteomics, which allows a protein to be labeled with a mass tag. Proteomic technologies must be sensitive and robust, it is for these reasons, the mass spectrometer has been the workhorse of protein analysis. The high precision of mass spectrometry can distinguish between closely related species and species of interest can be isolated and fragmented within the instrument. Its applications to protein analysis was only possible in the late 1980s with the development of protein and peptide ionization with minimal fragmentation. These breakthroughs were ESI and MALDI. Mass spectrometry technologies are modular and can be chosen or optimized to the system of interest.

Chemical biologists are poised to impact proteomics through the development of techniques, probes and assays with synthetic chemistry for the characterization of protein samples of high complexity. These approaches include the development of enrichment strategies, chemical affinity tags and probes.

Enrichment Techniques

Samples for Proteomics contain a myriad of peptide sequences, the sequence of interest may be highly represented or of low abundance. However, for successful MS analysis the peptide should be enriched within the sample. Reduction of sample complexity is achieved through selective enrichment using affinity chromatography techniques. This involves targeting a peptide with a distinguishing feature like a biotin label or a post translational modification. Interesting methods have been developed that include the use of antibodies, lectins to capture glycoproteins, immobilized metal ions to capture phosphorylated peptides and suicide enzyme substrates to capture specific enzymes. Here, chemical biologists can develop reagents to interact with substrates, specifically and tightly, to profile a targeted functional group on a proteome scale. Development of new enrichment strategies is needed in areas like non-ser/thr/tyr phosphorylation sites and other post translational modifications. Other methods of decomplexing samples relies on upstream chromatographic separations.

Affinity Tags

Chemical synthesis of affinity tags has been crucial to the maturation of quantitative proteomics. iTRAQ, Tandem mass tags (TMT) and Isotope-coded affinity tag (ICAT) are protein mass-tags that consist of a covalently attaching group, a mass (isobaric or isotopic) encoded linker and a handle for isolation. Varying mass-tags bind to different proteins as a sort of footprint such that when analyzing cells of differing perturbations, the levels of each protein can be compared relatively after enrichment by the introduced handle. Other methods include SILAC and heavy isotope labeling. These methods have been adapted to identify complexing proteins by labeling a bait protein, pulling it down and analyzing the proteins it has complexed. Another method creates an internal tag by introducing novel amino acids that are genetically encoded in prokaryotic and eukaryotic organisms. These modifications create a new level of control and can facilitate photocrosslinking to probe protein–protein interactions. In addition, keto, acetylene, azide, thioester, boronate, and dehydroalanine- containing amino acids can be used to selectively introduce tags, and novel chemical functional groups into proteins.

Enzyme Probes

To investigate enzymatic activity as opposed to total protein, activity-based reagents have been developed to label the enzymatically active form of proteins (Activity-based proteomics). For example, serine hydrolase- and cysteine protease-inhibitors have been converted to suicide inhibitors. This strategy enhances the ability to selectively analyze low abundance constituents through direct targeting. Structures that mimic these inhibitors could be introduced with modifications that will aid proteomic analysis- like an identification handle or mass tag. Enzyme activity can also be monitored through converted substrate. This strategy relies on using synthetic substrate conjugates that contain moieties that are acted upon by specific enzymes. The product conjugates are then captured by an affinity reagent and analyzed. The measured concentration of product conjugate allow the determination of the enzyme velocity. Identification of enzyme substrates (of which there may be hundreds or thousands, many of which unknown) is a problem of significant difficulty in proteomics and is vital to the understanding of signal transduction pathways in cells; techniques for labelling cellular substrates of enzymes is an area chemical biologists can address. A method that has been developed uses "analog-sensitive" kinases to label substrates using an unnatural ATP analog, facilitating visualization and identification through a unique handle.

Glycobiology

While DNA, RNA and proteins are all encoded at the genetic level, there exists a separate system of trafficked molecules in the cell that are not encoded directly at any direct level: sugars. Thus, glycobiology is an area of dense research for chemical biologists. For instance, live cells can be supplied with synthetic variants of natural sugars in order to probe the function of the sugars in vivo. Carolyn Bertozzi at University of California, Berkeley has developed a method for site-specifically reacting molecules the surface of cells that have been labeled with synthetic sugars.

Combinatorial Chemistry

Chemical biologists used automated synthesis of many diverse compounds in order to experiment with effects of small molecules on biological processes. More specifically, they observe changes in the behaviors of proteins when small molecules bind to them. Such experiments may supposedly lead to discovery of small molecules with antibiotic or chemotherapeutic properties. These approaches are identical to those employed in the discipline of pharmacology.

Molecular Sensing

Chemical biologists are also interested in developing new small-molecule and biomolecule-based tools to study biological processes, often by molecular imaging techniques. The field of molecular sensing was popularized by Roger Tsien's work developing calcium-sensing fluorescent compounds as well as pioneering the use of GFP, for which he was awarded the 2008 Nobel Prize in Chemistry. Today, researchers continue to utilize basic chemical principles to develop new compounds for the study of biological metabolites and processes.

siRNA-A Tool in Chemical Biology

siRNA or small interfering RNAs owe their origins to the difficulties the scientific community faced utilizing classical and reverse genetics methods in studying gene expression. Disrupting genes to study their functions is not always optimal; neither is mapping mutations back to their genes easy. The whole process is expensive as well as time-consuming, which is why a lot of effort has been devoted to develop methods to silence gene expression in sequence specific manner using nucleic acids. They have the potential to be powerful tools in the field of chemical biology to study the chemistry of gene expression in therapeutic targets of bacteria and viruses.

A number of different types of nucleic acid molecules have already gained prominence because of their potential as therapeutics. They target mRNAs to silence the genes in a sequence specific manner. Oligodeoxyribonucleic acids, ODNs utilize steric interaction to silence gene expression. They can also form triple helices in conjunction with the DNA duplex. Whereas ribozymes can be chemically designed to target specific genes and cleave them in a sequence specific manner. The most promising of these methods however is utilization of short interfering RNA or siRNA to silence gene expression.

siRNA

siRNA or short interfering RNAs exist in nature as a means for the express purpose of controlling gene expression. It was discovered in petunia as a post-transcriptional gene silencing measure. It is the resultant product when a long double-strand RNA of 20 -25 nucleotides length was processed in the cells by the enzyme DICER. The newly synthesized siRNA assemble into endoribonuclease-containing complexes known as RNA-induced silencing complexes (RISCs), unwinding in the process. The activated RISC then binds to the complementary RNA molecules by base pairing interactions between the siRNA strand and the mRNA, which is then cleaved. This mechanism is known as RNA interference or RNAi.

Designing and Synthesizing siRNAs

It is now possible to order siRNAs designed and synthesized with the express purpose of targeting a particular sequence. The ambion website has a lot of information on the optimal design of siRNAs.

siRNAs can be synthesized chemically, or enzymatically. RNase III or DICER can be used to cleave the long dsRNAs to produce siRNAs. However the most expedient method is the use of plasmids to express them in vivo by delivering them into the target cell using vectors. This method allows the siRNAs to be expressed in the target cell stably, over a period of time and overcomes the drawbacks of the transience of their effect. Numerous strategies have been developed in order to deliver the siRNA into the cell efficiently:

1. Electroporation
2. Local and systemic injection: This method was the first success scientists had in silencing genes using siRNAs. They were successfully delivered into highly vascularized tissue in mice through using high-pressure tail vein injection. Greater than 90% loss in gene expression was observed in the targets.

3. siRNA producing viruses: This method shows great promise in gene therapy, and research is progressing in order to generate recombinant viruses that can produce siRNA in target cells.

4. Small molecules that enhance transdermal penetration: Research in this field is moving at a fast pace in order to synthesize small organic molecules that, if injected in conjunction with siRNAs, can help them penetrate into the target cells.

Biological Uses of the RNAi Approach

The principal purpose of studying siRNA mediated RNA interference is probably to investigate gene function.

It is so much easier to make genetic knock-outs by simply introducing sequence-specific siRNAs into cells; multi-copy genes can be silenced in one fell swoop by this method. Creation of double-knockout mutants is also easier and consumes much less time. Using local injections in specific regions of the model organisms also help in creating spatially separated and restricted knockout. siRNAs are also being successfully used to screen whole genomes in organisms such as C. elegans and Drosophilla melanogaster. Even in mammalian systems such as Danio rerio (zebrafish) that usually prove intractable to all gene silencing methods, even dsRNA injection, siRNA can do the job. It is paving a new way in development of therapeutics by identifying human gene orthologs in other species in a remarkably short period of time.

Numerous high-throughput screening approaches are being developed to screen large libraries of cells rapidly in order to identify drug targets. A brief description of few of the screening techniques:

1. Pooled Format Screening: A reagent library of RNAi has to be introduced to the cells so that a particular cell is in one particular reagent. The primary hits are then identified and their identity elucidate by sequencing techniques.

2. Arrayed Format Screening: Each RNAi reagent is placed in separate wells in a plate and multiple manipulations can be done to identify their targets, which are then detected by fluorescence readouts, imaging techniques and other methods as well. Thus the identity of the target cell can be determined through the identity of the reagent in the database.

3. Multiplexed methods: A combination of various assays can be used for high-throughput screening of candidate drug targets. For example, candidate genes can be identified through informatics based methods and then screened against a library of reagents. Many other such methods are being developed in order to make the job of screening therapeutic targets easier.

siRNA Based Therapeutics

The future in this field rests in the development of siRNA-based drugs.

This could prove to be a powerful tool in gene based therapy. Research is now concentrated on developing strategies to design siRNA therapeutics for clinical use. A brief description of some novel strategies for siRNA drug development is provided here:

1. Direct Mutation Targeting: The siRNAs are designed to perfectly match mutant alleles but contain one or more mismatches with wild-type alleles, leading to specific degradation of the matching, mutant transcripts.
2. Indirect Mutation Targeting: The siRNA approach will not work if the mutant alleles are too similar to wild type. So an indirect approach is taken in which siRNAs are designed against disease linked markers such as SNP variations. The ones that are screened as positive are targeted for degradation.
3. Exon-specific targeting: siRNAs are designed to target expressed regions (exons) of the gene.
4. Targeting exon skipped transcripts: If the problem in the gene lies in aberrant splicing post-transcription, siRNA can be designed to target the unnatural exon-exon interface arising as a result of such alternative splicing.

Employing Biology

Many research programs are also focused on employing natural biomolecules to perform a task or act as support for a new chemical method or material. In this regard, researchers have shown that DNA can serve as a template for synthetic chemistry, self-assembling proteins can serve as a structural scaffold for new materials, and RNA can be evolved *in vitro* to produce new catalytic function.

Protein Misfolding and Aggregation as a Cause of Disease

A common form of aggregation is long, ordered spindles called amyloid fibrils that are implicated in Alzheimer's disease and that have been shown to consist of cross-linked beta sheet regions perpendicular to the backbone of the polypeptide. Another form of aggregation occurs with prion proteins, the glycoproteins found with Creutzfeldt–Jakob disease and bovine spongiform encephalopathy. In both structures, aggregation occurs through hydrophobic interactions and water must be excluded from the binding surface before aggregation can occur. A movie of this process can be seen in "Chemical and Engineering News". The diseases associated with misfolded proteins are life-threatening and extremely debilitating, which makes them an important target for chemical biology research.

Through the transcription and translation process, DNA encodes for specific sequences of amino acids. The resulting polypeptides fold into more complex secondary, tertiary, and quaternary structures to form proteins. Based on both the sequence and the structure, a particular protein is conferred its cellular function. However, sometimes the folding process fails due to mutations in the genetic code and thus the amino acid sequence or due to changes in the cell environment (e.g. pH, temperature, reduction potential, etc.). Misfolding occurs more often in aged individuals or in cells exposed to a high degree of oxidative stress, but a fraction of all proteins misfold at some point even in the healthiest of cells.

Normally when a protein does not fold correctly, molecular chaperones in the cell can encourage refolding back into its active form. When refolding is not an option, the cell can also target the protein for degradation back into its component amino acids via proteolytic, lysosomal, or autophagic mechanisms. However, under certain conditions or with certain mutations, the cells can no longer cope with the misfolded protein(s) and a disease state results. Either the protein has a loss-of-function, such as in cystic fibrosis, in which it loses activity or cannot reach its target, or the protein has a gain-of-function, such as with Alzheimer's disease, in which the protein begins to aggregate causing it to become insoluble and non-functional.

Protein misfolding has previously been studied using both computational approaches as well as *in vivo* biological assays in model organisms such as *Drosophila melanogaster* and *C. elegans.* Computational models use a *de novo* process to calculate possible protein structures based on input parameters such as amino acid sequence, solvent effects, and mutations. This method has the shortcoming that the cell environment has been drastically simplified, which limits the factors that influence folding and stability. On the other hand, biological assays can be quite complicated to perform *in vivo* with high-throughput like efficiency and there always remains the question of how well lower organism systems approximate human systems.

Dobson et al. propose combining these two approaches such that computational models based on the organism studies can begin to predict what factors will lead to protein misfolding. Several experiments have already been performed based on this strategy. In experiments on *Drosophila,* different mutations of beta amyloid peptides were evaluated based on the survival rates of the flies as well as their motile ability. The findings from the study show that the more a protein aggregates, the more detrimental the neurological dysfunction. Further studies using transthyretin, a component of cerebrospinal fluid that binds to beta amyloid peptide deterring aggregation but can itself aggregate especially when mutated, indicate that aggregation prone proteins may not aggregate where they are secreted and rather are deposited in specific organs or tissues based on each mutation. Kelly et al. have shown that the more stable, both kinetically and thermodynamically, a misfolded protein is the more likely the cell is to secrete it from the endoplasmic reticulum rather than targeting the protein for degradation. In addition, the more stress that a cell feels from misfolded proteins the more probable new proteins will misfold. These experiments as well as others having begun to elucidate both the intrinsic and extrinsic causes of misfolding as well as how the cell recognizes if proteins have folded correctly.

As more information is obtained on how the cell copes with misfolded proteins, new therapeutic strategies begin to emerge. An obvious path would be prevention of misfolding. However, if protein misfolding cannot be avoided, perhaps the cell's natural mechanisms for degradation can be bolstered to better deal with the proteins before they begin to aggregate. Before these ideas can be realized, many more experiments need to be done to understand the folding and degradation machinery as well as what factors lead to misfolding. More information about protein misfolding and how it relates to disease can be found in the recently published book by Dobson, Kelly, and Ramerez-Alvarado entitled Protein Misfolding Diseases Current and Emerging Principles and Therapies.

Chemical Synthesis of Peptides

In contrast to the traditional biotechnological practice of obtaining peptides or proteins by isolation from cellular hosts through cellular protein production, advances in chemical techniques for the synthesis and ligation of peptides has allowed for the total synthesis of some peptides and proteins. Chemical synthesis of proteins is a valuable tool in chemical biology as it allows for the introduction of non-natural amino acids as well as residue specific incorporation of "posttranslational modifications" such as phosphorylation, glycosylation, acetylation, and even ubiquitination. These capabilities are valuable for chemical biologists as non-natural amino acids can be used to probe and alter the functionality of proteins, while post translational modifications are widely known to regulate the structure and activity of proteins. Although strictly biological techniques have been developed to achieve these ends, the chemical synthesis of peptides often has a lower technical and practical barrier to obtaining small amounts of the desired protein. Given the widely recognized importance of proteins as cellular catalysts and recognition elements, the ability to precisely control the composition and connectivity of polypeptides is a valued tool in the chemical biology community and is an area of active research.

While chemists have been making peptides for over 100 years, the ability to efficiently and quickly synthesize short peptides came of age with the development of Bruce Merrifield's solid phase peptide synthesis (SPPS). Prior to the development of SPPS, the concept of step-by-step polymer synthesis on an insoluble support was without chemical precedent. The use of a covalently bound insoluble polymeric support greatly simplified the process of peptide synthesis by reducing purification to a simple "filtration and wash" procedure and facilitated a boom in the field of peptide chemistry. The development and "optimization" of SPPS took peptide synthesis from the hands of the specialized peptide synthesis community and put it into the hands of the broader chemistry, biochemistry, and now chemical biology community. SPPS is still the method of choice for linear synthesis of polypeptides up to 50 residues in length and has been implemented in commercially available automated peptide synthesizers. One inherent shortcoming in any procedure that calls for repeated coupling reactions is the buildup of side products resulting from incomplete couplings and side reactions. This places the upper bound for the synthesis of linear polypeptide lengths at around 50 amino acids, while the "average" protein consists of 250 amino acids. Clearly, there was a need for development of "non-linear" methods to allow synthetic access to the average protein.

Although the shortcomings of linear SPPS were recognized not long after its inception, it took until the early 1990s for effective methodology to be developed to ligate small peptide fragments made by SPPS, into protein sized polypeptide chains (for recent review of peptide ligation strategies. The oldest and best developed of these methods is termed native chemical ligation. Native chemical ligation was unveiled in a 1994 paper from the laboratory of Stephen B. H. Kent. Native chemical ligation involves the coupling of a C-terminal thioester and an N-terminal cysteine residue, ultimately resulting in formation of a "native" amide bond. Further refinements in native chemical ligation have allowed for kinetically controlled coupling of multiple peptide fragments, allowing access to moderately sized peptides such as an HIV-protease dimer and human lysozyme. Even with the successes and attractive features of native chemical ligation, there are still some drawbacks in the utilization of this technique. Some of these drawbacks include the installation and preservation of a reactive C-terminal thioester, the requirement of an N-terminal cysteine residue (which is the second-least-common amino acid in proteins), and the requirement for a sterically unincumbering C-terminal residue.

Other strategies that have been used for the ligation of peptide fragments using the acyl transfer chemistry first introduced with native chemical ligation include expressed protein ligation, sulfurization/desulfurization techniques, and use of removable thiol auxiliaries.

Expressed protein ligation allows for the biotechnological installation of a C-terminal thioester using intein biochemistry, thereby allowing the appendage of a synthetic N-terminal peptide to the recombinantly produced C-terminal portion. This technique allows for access to much larger proteins, as only the N-terminal portion of the resulting protein has to be chemically synthesized. Both sulfurization/desulfurization techniques and the use of removable thiol auxiliaries involve the installation of a synthetic thiol moiety to carry out the standard native chemical ligation chemistry, followed by removal of the auxiliary/thiol. These techniques help to overcome the requirement of an N-terminal cysteine needed for standard native chemical ligation, although the steric requirements for the C-terminal residue are still limiting.

A final category of peptide ligation strategies include those methods not based on native chemical ligation type chemistry. Methods that fall in this category include the traceless Staudinger ligation, azide-alkyne dipolar cycloadditions, and imine ligations.

Major contributors in this field today include Stephen B. H. Kent, Philip E. Dawson, and Tom W. Muir, as well as many others involved in methodology development and applications of these strategies to biological problems.

Protein Design by Directed Evolution

One of the primary goals of protein engineering is the design of novel peptides or proteins with a desired structure and chemical activity. Because our knowledge of the relationship between primary sequence, structure, and function of proteins is limited, rational design of new proteins with enzymatic activity is extremely challenging. Directed evolution, repeated cycles of genetic diversification followed by a screening or selection process, can be used to mimic Darwinian evolution in the laboratory to design new proteins with a desired activity.

Several methods exist for creating large libraries of sequence variants. Among the most widely used are subjecting DNA to UV radiation or chemical mutagens, error-prone PCR, degenerate codons, or recombination. Once a large library of variants is created, selection or screening techniques are used to find mutants with a desired attribute. Common selection/screening techniques include fluorescence-activated cell sorting (FACS), mRNA display, phage display, or *in vitro* compartmentalization. Once useful variants are found, their DNA sequence is amplified and subjected to further rounds of diversification and selection. Since only proteins with the desired activity are selected, multiple rounds of directed evolution lead to proteins with an accumulation beneficial traits.

There are two general strategies for choosing the starting sequence for a directed evolution experiment: *de novo* design and redesign. In a protein design experiment, an initial sequence is chosen at random and subjected to multiple rounds of directed evolution. For example, this has been employed successfully to create a family of ATP-binding proteins with a new folding pattern not found in nature. Random sequences can also be biased towards specific folds by specifying the characteristics (such as polar vs. nonpolar) but not the specific identity of each amino acid in a sequence. Among other things, this strategy has been used to successfully design four-helix bundle proteins. Because it is often thought that a well-defined structure is required for activity, biasing a designed protein towards adopting a specific folded structure is likely to increase the frequency of desirable variants in constructed libraries.

In a protein redesign experiment, an existing sequence serves as the starting point for directed evolution. In this way, old proteins can be redesigned for increased activity or new functions. Protein redesign has been used for protein simplification, creation of new quaternary structures, and topological redesign of a chorismate mutase. To develop enzymes with new activities, one can take advantage of promiscuous enzymes or enzymes with significant side reactions. In this regard, directed evolution has been used on γ-humulene synthase, an enzyme that creates over 50 different sesquiterpenes, to create enzymes that selectively synthesize individual products. Similarly, completely new functions can be selected for from existing protein scaffolds. In one example of this, an RNA ligase was created from a zinc finger scaffold after 17 rounds of directed evolution. This new enzyme catalyzes a chemical reaction not known to be catalyzed by any natural enzyme.

Computational methods, when combined with experimental approaches, can significantly assist both the design and redesign of new proteins through directed evolution. Computation has been used to design proteins with unnatural folds, such as a right-handed coiled coil. These computational approaches could also be used to redesign proteins to selectively bind specific target molecules. By identifying lead sequences using computational methods, the occurrence of functional proteins in libraries can be dramatically increased before any directed evolution experiments in the laboratory.

Manfred T. Reetz, Frances Arnold, Donald Hilvert, and Jack W. Szostak are significant researchers in this field.

Biocompatible Click Cycloaddition Reactions in Chemical Biology

Recent advances in technology have allowed scientists to view substructures of cells at levels of unprecedented detail. Unfortunately these "aerial" pictures offer little information about the mechanics of the biological system in question. To be fully effective, precise imaging systems require a complementary technique that better elucidates the machinery of a cell. By attaching tracking devices (optical probes) to biomolecules *in vivo*, one can learn far more about cell metabolism, molecular transport, cell-cell interactions and many other processes

Bioorthogonal Reactions

Successful labeling of a molecule of interest requires specific functionalization of that molecule to react chemospecifically with an optical probe. For a labeling experiment to be considered robust, that functionalization must minimally perturb the system. Unfortunately, these requirements can often be extremely hard to meet. Many of the reactions normally available to organic chemists in the laboratory are unavailable in living systems. Water- and redox- sensitive reactions would not proceed, reagents prone to nucleophilic attack would offer no chemospecificity, and any reactions with large kinetic barriers would not find enough energy in the relatively low-heat environment of a living cell. Thus, chemists have recently developed a panel of bioorthogonal chemistry that proceed chemospecifically, despite the milieu of distracting reactive materials *in vivo*.

Design of Bioorthogonal Reagents and Bioorthogonal Chemical Reporters

The coupling of an optical probe to a molecule of interest must occur within a reasonably short time frame; therefore, the kinetics of the coupling reaction should be highly favorable. Click chemistry is well suited to fill this niche, since click reactions are, by definition, rapid, spontaneous, selective, and high-yielding. Unfortunately, the most famous "click reaction," a [3+2] cycloaddition between an azide and an acyclic alkyne, is copper-catalyzed, posing a serious problem for use *in vivo* due to copper's toxicity.

The issue of copper toxicity can be alleviated using copper-chelating ligands, enabling copper-catalyzed labeling of the surface of live cells.

To bypass the necessity for a catalyst, the lab of Dr. Carolyn Bertozzi introduced inherent strain into the alkyne species by using a cyclic alkyne. In particular, cyclooctyne reacts with azido-molecules with distinctive vigor. Further optimization of the reaction led to the use of difluorinated cyclooctynes (DIFOs), which increased yield and reaction rate. Other coupling partners discovered by separate labs to be analogous to cyclooctynes include trans cyclooctene, norbornene, and a cyclobutene-functionalized molecule.

Use in Biological Systems

As mentioned above, the use of bioorthogonal reactions to tag biomolecules requires that one half of the reactive "click" pair is installed in the target molecule, while the other is attached to an optical probe. When the probe is added to a biological system, it will selectively conjugate with the target molecule.

The most common method of installing bioorthogonal reactivity into a target biomolecule is through metabolic labeling. Cells are immersed in a medium where access to nutrients is limited to synthetically modified analogues of standard fuels such as sugars. As a consequence, these altered biomolecules are incorporated into the cells in the same manner as their wild-type brethren. The optical probe is then incorporated into the system to image the fate of the altered biomolecules. Other methods of functionalization include enzymatically inserting azides into proteins, and synthesizing phospholipids conjugated to cyclooctynes.

Future Directions

As these bioorthogonal reactions are further optimized, they will likely be used for increasingly complex interactions involving multiple different classes of biomolecules. More complex interactions have a smaller margin for error, so increased reaction efficiency is paramount to continued success in optically probing cellular machinery. Also, by minimizing side reactions, the experimental design of a minimally perturbed living system is closer to being realized.

Discovery of Biomolecules through Metagenomics

The advances in modern sequencing technologies in the late 1990s allowed scientists to investigate DNA of communities of organisms in their natural environments, so-called "eDNA", without culturing individual species in the lab. This metagenomic approach enabled scientists to study a wide selection of organisms that were previously not characterized due in part to an incompetent growth condition. These sources of eDNA include, but are not limited to, soils, ocean, subsurface, hot springs, hydrothermal vents, polar ice caps, hypersaline habitats, and extreme pH environments. of the many applications of metagenomics, chemical biologists and microbiologists such as Jo Handelsman, Jon Clardy, and Robert M. Goodman who are pioneers of metagenomics, explored metagenomic approaches toward the discovery of biologically active molecules such as antibiotics.

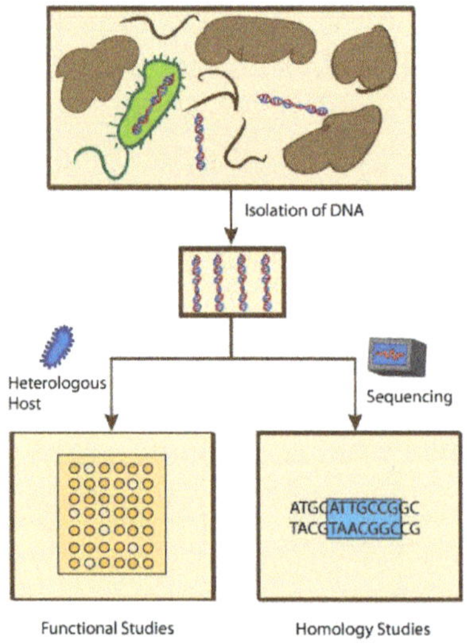

Overview of metagenomic methods

Functional or homology screening strategies have been used to identify genes that produce small bioactive molecules. Functional metagenomic studies are designed to search for specific phenotypes that are associated with molecules with specific characteristics. Homology metagenomic studies, on the other hand, are designed to examine genes to identify conserved sequences that are previously associated with the expression of biologically active molecules.

Functional metagenomic studies enable scientists to discover novel genes that encode biologically active molecules. These assays include top agar overlay assays where antibiotics generate zones of growth inhibition against test microbes, and pH assays that can screen for pH change due to newly synthesized molecules using pH indicator on an agar plate. Substrate-induced gene expression screening (SIGEX), a method to screen for the expression of genes that are induced by chemical compounds, has also been used to search genes with specific functions. These led to the discovery and isolation of several novel proteins and small molecules. For example, the Schipper group identified three eDNA derived AHL lactonases that inhibit biofilm formation of Pseudomonas aeruginosa via functional metagenomic assays. However, these functional screening methods require a good design of probes that detect molecules being synthesized and depend on the ability to express metagenomes in a host organism system.

In contrast, homology metagenomic studies led to a faster discovery of genes that have homologous sequences as the previously known genes that are responsible for the biosynthesis of biologically active molecules. As soon as the genes are sequenced, scientists can compare thousands of bacterial genomes simultaneously. The advantage over functional metagenomic assays is that homology metagenomic studies do not require a host organism system to express the metagenomes, thus this method can potentially save the time spent on analyzing nonfunctional genomes. These also led to the discovery of several novel proteins and small molecules. For example, Banik et al. screened for clones containing genes associated with the synthesis of teicoplanin and vancomycin-like glycopeptide antibiotics and found two new biosynthetic gene clusters. In addition, an *in silico* examination from the Global Ocean Metagenomic Survey found 20 new lantibiotic cyclases.

There are challenges to metagenomic approaches to discover new biologically active molecules. Only 40% of enzymatic activities present in a sample can be expressed in *E. coli.*. In addition, the purification and isolation of eDNA is essential but difficult when the sources of obtained samples are poorly understood. However, collaborative efforts from individuals from diverse fields including bacterial genetics, molecular biology, genomics, bioinformatics, robots, synthetic biology, and chemistry can solve this problem together and potentially lead to the discovery of many important biologically active molecules.

Protein Phosphorylation

Posttranslational modification of proteins with phosphate groups has proven to be a key regulatory step throughout all biological systems. Phosphorylation events, either phosphorylation by protein kinases or dephosphorylation by phosphatases, result in protein activation or deactivation. These events have an immense impact on the regulation of physiological pathways, which makes the ability to dissect and study these pathways integral to understanding the details of cellular processes. There exist a number of challenges—namely the sheer size of the phosphoproteome, the fleeting nature of phosphorylation events and related physical limitations of classical biological and biochemical techniques—that have limited the advancement of knowledge in this area. A recent review provides a detailed examination of the impact of newly developed chemical approaches to dissecting and studying biological systems both in vitro and in vivo.

Through the use of a number of classes of small molecule modulators of protein kinases, chemical biologists have been able to gain a better understanding of the effects of protein phosphorylation. For example, nonselective and selective kinase inhibitors, such as a class of pyridinylimidazole compounds described by Wilson, et al., are potent inhibitors useful in the dissection of MAP kinase signaling pathways. These pyridinylimidazole compounds function by targeting the ATP binding pocket. Although this approach, as well as related approaches, with slight modifications, has proven effective in a number of cases, these compounds lack adequate specificity for more general applications. Another class of compounds, mechanism-based inhibitors, combines detailed knowledge of the chemical mechanism of kinase action with previously utilized inhibition motifs. For example, Parang, et al. describe the development of a "bisubstrate analog" that inhibits kinase action by binding both the conserved ATP binding pocket and a protein/peptide recognition site on the specific kinase. While there is no published in vivo data on compounds of this type, the structural data acquired from in vitro studies have expanded the current understanding of how a number of important kinases recognize target substrates. Interestingly, many research groups utilized ATP analogs as a chemical probe to study kinases and identify their substrates.

The development of novel chemical means of incorporating phosphomimetics into proteins has provided important insight into the effects of phosphorylation events. Historically, phosphorylation events have been studied by mutating an identified phosphorylation site (serine, threonine or tyrosine) to an amino acid, such as alanine, that cannot be phosphorylated. While this approach has been successful in some cases, mutations are permanent in vivo and can have potentially detrimental effects on protein folding and stability. Thus, chemical biologists have developed new ways of investigating protein phosphorylation. By installing phospho-serine, phospho-threonine or analogous phosphonate mimics into native proteins, researchers are able to perform in vivo studies to investigate the effects of phosphorylation by extending the amount of time a phosphorylation event occurs while minimizing the often-unfavorable effects of mutations. Protein semisynthesis, or more specifically expressed protein ligation (EPL), has proven to be successful techniques for synthetically producing proteins that contain phosphomimetic molecules at either the C- or the N-terminus. In addition, researchers have built upon an established technique in which one can insert an unnatural amino acid into a peptide sequence by charging synthetic tRNA that recognizes a nonsense codon with an unnatural amino acid. Recent developments indicate that this technique can also be employed in vivo, although, due to permeability issues, these in vivo experiments using phosphomimetic molecules have not yet been possible.

Advances in chemical biology have also improved upon classical techniques of imaging kinase action. For example, the development of peptide biosensors—peptides containing incorporated fluorophore molecules—allowed for improved temporal resolution in in vitro binding assays. Experimental limitations, however, prevent this technique from being effectively used in vivo. One of the most useful techniques to study kinase action is Fluorescence Resonance Energy Transfer (FRET). To utilize FRET for phosphorylation studies, fluorescent proteins are coupled to both a phosphoamino acid binding domain and a peptide that can by phosphorylated. Upon phosphorylation or dephosphorylation of a substrate peptide, a conformational change occurs that results in a change in fluorescence. FRET has also been used in tandem with Fluorescence Lifetime Imaging Microscopy (FLIM) or fluorescently conjugated antibodies and flow cytometry to provide a detailed, specific, quantitative results with excellent temporal and spatial resolution.

Through the augmentation of classical biochemical methods as well as the development of new tools and techniques, chemical biologists have improved accuracy and precision in the study of protein phosphorylation.

Metal Complexes in Medicine

Metal complexes have many characteristics that can be advantageous in drug design. In comparison to organic-based medicines, metal complexes have many more Coordination Numbers, geometries, and oxidation/reduction states that can be used to make structures that interact with targets in unique ways unavailable to most organic molecules. In addition, the cationic metal is advantageous in complexing with charged targets within biological systems like the phosphate backbone of DNA. Targets of metal-based medicines include DNA, proteins, and enzymes. Each target tupe is described in turn below.

Metal Complexes Targeting DNA

DNA has been the primary target of metal complexes due to the ability of cationic metal interacting with the anionic backbone of DNA. The anticancer chemotherapy drug cisplatin covalently binds to DNA, which disrupts transcription and leads to programmed cell death. Assuming early detection, cisplatin cures almost all cases of testicular cancer. This drug, however, has severe side effects and great effort is being made to improve drug delivery including attachment to single-walled carbon nanotubes, encapsulation in proteins cages, among other clever strategies.

Another major effort for anticancer metal-based drugs centers around stabilization of the G-quadruplex of DNA. In general, these drugs have a non-covalent interaction with the G-quadruplex as well as a planar positively charged structure.

Metal Complexes Targeting Enzymes and Proteins

Though DNA has been a primary target for inorganic medicines, enzymes, and proteins also can be modulated through interactions with these compounds. Metal complexes can interact with the amino acids with the highest reduction potential (histidine, cysteine, and selenocysteine). Metals used in such complexes include gold, platinum, ruthenium, vanadium, cobalt and others. Several new potential therapeutic complexes are currently in the process of discovery and investigation.

Gold

Some gold complexes are showing potential as medicines. A rheumatoid arthritis drug (auranofin, a gold(I) phosphine complex) has shown value in treating parasitic disease through inhibiting thioredoxin glutathione reductase.

Platinum

Along with cisplatin, many other platinum complexes are potential therapeutics. Like auranofin, terpyridine platinum inhibits thioredoxin reductase with nanomolar IC50. This complex also is an inhibitor of the common target enzyme topoisomerase I. Yet another family of complexes with potential anticancer properties are dichloro(SMP)-platinum(II) complexes. These complexes target the matrix metalloproteinase, where the complex coordinates with amino acids of the enzyme in the coordination sites previously held by chlorides, and through the smp ligand. As seen by these few examples, platinum complexes are a particularly active area of research for metal-based medicines.

Ruthenium

Ruthenium complexes have anticancer activity. A library of glutathione transferase inhibitors were created through a combination of ethacrynic acid (a known inhibitor of the enzyme) and ruthenium complexes.

Vanadium

Vanadium complexes have been used in multiple therapeutic settings. A new area in which vanadium may have a great medicinal impact is through the oxovanadium porphyrin complexes. These complexes have demonstrated HIV-1 reverse transcriptase inhibition in vitro.

Issues and Outlook

Though there is currently much excitement in the field of metal-based medicines, many challenges still face researchers. One such challenge is selectivity of complexes in vivo. Many of these complexes can bind to common proteins like serum albumin in addition to other proteins with amino acids that are common in protein–metal complex interactions like histidine, cysteine, and selenocysteine. Along with selectivity issues, much is yet unknown about mechanisms through which metal complexes interact with proteins. How complexing between a given metal complex and target protein or enzyme occurs is often unknown or unclear and requires much more elucidation before truly effective metal complexes can be designed and delivered. Currently, physicians utilize very few metal-based medicines in the clinics. For example, none of the 21 drugs approved by the U.S. Food and Drug Administration (FDA) in 2008 were inorganic. However, with the success of cisplatin in cancer treatment, it is not unreasonable to anticipate more metal complexes will be actively used in the treatment of diseases.

Synthetic Biology

Synthetic biology focuses on the manipulation of biological components to form new systems or the generation of living systems with synthetic parts. The canonical idea of synthetic biology is the creation of new life, but recently it has come to include bioengineering in terms of the use of interchangeable components to give novel outputs. In the search for modular parts, it is most facile if the building blocks contribute independently to the function of the whole unit so that the modules can be recombined in predictable ways. It is useful for synthetic biologists to define "life": in this context, to be alive an organism must be capable of Darwinian evolution – genetic mutation, self-replication and inheritance of mutations.

Synthetic Cells

J. Craig Venter's group has created the first "synthetic" cell – the first cells to exist with fully synthetic DNA. Venter was able to manipulate the synthetic genome to dictate the proteins expressed in the organism. Note that these were not fully synthetic cells but that the synthetic DNA was able to take over all metabolic processes necessary for cell survival and proliferation.

DNA as Interchangeable Parts

DNA is composed of repeating modular units consisting of an anion phosphate group that forms the polyanion backbone, and nucleotide base pairs that engage in Watson-Crick base pairing to form the double strand. Because the molecular recognition of DNA is based mostly on the polyanion backbone, the nucleotides can be modified without altering the structural integrity of the DNA. Steven Benner's group has generated an artificial genetic alphabet of eight new base pairs that can be amplified by polymerase chain reaction; this indicates that these base pairs can be used in systems that undergo Darwinian evolution.

Proteins as Interchangeable Parts

Amino Acids

Amino acids are poor modular building-blocks because they do not act independently and there is a fundamental lack of understanding about the relationship between linear amino acid sequences and the folding and functionality of proteins. Chemical biologists have been able to model, design, and synthesize peptides and evaluate their function.

Protein Secondary Structure

Modules consisting of protein secondary structure can be designed to perform specific functions; for example, it has been demonstrated that alpha helices can be used as functional peptide catalysts. The Ghadiri group has created a template peptide that promotes the ligation of two modified helices by bringing the helices into close proximity by specifically designed hydrophobic interactions of the helices with the template.

Folded Proteins

Fully folded proteins can be combined in novel ways to generate specific non-natural outcomes. This is highly useful commercially from drug development to the production of polymers – one can imagine the economic benefits if scientists can design systems in which proteins catalyze reactions without the necessity of excessive human intervention to produce commercially relevant materials. For example, the Keasling group has developed a series of proteins that catalyze conversion of acetyl CoA, a common cellular metabolite, into a precursor for the potent antimalarial drug artemisinin.

Modifying Molecular Switches

Signaling pathways can be modified to be turned on or off by non-natural ligands or inputs to the system. For instance, systems can be modified so that they are autoinhibited by non-natural proteins that release their inhibition upon binding with a specific molecule that is different from the natural signaling molecule of the path. This allows new approaches to studying signal circuits specifically and with user-designed inputs.

Chemical Approaches to Stem-cell Biology

Advances in stem-cell biology have typically been driven by discoveries in molecular biology and genetics. These have included optimization of culture conditions for the maintenance and differentiation of pluripotent and multipotent stem-cells and the deciphering of signaling circuits that control stem-cell fate. However, chemical approaches to stem-cell biology have recently received increased attention due to the identification of several small molecules capable of modulating stem-cell fate in vitro. A small molecule approach offers particular advantages over traditional methods in that it allows a high degree of temporal control, since compounds can be added or removed at will, and tandem inhibition/activation of multiple cellular targets.

Small molecules that modulate stem-cell behavior are commonly identified in high-throughput screens. Libraries of compounds are screened for the induction of a desired phenotypic change in cultured stem-cells. This is usually observed through activation or repression of a fluorescent reporter or by detection of specific cell surface markers by FACS or immunohistochemistry. Hits are then structurally optimized for activity by the synthesis and screening of secondary libraries. The cellular targets of the small molecule can then be identified by affinity chromatography, mass spectrometry, or DNA microarray.

A trademark of pluripotent stem-cells, such as embryonic stem-cells (ESCs), is the ability to self-renew indefinitely. The conventional use of feeder cells and various exogenous growth factors in the culture of ESCs presents a problem in that the resulting highly variable culture conditions make the long-term expansion of un-differentiated ESCs challenging. Ideally, chemically defined culture conditions could be developed to maintain ESCs in a pluripotent state indefinitely. Toward this goal, the Schultz and Ding labs at the Scripps Research Institute identified a small molecule that can preserve the long-term self-renewal of ESCs in the absence of feeder cells and other exogenous growth factors. This novel molecule, called pluripotin, was found to simultaneously inhibit multiple differentiation inducing pathways.

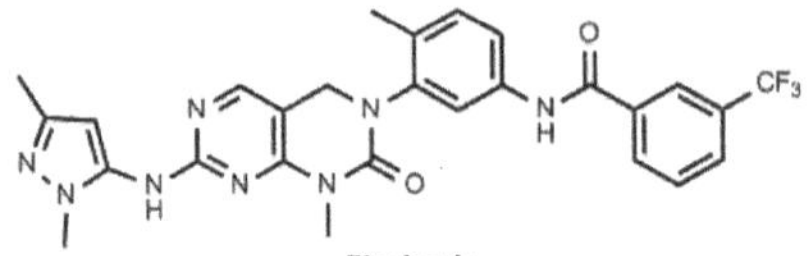

Pluripotin

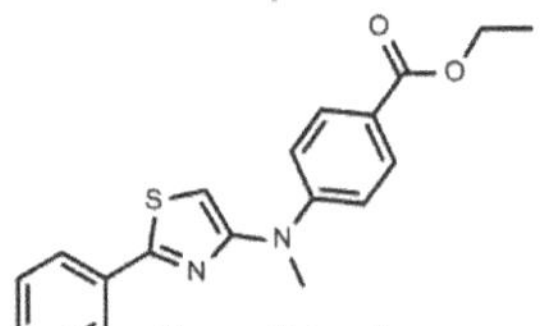

Neuropathioazol

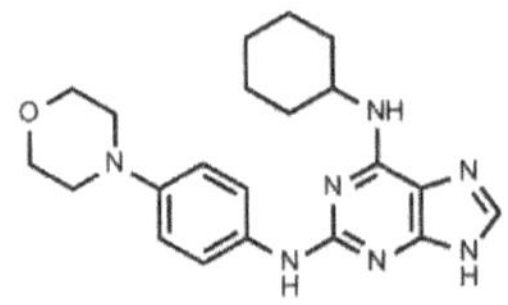

Reversine

Small molecule modulators of stem-cell fate.

The utility of stem-cells is in their ability to differentiate into all cell types that make up an organism. Differentiation can be achieved in vitro by favoring development toward a particular cell type through the addition of lineage specific growth factors, but this process is typically non-specific and generates low yields of the desired phenotype. Alternatively, inducing differentiation by small molecules is advantageous in that it allows for the development of completely chemically defined conditions for the generation of one specific cell type. A small molecule, neuropathiazol, has been identified which can specifically direct differentiation of multipotent neural stem cells into neurons. Neuropathiazol is so potent that neurons develop even in conditions that normally favor the formation of glial cells, a powerful demonstration of controlling differentiation by chemical means.

Because of the ethical issues surrounding ESC research, the generation of pluripotent cells by reprogramming existing somatic cells into a more "stem-like" state is a promising alternative to the use of standard ESCs. By genetic approaches, this has recently been achieved in the creation of ESCs by somatic cell nuclear transfer and the generation of induced pluripotent stem-cells by viral transduction of specific genes. From a therapeutic perspective, reprogramming by chemical means would be safer than genetic methods because induced stem-cells would be free of potentially dangerous transgenes. Several examples of small molecules that can de-differentiate somatic cells have been identified. In one report, lineage-committed myoblasts were treated with a compound, named reversine, and observed to revert to a more stem-like phenotype. These cells were then shown to be capable of differentiating into osteoblasts and adipocytes under appropriate conditions.

Stem-cell therapies are currently the most promising treatment for many degenerative diseases. Chemical approaches to stem-cell biology support the development of cell-based therapies by enhancing stem-cell growth, maintenance, and differentiation in vitro. Small molecules that have been shown to modulate stem-cell fate are potential therapeutic candidates and provide a natural lean-in to pre-clinical drug development. Small molecule drugs could promote endogenous stem-cells to differentiate, replacing previously damaged tissues and thereby enhancing the body's own regenerative ability. Further investigation of molecules that modulate stem-cell behavior will only unveil new therapeutic targets.

Fluorescence for Assessing Protein Location and Function

Fluorophores and Techniques to Tag Proteins

Organisms are composed of cells that, in turn, are composed of macromolecules, e.g. proteins, ribosomes, etc. These macromolecules interact with each other, changing their concentration and suffering chemical modifications. The main goal of many biologists is to understand these interactions, using MRI, ESR, electrochemistry, and fluorescence among others. The advantages of fluorescence reside in its high sensitivity, non-invasiveness, safe detection, and ability to modulate the fluorescence signal. Fluorescence was observed mainly from small organic dyes attached to antibodies to the protein of interest. Later, fluorophores could directly recognize organelles, nucleic acids, and important ions in living cells. In the past decade, the discovery of green fluorescent protein (GFP), by Roger Y. Tsien, hybrid system and quantum dots have enable assessing protein location and function more precisely. Three main types of fluorophores are used: small organic dyes, green fluorescent proteins, and quantum dots. Small organic dyes usually are less than 1 kD, and have been modified to increase photostability, enhance brightness, and reduce self-quenching. Quantum dots have very sharp wavelength, high molar absorptivity and quantum yield. Both organic dyes and quantum dyes do not have the ability to recognize the protein of interest without the aid of antibodies, hence they must use immunolabeling. Since the size of the fluorophore-targeting complex typically exceeds 200 kD, it might interfere with multiprotein recognition in protein complexes, and other methods should be use in parallel. An advantage includes diversity of properties and a limitation is the ability of targeting in live cells. Green fluorescent proteins are genetically encoded and can be covalently fused to your protein of interest. A more developed genetic tagging technique is the tetracysteine biarsenical system, which requires modification of the targeted sequence that includes four cysteines, which binds membrane-permeable biarsenical molecules, the green and the red dyes "FlAsH" and "ReAsH", with picomolar affinity. Both fluorescent proteins and biarsenical tetracysteine can be expressed in live cells, but present major limitations in ectopic expression and might cause lose of function. Giepmans shows parallel applications of targeting methods and fluorophores using GFP and tetracysteine with ReAsH for α-tubulin and β-actin, respectively. After fixation, cells were immunolabeled for the Golgi matrix with QD and for the mitochondrial enzyme cytochrome with Cy5.

Protein Dynamics

Fluorescent techniques have been used assess a number of protein dynamics including protein tracking, conformational changes, protein–protein interactions, protein synthesis and turnover, and enzyme activity, among others.

Three general approaches for measuring protein net redistribution and diffusion are single-particle tracking, correlation spectroscopy and photomarking methods. In single-particle tracking, the individual molecule must be both bright and sparse enough to be tracked from one video to the other. Correlation spectroscopy analyzes the intensity fluctuations resulting from migration of fluorescent objects into and out of a small volume at the focus of a laser. In photomarking, a fluorescent protein can be dequenched in a subcellular area with the use of intense local illumination and the fate of the marked molecule can be imaged directly. Michalet and coworkers used quantum dots for single-particle tracking using biotin-quantum dots in HeLa cells.

One of the best ways to detect conformational changes in proteins is to sandwich said protein between two fluorophores. FRET will respond to internal conformational changes result from reorientation of the fluorophore with respect to the other. Dumbrepatil sandwiched an estrogen receptor between a CFP (cyan fluorescent protein) and a YFP (yellow fluorescent protein) to study conformational changes of the receptor upon binding of a ligand.

Fluorophores of different colors can be applied to detect their respective antigens within the cell. If antigens are located close enough to each other, they will appear colocalized and this phenomenon is known as colocalization. Specialized computer software, such as CoLocalizer Pro, can be used to confirm and characterize the degree of colocalization.

FRET can detect dynamic protein–protein interaction in live cells providing the fluorophores get close enough. Galperin *et al.* used three fluorescent proteins to study multiprotein interactions in live cells.

Tetracysteine biarsenical systems can be used to study protein synthesis and turnover, which requires discrimination of old copies from new copies. In principle, a tetracysteine-tagged protein is labeled with FlAsH for a short time, leaving green labeled proteins. The protein synthesis is then carried out in the presence of ReAsH, labeling the new proteins as red.

One can also use fluorescence to see endogenous enzyme activity, typically by using a quenched activity based proteomics (qABP). Covalent binding of a qABP to the active site of the targeted enzyme will provide direct evidence concerning if the enzyme is responsible for the signal upon release of the quencher and regain of fluorescence.

The unique combination of high spatial and temporal resolution, nondestructive compatibility with living cells and organisms, and molecular specificity insure that fluorescence techniques will remain central in the analysis of protein networks and systems biology.

Applications of DNA Microarrays in Chemical Biology

Planar surfaces functionalized with single- or double-strand nucleic acids have enabled researchers to address a variety of salient biological and biochemical questions in recent years. The general architecture of modern DNA microarrays reflects the historical progression from the sequence-specific probing of whole chromosomes immobilized on glass slides (as early as 1961 with fluorescent *in situ* hybridization) and the low-density porous membrane arrays available since the early 1990s, to the high-density (10^2-10^4 features/mm^2) solid support platforms that exist today. The massively parallel processing capabilities of these picomolar-range contemporary arrays provide for the generation of large data sets and multiplexed analysis. Furthermore, several top-down and bottom-up assembly methodologies provide researchers with the option for "in-house" production of arrays from custom oligonucleotide libraries or the use of commercial genome chips, notably those developed by Affymetrix and Agilent Technologies.

DNA microarrays can be used to conduct several general types of experiments, most of which relying on the hybridization of fluorescently labeled single-strand DNA molecules isolated from a biological sample to their single-strand complement probes presented on an array. One of the earliest conceived applications for DNA microarrays was for single-nucleotide polymorphism (SNP) genotyping. Since SNPs are a "quick and dirty" approach to detect genetic indicators of pathologies and lineages, arrays in theory provide a facile method for diagnosis; this was confirmed experimentally in the late 1990s in the successful SNP analysis of human tumors. Although there are currently commercially available arrays (e.g. bovine mapping chips) to characterize SNPs, it seems likely that the nascent availability of high-throughput and low-cost pyrosequencing will become the preferred method of recognition, or replace the need for SNP detection altogether with rapid whole-genome sequencing.

A different application of microarray technology that has become the gold standard for RNA analysis in recent years is the widespread utilization of expression microarrays, or "gene chips". Gene chip preparation calls for the quantitative reverse transcription of the total cellular RNA pool into labeled and fragmented single-strand DNA prior to hybridization-based capture. Up- and down-regulation of genes in response to stressors or disease states are quantitatively compared in cell lines and organisms. Coupled expression microarray and quantitative proteomics experiments have allowed for the in-depth exploration of the oftentimes non-linear relationship between the abundance of a particular transcribed message and that of its corresponding translated protein. These integrative studies, partially enabled by quantitative DNA microarray technology, have been successfully applied to a variety of biological systems, including yeast, bovine, mouse, bacterial, and human. The expression analysis community has amassed such a significant amount of expression microarray data that they are freely available in public databases.

These types of surfaces can also be used to analyze DNA-protein interactions on a genome-wide scale via chromatin immunoprecipitation, followed by an array-based analysis of the DNA (ChIP-chip). ChIP-chip experiments are enabled by the co-purification of a DNA-binding protein of interest with its corresponding genomic loci when a cross-linked chromatin extract is probed with an antibody to said protein. After purification, amplification and labeling, the DNA is applied to a microarray representing the entire genome; the data are plotted as a histogram that resolves the specific genomic regions associated with that protein. ChIP-chip experiments have provided the scientific community with a wealth of information about the steady-state genomic locations of DNA-binding proteins, such as histones, transcription factors, and polymerase machinery, and have also been successfully applied to studies on the dynamics of transcription factor binding. The data from these experiments may be further manipulated to computationally derive consensus binding sequences for some transcription factors, giving the opportunity for insight into the in vivo behavior of the factor, deeper than simple information about localization.

DNA microarrays are also amenable to the direct analysis of protein–DNA interactions in kinetic binding assays as analyzed by surface plasmon resonance (SPR). This experimental approach also relies on single-strand DNA immobilized on a high-density array; however, the quantitative readout is based on a change in the optical properties of the DNA-functionalized surface when a protein flowed over the surface binds to the sequence in a particular surface feature. DNA-functionalized arrays analyzed with SPR in this way have yielded kinetic data regarding fundamental molecular biological processes. Recently, SPR analysis of a DNA microarray and components of the DNA replication machinery helped to elucidate the biochemical nuances of the replication fork.

High-density DNA microarrays have emerged as an important component of the chemical biology toolkit. The existing technology allows for the construction of customizable, as well as general, arrays and provides researchers with the opportunity to generate robust data from many different types of biological inputs. Considering the relatively recent shift in the scientific community away from binary perturbation/readout studies and toward "big science" and large data sets, it seems likely that DNA microarrays will continue to enable pertinent biological research for many years to come.

Applications of Chemical Biology in Drug Discovery

Chemical biology approaches can help answer important questions of relevance to small molecule drug discovery projects. This includes questions related to the characterization of protein targets and the molecular pharmacology of small molecule drugs that modulate target function.

Chemical Biology Approaches to Characterize Protein Targets

Protein Target Characterization		
Question	**Why ask?**	**Chemical Biology Technique**
What is my target?	Enables target-based drug discovery on hits/mechanisms from phenotypic screens.	Affinity chemoproteomics; Mutagenesis; Phenotypic screening; Chemogenomics; Proteomics
What is the subcellular distribution of my target?	Location of target may influence screening assays or inhibitor design; active and inactive species of the target may localize to different cellular regions; target may be activated by particular environment owing to localization to a particular organelle (for example, acidic lysosome)	Imaging; Microscopy
Does my target exist in multiple forms, and does this vary across tissues and species?	May reveal species differences, splice variants, relevance of full length target vs. catalytic domains for screening; splice variants of the target may have different protein domains, activity, cellular location, tissue distribution, and affinity for substrate; knowing correct sequence cDNA enables potential to express recombinant protein or to generate overexpression cell line which can inform choice of primary assay and screening sequence.	Computational biology; Genotype-tissue-expression analysis; Proteomics
What is the endogenous ligand for my target and its concentration in the diseased state?	Allows one to understand what biological pathways are being modulated (one of the keys for establishing a physiologically relevant assay); allows one to theorize what will happen if endogenous substrate levels are increased by inhibiting the desired target.	Immunoprecipitation; Metabolomics; Peptide Microarrays; Peptidomics
How well characterized is the interaction of the endogenous substrate with my target?	Informs chemical feasibility, screening strategy and medicinal chemistry approach (for example, substrate concentration and Km for enzyme target)	Biochemical enzyme and cellular activity assays
Is my target post-translationally modified?	Can rationalize differences in affinity/efficacy between biochemical in vitro and cellular/in vivo systems; informs primary assay choice and screen sequence.	Chemical Probe; Immunodetection; Metabolomics; Proteomics; Phosphoproteomics

What other proteins are influential in regulating substrate concentrations?	Can represent alternate strategies for modulating the desired pathway. Can lead to complimentary pharmacology.	Computational Biology; RNAi
What is the turnover of my target and is this affected by my compound?	Especially important to understand the cellular efficiencies of covalent modalities.	Chemical probe; Immunoprecipitation; Pulse/chase; SILAC/MS
What is the abundance of my target, and does it vary?	Abundance of a protein might vary by tissue, disease state, or as a response to drug action.	Targeted quantitative proteomics coupled with immunocapture
Does my target interact with other proteins and what is the consequence of these interactions?	Can lead to a better understanding of signaling pathway; informs screening assay; could be important that protein forms hetero or homodimers.	Immunoprecipitation
What potential off-targets are most closely related to my target sequence and function?	Informs screening sequence design; important to consider not only targets that are closely related in terms of binding site sequence, but also those most closely related in a chemogenomic sense; understand which tissues express off-targets.	Computational Biology

Protein Target Characterization

- What is my target?

  - *Affinity chemoproteomics* can be used to identify the targets of compounds that have shown activity in a phenotypic or pathway screen. For example, this approach was used to identify of MTH1 as a potential anticancer target of the (S) form of Crizotinib. Affinity chemoproteomics technology was also used to identify the BET bromodomains as the molecular targets of a series of small molecule modulators of Apolipoprotein A1 (ApoA1).

  - *Mutagensis* is important target validation approach is to generate a catalytically-dead target protein to tease apart the role of the enzymatic and scaffolding functions. This method is often used in kinase validation experiments where the pharmacological performance of a putative inhibitor is tested in cells with the kinase knocked out and replaced through transfection of the recombinant wild-type protein or its kinase dead mutant. This approach was used recently to confirm DCLK1 as the functional anti-proliferative target of a previously developed LRRK2 inhibitor for Parkinson's disease

- Does my target exist in multiple related forms and does this vary across different tissues and species?

  - *Genotype-tissue expression project (GTEx)* has generated a database of mRNA expression levels across multiple tissues. An illustration of the use of this resource shows that PDE4B is expressed in multiple forms that vary across human tissues. Being aware of these different forms has enabled development of a selective inhibitor targeting PDE4B in the brain through taking advantage of differences in sequences in the long and short forms outside the active site.

- What is the endogenous ligand for my target and its concentration in the diseased state?

- *Metabolomics* in combination with clickable chemical reporters can enable chemoproteomic methods to be applied to substrate identification, as shown recently for protein lipidation.

- Is my target post-translationally modified?

  - Dimedone *chemical probes* that chemoselectively react with sulfenic acid residues in EGFR have shown how cysteine oxidation can be use for enzymatic regulation.

- What is the turnover of my target and is this affected by my compound?

  - *Chemical probes* of BTK were used to confirm its slow turnover rate.

  - *Immunoprecipitation* of EGFR has been used to confirm increased protein half-livey, likely due to a concomitant decrease in binding to the Cbl ubiquitin ligase responsible for regulating the degradation of EGFR. This increased half-life could result in a need to change the projected dose or frequency. In addition to mutations altering the turnover rate of a protein, small molecule drugs can also change the turnover and thus the amount of protein in the cell. For instance, many kinases are known to be clients of the Hsp90-Cdc37 chaperone system and recent studies have found that some ATP-competitive inhibitors disrupt the ability of Cdc37 to bind to the target kinase and recruit it to Hsp9015.

- What potential off- targets are most closely related to my target sequence and function?

  - *Computational biology* and structural bioinformatic analyses of protein sequences and binding pocket topology can be used to delineate which proteins are similar to a desired target's biological sequence as well as its binding site shape. Additionally, chemoinformatic approaches using in silico tools to predict potential off-targets based on the structural similarity between a novel small molecule and previously known compounds with different pharmacology are also useful.

Chemical Biology Approaches to Characterize Molecular Pharmacology

Molecular Pharmacology Characterization		
Question	**Why ask?**	**Chemical Biology Technique**
What target(s) and off-targets does my molecule bind?	Key to understanding what drives molecule efficacy and potential safety liabilities; allows team to understand mechanism of action, pathways affected and target that drives efficacy; enables target-based drug discovery and understanding or improvement of potency and selectivity.	Activity probes; Affinity capture; ABPP; DARTS; Photoaffinity; Proteomics; Protein microarrays; SPROX, Thermal aggregation; Y3H assays
Where does my molecule bind?	Characterization of binding kinetics and binding site may be key to functional translation and pharmacokinetic-pharmacodynamic relationships.	Hydrogen-deuterium exchange
How does my molecule bind?	Characterization of binding kinetics and binding site may be key to functional translation and PK/PD.	Kinetics
What is the tissue distribution of my molecule?	Assuming plasma exposure reflects target tissue exposure is often incorrect. Affects Pillar 1.	MALDI MS; PK: Protein binding; Tissue distribution

What are the functional consequences to my target when my molecule binds?	Provides an understanding of how the molecule works; binding could modulate target degradation, stabilization, translocation or interactions with other proteins.	Immunodetection; Proteomics
How much target occupancy do I need to drive my relevant biological phenotype?	Allows team to develop Pillar II confidence and link to Pillar III. Key enabler to help define Ceff.	Occupancy probes; "Three pillars in a tube"

Molecular Pharmacology Characterization

- What target(s) and off-targets does my molecule bind?

  - *Activity probes* such as nucleotide acyl phosphates (KiNativ™) probes that react chemoselectively with the catalytic lysine of >80% of all known kinases can be used to determine many of the kinase targets of a molecule.

  - *ABPP* using probes derived from covalent kinase inhibitors, have been used to determine the cellular targets of these covalent inhibitors in living cells.

  - *Chemoproteomics* was used to assess the multiple binding partners of HDAC inhibitors in protein complexes scaffolded by ELM-SANT domain subunits. *Thermal proteomics*, where one looks for proteins that show increased thermal stability at elevated temperatures due to compound binding, has been used to identify off-targets of several kinase inhibitors including the BRAF inhibitor Vemurafenib. *Affinity capture* where a chemical probe is tethered to a solid support, was used to identify cereblon (CRBN) as a thalidomide binding protein.

  - *DARTS (Drug Affinity Responsive Target Stability);* takes advantage of a reduction in the protease susceptibility of the target protein upon drug binding. This technique was used to confirm binding biological binding partners for a number of molecules: rapamycin and FKBP with FKBP12 and resveratrol with eIF4A.

  - *SPROX (Stability of Proteins from Rates of Oxidation);* utilizes the chemical denaturant-dependent oxidation rates of methionine residues in a protein to report on the thermodynamic properties of its global and/or subglobal unfolding/refolding reactions.

- What is the tissue distribution of my molecule?

  - Quantitative *mass spectrometry (MS)* was recently used to determine the concentrations of the chemotherapeutic agent YM155 in various cancer cells.

- What are the functional consequences to my target when my molecule binds?

  - The use of shRNA gene silencing combined with *immunoblotting* showed that BRAF inhibitors GDC-0879 and PLX4720 block MAPK signaling in BRAF (V600E) tumors while these same compounds activate the RAF-MEK-ERK pathway in KRAS mutant and RAS/RAF wild-type tumors.

- How much target occupancy do I need to drive my relevant biological phenotype?

- A clickable covalent *occupancy probe* of fatty acid amide hydrolase (FAAH) has been used to relate target engagement to anandamide elevation and efficacy in models of rodent inflammatory and neuropathic pain. Target occupancy quantification can also be used to validate the relevance of a drug target identified from phenotypic screening. A clickable covalent probe of the mRNA decapping enzyme DcpS, which used a sulfonyl fluoride warhead to target a reactive tyrosine in the binding site of the enzyme, enabled the assessment of DcpS target engagement of a diaminoquinazoline inhibitor (developed from a phenotypic screen for the treatment of spinal muscular atrophy).

Organometallic Chemistry

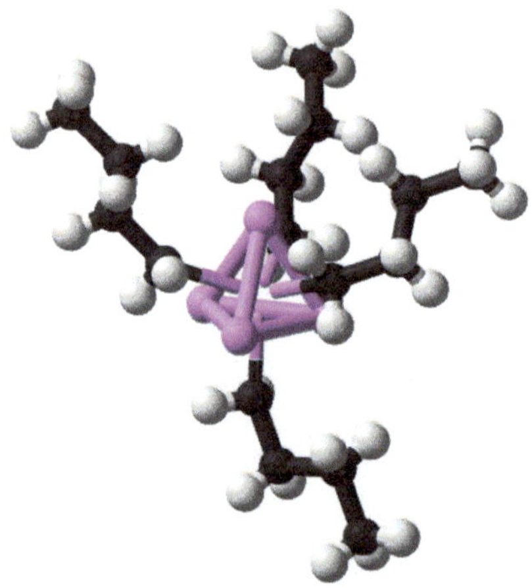

n-Butyllithium, an organometallic compound. Four lithium atoms (in purple) form a tetrahedron, with four butyl groups attached to the faces (carbon is black, hydrogen is white).

Organometallic chemistry is the study of chemical compounds containing at least one bond between a carbon atom of an organic compound and a metal, including alkaline, alkaline earth, transition metal, and other cases. Moreover, some related compounds such as transition metal hydrides and metal phosphine complexes are often included in discussions of organometallic compounds. The field of organometallic chemistry combines aspects of traditional inorganic and organic chemistry.

Organometallic compounds are widely used both stoichiometrically in research and industrial chemical reactions, as well as in the role of catalysts to increase the rates of such reactions (e.g., as in uses of homogeneous catalysis), where target molecules include polymers, pharmaceuticals, and many other types of practical products.

Organometallic Compounds

Organometallic compounds are distinguished by the prefix "organo-" e.g. organopalladium compounds. Examples of such organometallic compounds include all Gilman reagents, which contain lithium and copper. Tetracarbonyl nickel, and ferrocene are examples of organometallic compounds containing transition metals. Other examples include organomagnesium compounds like iodo(methyl)magnesium MeMgI, dimethylmagnesium (Me_2Mg), and all Grignard reagents; organolithium compounds such as n-butyllithium (n-BuLi), organozinc compounds such as diethylzinc (Et_2Zn) and chloro(ethoxycarbonylmethyl)zinc ($ClZnCH_2C(=O)OEt$); and organocopper compounds such as lithium dimethylcuprate ($Li^+[CuMe_2]^-$).

The term "metalorganics" usually refers to metal-containing compounds lacking direct metal-carbon bonds but which contain organic ligands. Metal beta-diketonates, alkoxides, and dialkylamides are representative members of this class.

In addition to the traditional metals, lanthanides, actinides, and semimetals, elements such as boron, silicon, arsenic, and selenium are considered to form organometallic compounds, e.g. organoborane compounds such as triethylborane (Et_3B).

- Representative Organometallic Compounds

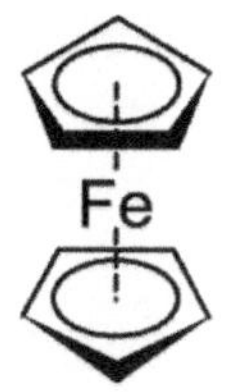

Ferrocene is an archetypal organoiron complex. It is an air-stable, sublimable compound.

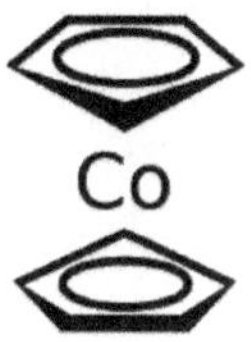

Cobaltocene is a structural analogue of ferrocene, but is highly reactive toward air.

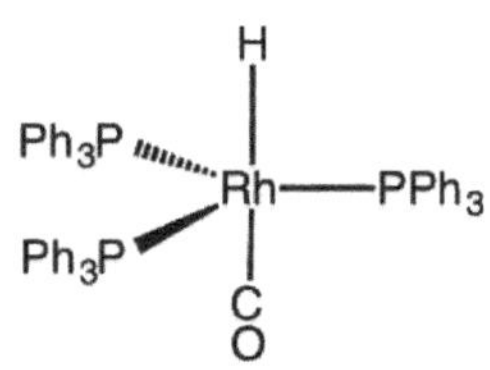

Tris(triphenylphosphine)rhodium carbonyl hydride is used in the commercial production of many aldehyde-based fragrances.

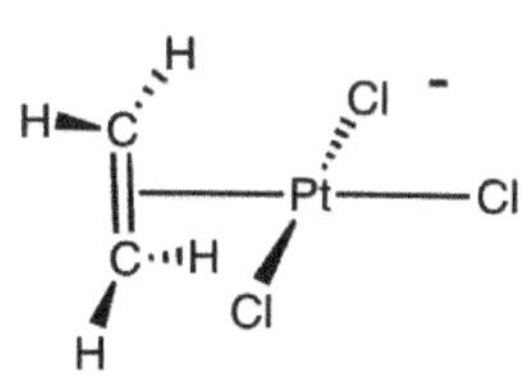

Zeise's salt is an example of a transition metal-alkene complex.

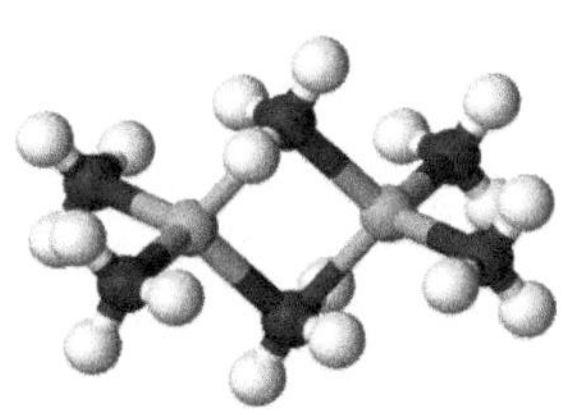

Trimethylaluminium is an organometallic compound with a bridging methyl group. It is used in the industrial production of some alcohols

Coordination Compounds with Organic Ligands

Many complexes feature coordination bonds between a metal and organic ligands. The organic ligands often bind the metal through a heteroatom such as oxygen or nitrogen, in which case such compounds are considered coordination compounds. However, if any of the ligands form a direct M-C bond, then complex is usually considered to be organometallic, e.g., $[(C_6H_6)Ru(H_2O)_3]^{2+}$. Furthermore, many lipophilic compounds such as metal acetylacetonates and metal alkoxides are called "metalorganics."

Many organic coordination compounds occur naturally. For example, hemoglobin and myoglobin contain an iron center coordinated to the nitrogen atoms of a porphyrin ring; magnesium is the center of a chlorin ring in chlorophyll. The field of such inorganic compounds is known as bioinorganic chemistry. In contrast to these coordination compounds, methylcobalamin (a form of Vitamin B_{12}), with a cobalt-methyl bond, is a true organometallic complex, one of the few known in biology. This subset of complexes are often discussed within the subfield of bioorganometallic chemistry. Illustrative of the many functions of the B_{12}-dependent enzymes, the MTR enzyme catalyzes the transfer of a methyl group from a nitrogen on N5-methyl-tetrahydrofolate to the sulfur of homocysteine to produce methionine.

The status of compounds in which the canonical anion has a delocalized structure in which the negative charge is shared with an atom more electronegative than carbon, as in enolates, may vary with the nature of the anionic moiety, the metal ion, and possibly the medium; in the absence of direct structural evidence for a carbon–metal bond, such compounds are not considered to be organometallic.

Structure and Properties

The metal-carbon bond in organometallic compounds are generally highly covalent. For highly electropositive elements, such as lithium and sodium, the carbon ligand exhibits carbanionic character, but free carbon-based anions are extremely rare, an example being cyanide.

Concepts and Techniques

As in other areas of chemistry, electron counting is useful for organizing organometallic chemistry. The 18-electron rule is helpful in predicting the stabilities of metal carbonyls and related compounds. Most organometallic compounds do not however follow the 18e rule. Chemical bonding and reactivity in organometallic compounds is often discussed from the perspective of the isolobal principle.

As well as X-ray diffraction, NMR and infrared spectroscopy are common techniques used to determine structure. The dynamic properties of organometallic compounds is often probed with variable-temperature NMR and chemical kinetics.

- oxidative addition and reductive elimination
- transmetalation
- carbometalation

- hydrometalation
- electron transfer
- beta-hydride elimination
- organometallic substitution reaction
- carbon-hydrogen bond activation
- cyclometalation
- migratory insertion
- nucleophilic abstraction-->

History

Early developments in organometallic chemistry include Louis Claude Cadet's synthesis of methyl arsenic compounds related to cacodyl, William Christopher Zeise's platinum-ethylene complex, Edward Frankland's discovery of dimethyl zinc, Ludwig Mond's discovery of $Ni(CO)_4$, and Victor Grignard's organomagnesium compounds. The abundant and diverse products from coal and petroleum led to Ziegler-Natta, Fischer-Tropsch, hydroformylation catalysis which employ CO, H_2, and alkenes as feedstocks and ligands.

Recognition of organometallic chemistry as a distinct subfield culminated in the Nobel Prizes to Ernst Fischer and Geoffrey Wilkinson for work on metallocenes. In 2005, Yves Chauvin, Robert H. Grubbs and Richard R. Schrock shared the Nobel Prize for metal-catalyzed olefin metathesis.

Organometallic Chemistry Timeline

- 1760 Louis Claude Cadet de Gassicourt investigates inks based on cobalt salts and isolates cacodyl from cobalt mineral containing arsenic
- 1827 William Christopher Zeise produces Zeise's salt; the first platinum / olefin complex
- 1848 Edward Frankland discovers diethylzinc
- 1863 Charles Friedel and James Crafts prepare organochlorosilanes
- 1890 Ludwig Mond discovers nickel carbonyl
- 1899 Introduction of Grignard reaction
- 1899 John Ulric Nef discovers alkynation using sodium acetylides.
- 1900 Paul Sabatier works on hydrogenation organic compounds with metal catalysts. Hydrogenation of fats kicks off advances in food industry, see margarine
- 1909 Paul Ehrlich introduces Salvarsan for the treatment of syphilis, an early arsenic based organometallic compound

- 1912 Nobel Prize Victor Grignard and Paul Sabatier
- 1930 Henry Gilman works on lithium cuprates, see Gilman reagent
- 1951 Walter Hieber was awarded the Alfred Stock prize for his work with metal carbonyl chemistry.
- 1951 Ferrocene is discovered
- 1963 Nobel prize for Karl Ziegler and Giulio Natta on Ziegler-Natta catalyst
- 1965 Discovery of cyclobutadieneiron tricarbonyl
- 1968 Heck reaction
- 1973 Nobel prize Geoffrey Wilkinson and Ernst Otto Fischer on sandwich compounds
- 1981 Nobel prize Roald Hoffmann and Kenichi Fukui for creation of the Woodward-Hoffman Rules
- 2001 Nobel prize W. S. Knowles, R. Noyori and Karl Barry Sharpless for asymmetric hydrogenation
- 2005 Nobel prize Yves Chauvin, Robert Grubbs, and Richard Schrock on metal-catalyzed alkene metathesis
- 2010 Nobel prize Richard F. Heck, Ei-ichi Negishi, Akira Suzuki for palladium catalyzed cross coupling reactions

Scope

Subspecialty areas of organometallic chemistry include:

- Period 2 elements: organolithium chemistry, organoberyllium chemistry, organoborane chemistry,
- Period 3 elements: organomagnesium chemistry, organoaluminum chemistry, organosilicon chemistry
- Period 4 elements: organotitanium chemistry, organochromium chemistry, organomanganese chemistry organoiron chemistry, organocobalt chemistry organonickel chemistry, organocopper chemistry, organozinc chemistry, organogallium chemistry, organogermanium chemistry
- Period 5 elements: organoruthenium chemistry, organopalladium chemistry, organosilver chemistry, organocadmium chemistry, organoindium chemistry, organotin chemistry
- Period 6 elements: organolanthanide chemistry, organoosmium chemistry, organoiridium chemistry, organoplatinum chemistry, organogold chemistry, organomercury chemistry, organothallium chemistry, organolead chemistry
- Period 7 elements: organouranium chemistry

The following is a presentation of elements of the periodic table with known compounds of carbon with other elements.

CH																	He
CLi	CBe											CB	CC	CN	CO	CF	Ne
CNa	CMg											CAl	CSi	CP	CS	CCl	CAr
CK	CCa	CSc	CTi	CV	CCr	CMn	CFe	CCo	CNi	CCu	CZn	CGa	CGe	CAs	CSe	CBr	CKr
CRb	CSr	CY	CZr	CNb	CMo	CTc	CRu	CRh	CPd	CAg	CCd	CIn	CSn	CSb	CTe	CI	CXe
CCs	CBa		CHf	CTa	CW	CRe	COs	CIr	CPt	CAu	CHg	CTl	CPb	CBi	CPo	CAt	Rn
Fr	CRa		Rf	Db	CSg	Bh	Hs	Mt	Ds	Rg	Cn	Uut	Fl	Uup	Lv	Uus	Uuo
		↓															
		CLa	CCe	CPr	CNd	CPm	CSm	CEu	CGd	CTb	CDy	CHo	CEr	CTm	CYb	CLu	
		Ac	CTh	CPa	CU	CNp	CPu	CAm	CCm	CBk	CCf	CEs	Fm	Md	No	Lr	

Chemical bonds to carbon	
Core organic chemistry	Many uses in chemistry
Academic research, but no widespread use	Bond unknown

Industrial Applications

Organometallic compounds find wide use in commercial reactions, both as homogeneous catalysis and as stoichiometric reagents For instance, organolithium, organomagnesium, and organoaluminium compounds, examples of which are highly basic and highly reducing, are useful stoichiometrically, but also catalyze many polymerization reactions.

Almost all processes involving carbon monoxide rely on catalysts, notable examples being described as carbonylations. The production of acetic acid from methanol and carbon monoxide is catalyzed via metal carbonyl complexes in the Monsanto process and Cativa process. Most synthetic aldehydes are produced via hydroformylation. The bulk of the synthetic alcohols, at least those larger than ethanol, are produced by hydrogenation of hydroformylation-derived aldehydes. Similarly, the Wacker process is used in the oxidation of ethylene to acetaldehyde.

Almost all industrial processes involving alkene-derived polymers rely on organometallic catalysts. The world's polyethylene and polypropylene are produced via both heterogeneously via Ziegler-Natta catalysis and homogeneously, e.g., via constrained geometry catalysts.

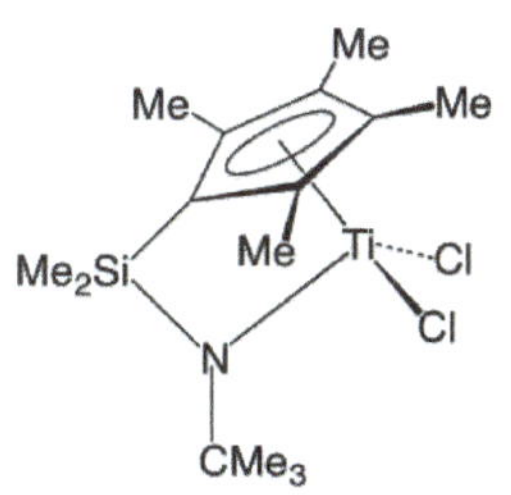

A constrained geometry organotitanium complex is a precatalyst for olefin polymerization.

Most processes involving hydrogen rely on metal-based catalysts. Whereas bulk hydrogenations, e.g. margarine production, rely on heterogeneous catalysts, For the production of fine chemicals, such hydrogenations rely on soluble organometallic complexes or involve organometallic intermediates. Organometallic complexes allow these hydrogenations to be effected asymmetrically.

Many semiconductors are produced from trimethylgallium, trimethylindium, trimethylaluminium, and trimethylantimony. These volatile compounds are decomposed along with ammonia, arsine, phosphine and related hydrides on a heated substrate via metalorganic vapor phase epitaxy (MOVPE) process in the production of light emitting diodes (LEDs).

Environmental Concerns

Natural and contaminant organometallic compounds are found in the environment. Some that are remnants of human use, such as organolead and organomercury compounds, are toxicity hazards. Tetraethyllead was prepared for use as a gasoline additive but has fallen into disuse because of lead's toxicity. Its replacements are other organometallic compounds, such as ferrocene and methylcyclopentadienyl manganese tricarbonyl (MMT). The organoarsenic compound roxarsone is a controversial animal feed additive. In 2006, approximately one million kilograms of t were produced in the U.S alone.

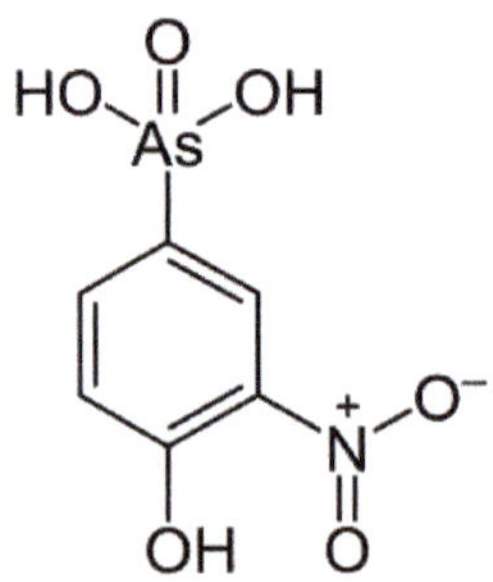

Roxarsone is an organoarsenic compound used as an animal feed.

References

- Theodore L. Brown, H. Eugene Lemay, Bruce Edward Bursten, H. Lemay. Chemistry: The Central Science. Prentice Hall; 8 edition (1999). ISBN 0-13-010310-1.
- Chemistry occupies an intermediate position in a hierarchy of the sciences by reductive level between physics and biology. Carsten Reinhardt. Chemical Sciences in the 20th Century: Bridging Boundaries. Wiley-VCH, 2001. ISBN 3-527-30271-9.
- Glaser, Christopher (1663). Traite de la chymie. Paris. as found in: Kim, Mi Gyung (2003). Affinity, That Elusive Dream - A Genealogy of the Chemical Revolution. The MIT Press. ISBN 0-262-11273-6.
- Nicolaou, K. C.; Sorensen, E. J. (1996). Classics in Total Synthesis: Targets, Strategies, Methods. Wiley. ISBN 978-3-527-29231-8.
- Allan, Barbara. Livesey, Brian (1994). How to Use Biological Abstracts, Chemical Abstracts and Index Chemicus. Gower. ISBN 978-0566075568
- Shriner, R.L.; Hermann, C.K.F.; Morrill, T.C.; Curtin, D.Y. and Fuson, R.C. (1997) The Systematic Identification of Organic Compounds. John Wiley & Sons, ISBN 0-471-59748-1
- Crabtree, Robert H. (2009). The Organometallic Chemistry of the Transition Metals (5th ed.). New York, NY: John Wiley and Sons. pp. 2, 560, and passim. ISBN 0470257628. Retrieved 23 May 2016.

- Oliveira, José; Elschenbroich, Christoph (2006). Organometallics (3., completely rev. and extended ed.). Weinheim: Wiley-VCH-Verl. ISBN 978-3-527-29390-2.
- Berg, Jeremy M.; Lippard, Stephen J. (1994). Principles of bioinorganic chemistry ([Pbk. ed.]. ed.). Mill Valley: University Science Books. ISBN 0-935702-73-3.
- Leeuwen, Piet W.N.M. van (2004). Homogeneous catalysis : understanding the art. Dordrecht: Springer. ISBN 978-1-4020-3176-2.
- Chemistry 412 course notes. "A Brief History of the Development of Periodic Table". Western Oregon University. Retrieved July 20, 2015.
- Savitski MM, et al. (2014). "Tracking cancer drugs in living cells by thermal profiling of the proteome.". Science. 346: 6205. doi:10.1126/science.1255784.
- "Ancients & Alchemists - Time line of achievement". Chemical Heritage Society. Archived from the original on 20 June 2010. Retrieved 23 March 2014.

6

Evolution of Medicine

Medicine has evolved over a period of thousands years. Prehistoric medicine integrated herbalism, animal organs and healing through faith. Modern medicine replaced ancient medicine through the discovery and promotion of scientific biomedical research. Newer trends in the field of medicine seek to eradicate and control disease through the development of pharmaceutical medicine. This chapter will provide an integrated understanding of the evolution of medicine.

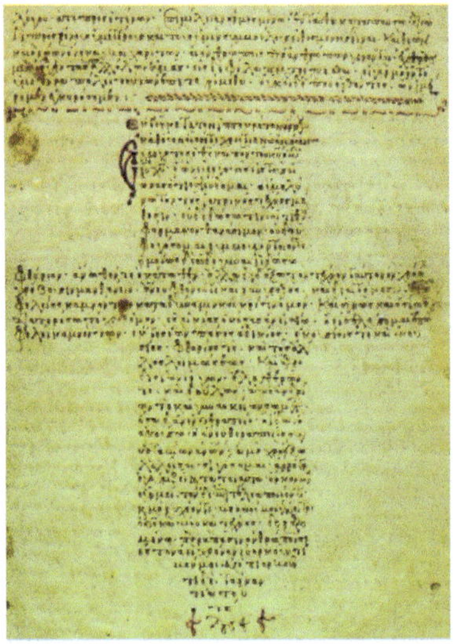

The *Hippocratic Corpus* is a collection of early medical works from ancient Greece that is strongly associated with the ancient Greek physician Hippocrates and his teachings.

This article deals with medicine as practiced by trained professionals from ancient times to the present. Early medical traditions include those of Babylon, China, Egypt and India. The Greeks went even further, introducing the concepts of medical diagnosis, prognosis, and advanced medical ethics. The Hippocratic Oath, still taken (although significantly changed from the original) by doctors up to today, was written in Greece in the 5th century BCE. In the medieval age, surgical practices inherited from the ancient masters were improved and then systematized in Rogerius's *The Practice of Surgery*. Universities began systematic training of physicians around the years 1220 in Italy. During the Renaissance, understanding of anatomy improved, and the microscope was invented. The germ theory of disease in the 19th century led to cures for many infectious diseases. Military doctors advanced the methods of trauma treatment and surgery. Public health measures were developed especially in the 19th century as the rapid growth of cities required systematic sanitary measures. Advanced research centers opened in the early 20th century, often connected with major hospitals. The mid-20th century was characterized by new biological treatments, such as antibiotics. These advancements, along with developments in chemistry, genetics, and lab technology (such as the x-ray) led to modern medicine. Medicine was heavily professionalized in the 20th century, and new careers opened to women as nurses (from the 1870s) and as physicians (especially after 1970). The 21st century is characterized by highly advanced research involving numerous fields of science.

Prehistoric Medicine

Although there is no record to establish when plants were first used for medicinal purposes (herbalism), the use of plants as healing agents, as well as clays and soils is ancient. Over time through emulation of the behavior of fauna a medicinal knowledge base developed and passed between generations. As tribal culture specialized specific castes, shamans and apothecaries fulfilled the role of healer.

The first known dentistry dates to about 7,000 B.C.E. in Baluchistan, where Neolithic dentists used flint-tipped drills and bowstrings.

The first known trepanning operation was carried out about 5,000 B.C.E. in Ensisheim, France.

The earliest known surgery, an amputation was carried out about 4,900 B.C.E. in Buthiers-Bulancourt, France.

Early Civilizations

The Edwin Smith Surgical Papyrus, written in the 17th century BC, contains the earliest recorded reference to the brain

Ancient Egypt developed a large, varied and fruitful medical tradition. Herodotus described the Egyptians as "the healthiest of all men, next to the Libyans", because of the dry climate and the notable public health system that they possessed. According to him, "the practice of medicine is so specialized among them that each physician is a healer of one disease and no more." Although Egyptian medicine, to a good extent, dealt with the supernatural, it eventually developed a practical use in the fields of anatomy, public health, and clinical diagnostics.

Medical information in the Edwin Smith Papyrus may date to a time as early as 3000 BC. Imhotep in the 3rd dynasty is sometimes credited with being the founder of ancient Egyptian medicine and with being the original author of the *Edwin Smith Papyrus*, detailing cures, ailments and anatomical observations. The *Edwin Smith Papyrus* is regarded as a copy of several earlier works and was written c. 1600 BC. It is an ancient textbook on surgery almost completely devoid of magical thinking and describes in exquisite detail the *examination, diagnosis, treatment,* and *prognosis* of numerous ailments.

The Kahun Gynaecological Papyrus treats women's complaints, including problems with conception. Thirty four cases detailing diagnosis and treatment survive, some of them fragmentarily. Dating to 1800 BCE, it is the oldest surviving medical text of any kind.

Medical institutions, referred to as *Houses of Life* are known to have been established in ancient Egypt as early as the 1st Dynasty.

The earliest known physician is also credited to ancient Egypt: Hesy-Ra, "Chief of Dentists and Physicians" for King Djoser in the 27th century BCE. Also, the earliest known woman physician,

Peseshet, practiced in Ancient Egypt at the time of the 4th dynasty. Her title was "Lady Overseer of the Lady Physicians." In addition to her supervisory role, Peseshet trained midwives at an ancient Egyptian medical school in Sais.

Babylonia

Babylonian cylinder seal and sketch depicting Nergal, god of plague, symbolized by the crooked stick. Wellcome Collection

The oldest Babylonian texts on medicine date back to the Old Babylonian period in the first half of the 2nd millennium BCE. The most extensive Babylonian medical text, however, is the *Diagnostic Handbook* written by the *ummânū*, or chief scholar, Esagil-kin-apli of Borsippa, during the reign of the Babylonian king Adad-apla-iddina (1069–1046 BCE).

Along with the Egyptians the Babylonians introduced the practice of diagnosis, prognosis, physical examination, and remedies. In addition, the *Diagnostic Handbook* introduced the methods of therapy and etiology. The text contains a list of medical symptoms and often detailed empirical observations along with logical rules used in combining observed symptoms on the body of a patient with its diagnosis and prognosis.

The *Diagnostic Handbook* was based on a logical set of axioms and assumptions, including the modern view that through the examination and inspection of the symptoms of a patient, it is possible to determine the patient's disease, its aetiology and future development, and the chances of the patient's recovery. The symptoms and diseases of a patient were treated through therapeutic means such as bandages, herbs and creams.

There was little development after the medieval era. Major European treatises on medicine took 200 years to reach the Middle East, where local rulers might consult Western doctors to get the latest treatments. Medical works in Arabic, Turkish, and Persian as late as 1800 were based on medieval Islamic medicine.

India

The Atharvaveda, a sacred text of Hinduism dating from the Early Iron Age, is one of the first Indian text dealing with medicine, like the medicine of the Ancient Near East based on concepts of the exorcism of demons and magic. The Atharvaveda also contain prescriptions of herbs for various ailments. The use of herbs to treat ailments would later form a large part of Ayurveda.

Ayurveda, meaning the "complete knowledge for long life" is another medical system of India. Its two most famous texts belong to the schools of Charaka and Sushruta. The earliest foundations of Ayurveda were built on a synthesis of traditional herbal practices together with a massive addition of theoretical conceptualizations, new nosologies and new therapies dating from about 600 BCE onwards, and coming out of the communities of thinkers who included the Buddha and others.

Ayurvedic herbal medicines

According to the compendium of Charaka, the Charakasamhitā, health and disease are not predetermined and life may be prolonged by human effort. The compendium of Suśruta, the Suśrutasamhitā defines the purpose of medicine to cure the diseases of the sick, protect the healthy, and to prolong life. Both these ancient compendia include details of the examination, diagnosis, treatment, and prognosis of numerous ailments. The Suśrutasamhitā is notable for describing procedures on various forms of surgery, including rhinoplasty, the repair of torn ear lobes, perineal lithotomy, cataract surgery, and several other excisions and other surgical procedures. Most remarkable is Sushruta's penchant for scientific classification: His medical treatise consists of 184 chapters, 1,120 conditions are listed, including injuries and illnesses relating to aging and mental illness. The Sushruta Samhita describe 125 surgical instrument, 300 surgical procedures and classifies human surgery in 8 categories.

The Ayurvedic classics mention eight branches of medicine: kāyācikitsā (internal medicine), śalyacikitsā (surgery including anatomy), śālākyacikitsā (eye, ear, nose, and throat diseases), kaumārabh□tya (pediatrics), bhūtavidyā (spirit medicine), and agada tantra (toxicology), rasāyana (science of rejuvenation), and vājīkara□a (Aphrodisiac). Apart from learning these, the student of Āyurveda was expected to know ten arts that were indispensable in the preparation and application of his medicines: distillation, operative skills, cooking, horticulture, metallurgy, sugar manufacture, pharmacy, analysis and separation of minerals, compounding of metals, and preparation of alkalis. The teaching of various subjects was done during the instruction of relevant clinical subjects. For example, teaching of anatomy was a part of the teaching of surgery, embryology was a part of training in pediatrics and obstetrics, and the knowledge of physiology and pathology was interwoven in the teaching of all the clinical disciplines. The normal length of the student's training appears to have been seven years. But the physician was to continue to learn.

As an alternative form of medicine in India, Unani medicine got deep roots and royal patronage during medieval times. It progressed during Indian sultanate and mughal periods. Unani medicine is very close to Ayurveda. Both are based on theory of the presence of the elements (in Unani, they are considered to be fire, water, earth and air) in the human body. According to followers of Unani medicine, these elements are present in different fluids and their balance leads to health and their imbalance leads to illness.

By the 18th century A.D., Sanskrit medical wisdom still dominated. Muslim rulers built large hospitals in 1595 in Hyderabad, and in Delhi in 1719, and numerous commentaries on ancient texts were written.

China

Assorted dried plant and animal parts used in traditional Chinese medicines, clockwise from top left corner: dried Lingzhi (lit. "spirit mushrooms"), ginseng, Luo Han Guo, turtle shell underbelly (plastron), and dried curled snakes

China also developed a large body of traditional medicine. Much of the philosophy of traditional Chinese medicine derived from empirical observations of disease and illness by Taoist physicians and reflects the classical Chinese belief that individual human experiences express causative principles effective in the environment at all scales. These causative principles, whether material, essential, or mystical, correlate as the expression of the natural order of the universe.

The foundational text of Chinese medicine is the Huangdi neijing, (or *Yellow Emperor's Inner Canon*), written 5th century to 3rd century BCE. Near the end of the 2nd century AD, during the Han dynasty, Zhang Zhongjing, wrote a *Treatise on Cold Damage*, which contains the earliest known reference to the *Neijing Suwen*. The Jin Dynasty practitioner and advocate of acupuncture and moxibustion, Huangfu Mi (215-282), also quotes the Yellow Emperor in his *Jiayi jing*, c. 265. During the Tang Dynasty, the *Suwen* was expanded and revised, and is now the best extant representation of the foundational roots of traditional Chinese medicine. Traditional Chinese Medicine that is based on the use of herbal medicine, acupuncture, massage and other forms of therapy has been practiced in China for thousands of years.

In the 18th century, during the Qing dynasty, there was a proliferation of popular books as well as more advanced encyclopedias on traditional medicine. Jesuit missionaries introduced Western science and medicine to the royal court, the Chinese physicians ignored them.

Finally in the 19th century, Western medicine was introduced at the local level by Christian medical missionaries from the London Missionary Society (Britain), the Methodist Church (Britain) and the Presbyterian Church (USA). Benjamin Hobson (1816–1873) in 1839, set up a highly successful Wai Ai Clinic in Guangzhou, China. The Hong Kong College of Medicine for Chinese was founded in 1887 by the London Missionary Society, with its first graduate (in 1892) being Sun Yat-sen, who later led the Chinese Revolution (1911). The Hong Kong College of Medicine for Chinese was the forerunner of the School of Medicine of the University of Hong Kong, which started in 1911.

Because of the social custom that men and women should not be near to one another, the women of China were reluctant to be treated by male doctors. The missionaries sent women doctors such as Dr. Mary Hannah Fulton (1854–1927). Supported by the Foreign Missions Board of the Presbyterian Church (USA) she in 1902 founded the first medical college for women in China, the Hackett Medical College for Women, in Guangzhou.

Greece and Roman Empire

Around 800 BCE Homer in The Iliad gives descriptions of wound treatment by the two sons of Asklepios, the admirable physicians Podaleirius and Machaon and one acting doctor, Patroclus. Because Machaon is wounded and Podaleirius is in combat Eurypylus asks Patroclus to *cut out this arrow from my thigh, wash off the blood with warm water and spread soothing ointment on the wound*. Askelpios like Imhotep becomes god of healing over time.

View of the *Askleipion* of Kos, the best preserved instance of an Asklepieion.

Temples dedicated to the healer-god Asclepius, known as *Asclepieia* functioned as centers of medical advice, prognosis, and healing. At these shrines, patients would enter a dream-like state of induced sleep known as *enkoimesis* not unlike anesthesia, in which they either received guidance from the deity in a dream or were cured by surgery. Asclepeia provided carefully controlled spaces conducive to healing and fulfilled several of the requirements of institutions created for healing. In the Asclepeion of Epidaurus, three large marble boards dated to 350 BCE preserve the names, case histories, complaints, and cures of about 70 patients who came to the temple with a problem and shed it there. Some of the surgical cures listed, such as the opening of an abdominal abscess or the removal of traumatic foreign material, are realistic enough to have taken place, but with the patient in a state of enkoimesis induced with the help of soporific substances such as opium. Alcmaeon of Croton wrote on medicine between 500 and 450 BCE. He argued that channels linked the sensory organs to the brain, and it is possible that he discovered one type of channel, the optic nerves, by dissection.

Hippocrates

A towering figure in the history of medicine was the physician Hippocrates of Kos (c. 460 – c. 370 BCE), considered the "father of Western medicine." The Hippocratic Corpus is a collection of around seventy early medical works from ancient Greece strongly associated with Hippocrates and his students. Most famously, Hippocrates invented the Hippocratic Oath for physicians, which is still relevant and in use today.

Hippocrates and his followers were first to describe many diseases and medical conditions. He is given credit for the first description of clubbing of the fingers, an important diagnostic sign in chronic suppurative lung disease, lung cancer and cyanotic heart disease. For this reason, clubbed fingers are sometimes referred to as "Hippocratic fingers". Hippocrates was also the first physician to describe Hippocratic face in *Prognosis*. Shakespeare famously alludes to this description when writing of Falstaff's death in Act II, Scene iii. of *Henry V*.

Hippocrates began to categorize illnesses as acute, chronic, endemic and epidemic, and use terms such as, "exacerbation, relapse, resolution, crisis, paroxysm, peak, and convalescence."

Another of Hippocrates's major contributions may be found in his descriptions of the symptomatology, physical findings, surgical treatment and prognosis of thoracic empyema, i.e. suppuration of the lining of the chest cavity. His teachings remain relevant to present-day students of pulmonary medicine and surgery. Hippocrates was the first documented person to practise cardiothoracic surgery, and his findings are still valid.

Some of the techniques and theories developed by Hippocrates are now put into practice by the fields of Environmental and Integrative Medicine. These include recognizing the importance of taking a complete history which includes environmental exposures as well as foods eaten by the patient which might play a role in his or her illness.

Herophilus and Erasistratus

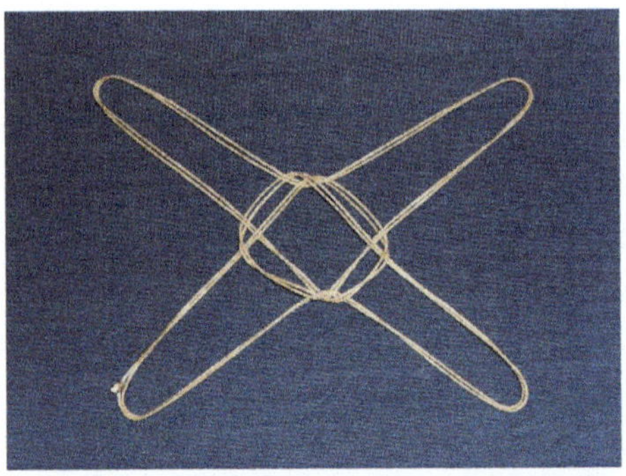

The *Plinthios Brokhos* as described by Greek physician Heraklas, a sling for binding a fractured jaw. These writings were preserved in one of Oribasius' collections.

Two great Alexandrians laid the foundations for the scientific study of anatomy and physiology, Herophilus of Chalcedon and Erasistratus of Ceos. Other Alexandrian surgeons gave us ligature (hemostasis), lithotomy, hernia operations, ophthalmic surgery, plastic surgery, methods of reduction of dislocations and fractures, tracheotomy, and mandrake as an anaesthetic. Some of what we know of them comes from Celsus and Galen of Pergamum.

Herophilus of Chalcedon, working at the medical school of Alexandria placed intelligence in the brain, and connected the nervous system to motion and sensation. Herophilus also distinguished between veins and arteries, noting that the latter pulse while the former do not. He and his contemporary, Erasistratus of Chios, researched the role of veins and nerves, mapping their courses across the body. Erasistratus connected the increased complexity of the surface of the human brain compared to other animals to its superior intelligence. He sometimes employed experiments to further his research, at one time repeatedly weighing a caged bird, and noting its weight loss between feeding times. In Erasistratus' physiology, air enters the body, is then drawn by the lungs into the heart, where it is transformed into vital spirit, and is then pumped by the arteries throughout the body. Some of this vital spirit reaches the brain, where it is transformed into animal spirit, which is then distributed by the nerves.

Galen

The Greek Galen (129–c. 216 CE) was one of the greatest physicians of the ancient world, studying and traveling widely in ancient Rome. He dissected animals to learn about the body, and performed many audacious operations—including brain and eye surgeries— that were not tried again for almost two millennia. In *Ars medica* ("Arts of Medicine"), he explained mental properties in terms of specific mixtures of the bodily parts.

Galen's medical works were regarded as authoritative until well into the Middle Ages. Galen left a physiological model of the human body that became the mainstay of the medieval physician's university anatomy curriculum, but it suffered greatly from stasis and intellectual stagnation because some of Galen's ideas were incorrect; he did not dissect a human body nor did the medieval lecturers.

The Renaissance rediscovered Galen. In 1523 Galen's *On the Natural Faculties* was published in London. In the 1530s Belgian anatomist and physician Andreas Vesalius launched a project to translate many of Galen's Greek texts into Latin. Vesalius's most famous work, *De humani corporis fabrica* was greatly influenced by Galenic writing and form.

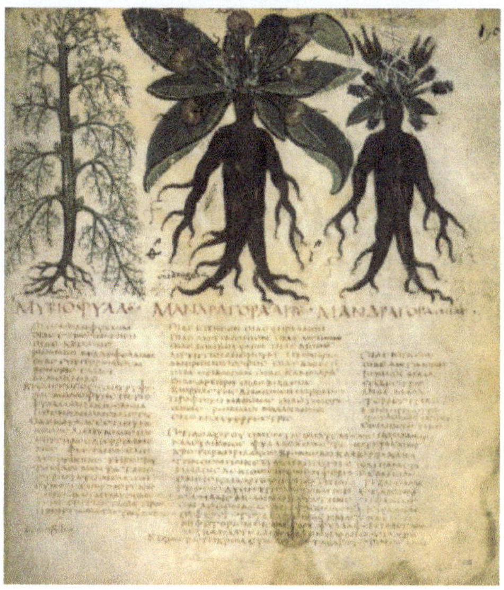

Mandrake (written 'ΜΑΝΔΡΑΓΟΡΑ' in Greek capitals). *Naples Dioscurides*, 7th century

Roman Contributions

The Romans invented numerous surgical instruments, including the first instruments unique to women, as well as the surgical uses of forceps, scalpels, cautery, cross-bladed scissors, the surgical needle, the sound, and speculas. Romans also performed cataract surgery.

The Roman army physician Dioscorides (c. 40–90 AD), was a Greek botanist and pharmacologist. He wrote the encyclopedia *De Materia Medica* describing over 600 herbal cures, forming an influential pharmacopoeia which was used extensively for the following 1,500 years.

The Middle Ages, AD 400 to 1400

Byzantine Empire

Islamic World

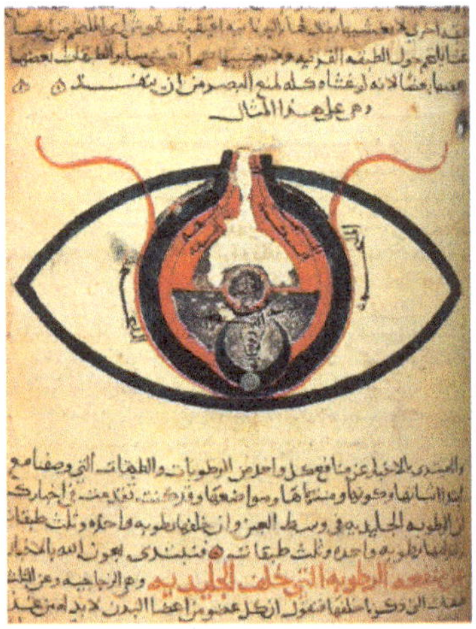

Arabic manuscript, *Anatomy of the Eye*, by al-Mutadibih, 1200 AD

The Islamic civilization rose to primacy in medical science as its physicians contributed significantly to the field of medicine, including anatomy, ophthalmology, pharmacology, pharmacy, physiology, surgery, and the pharmaceutical sciences. The Arabs were influenced by ancient Indian, Greek, Roman and Byzantine medical practices, and developed these further. Galen & Hippocrates were pre-eminent authorities. The translation of 129 of Galen's works into Arabic by the Nestorian Christian Hunayn ibn Ishaq and his assistants, and in particular Galen's insistence on a rational systematic approach to medicine, set the template for Islamic medicine, which rapidly spread throughout the Arab Empire.

Europe

After A.D. 400, the study and practice of medicine in the Western Roman Empire went into deep decline. Medical services were provided, especially for the poor, in the thousands of monastic hospitals that sprang up across Europe, but the care was rudimentary and mainly palliative. Most of the writings of Galen and Hippocrates were lost to the West, with the summaries and compendia of St. Isidore of Seville being the primary channel for transmitting Greek medical ideas. The Carolingian renaissance brought increased contact with Byzantium and a greater awareness of ancient medicine, but only with the twelfth century renaissance and the new translations coming from Muslim and Jewish sources in Spain, and the fifteenth century flood of resources after the fall of Constantinople did the West fully recover its acquaintance with classical antiquity.

Wallis identifies a prestige hierarchy with university educated physicians on top, followed by learned surgeons; craft-trained surgeons; barber surgeons; itinerant specialists such as dentist and oculists; empirics; and midwives.

Schools

The first medical schools were opened in the 9th century, most notably the Schola Medica Salernitana at Salerno in southern Italy. The cosmopolitan influences from Greek, Latin, Arabic, and

Hebrew sources gave it an international reputation as the Hippocratic City. Students from wealthy families came for three years of preliminary studies and five of medical studies. By the thirteenth century the medical school at Montpellier began to eclipse the Salernitan school. In the 12th century universities were founded in Italy, France and England which soon developed schools of medicine. The University of Montpellier in France and Italy's University of Padua and University of Bologna were leading schools. Nearly all the learning was from lectures and readings in Hippocrates, Galen, Avicenna and Aristotle. There was little clinical work or dissection.

A miniature depicting the Schola Medica Salernitana

Humours

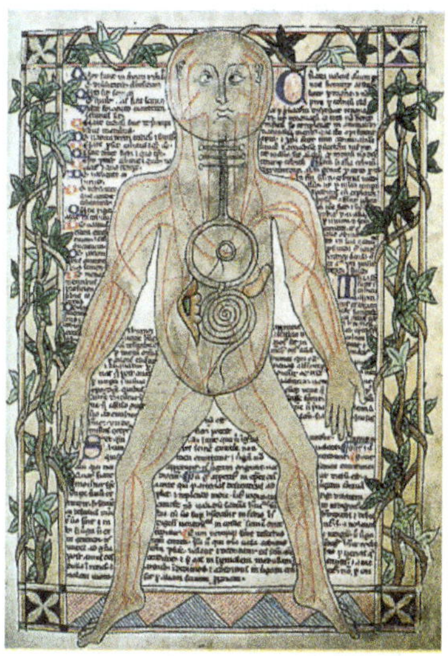

13th century illustration showing the veins.

The underlying principle of most medieval medicine was Galen's theory of humours. This was derived from the ancient medical works, and dominated all western medicine until the 19th century. The theory stated that within every individual there were four humours, or principal fluids - black bile, yellow bile, phlegm, and blood, these were produced by various organs in the body, and they had to be in balance for a person to remain healthy. Too much phlegm in the body, for example, caused lung problems; and the body tried to cough up the phlegm to restore a balance. The balance of humours in humans could be achieved by diet, medicines, and by blood-letting, using leeches. The four humours were also associated with the four seasons, black bile-autumn, yellow bile-summer, phlegm-winter and blood-spring.

Healing included both physical and spiritual therapeutics, such as the right herbs, a suitable diet, clean bedding, and the sense that care was always at hand. Other procedures used to help patients included the Mass, prayers, relics of saints, and music used to calm a troubled mind or quickened pulse.

Renaissance to Early Modern Period 16th-18th Century

The Renaissance brought an intense focus on scholarship to Christian Europe. A major effort to translate the Arabic and Greek scientific works into Latin emerged. Europeans gradually became experts not only the ancient writings of the Romans and Greeks, but in the contemporary writings of Islamic scientists. During the later centuries of the Renaissance came an increase in experimental investigation, particularly in the field of dissection and body examination, thus advancing our knowledge of human anatomy.

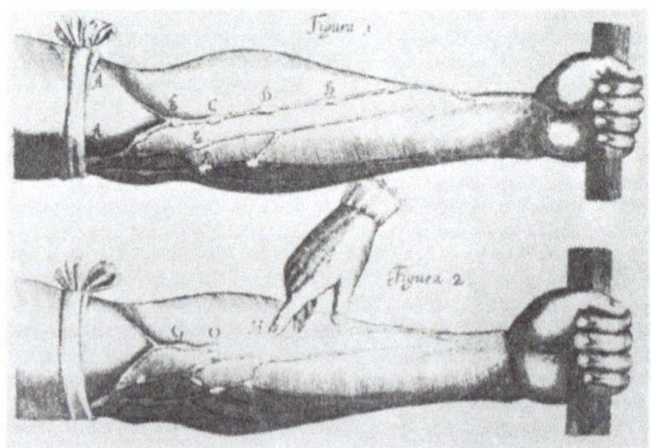

An experiment from William Harvey's *de Motu Cordis*, 1628

The development of modern neurology began in the 16th century with Vesalius, who described the anatomy of the brain and other organs; he had little knowledge of the brain's function, thinking that it resided mainly in the ventricles. Over his lifetime he corrected over 200 of Galen's mistakes. Understanding of medical sciences and diagnosis improved, but with little direct benefit to health care. Few effective drugs existed, beyond opium and quinine. Folklore cures and potentially poisonous metal-based compounds were popular treatments. Independently from Ibn al-Nafis, Michael Servetus rediscovered the pulmonary circulation, but this discovery did not reach the public because it was written down for the first time in the "Manuscript of Paris" in 1546, and later published in the theological work which he paid with his life in 1553. Later this was perfected by Renaldus Columbus and Andrea Cesalpino. Later William Harvey correctly described the circulatory system. The most useful tomes in medicine used both by students and expert physicians were *De Materia Medica* and Pharmacopoeia.

Replica of Antonie van Leeuwenhoek's microscope of the 1670s

Bacteria and protists were first observed with a microscope by Antonie van Leeuwenhoek in 1676, initiating the scientific field of microbiology.

Paracelsus

Paracelsus (1493–1541), was an erratic and abusive innovator who rejected Galen and bookish knowledge, calling for experimental research, with heavy doses of mysticism, alchemy and magic mixed in. The point is that he rejected sacred magic (miracles) under Church auspisces and looked for cures in nature. He preached but he also pioneered the use of chemicals and minerals in medicine. His hermetical views were that sickness and health in the body relied on the harmony of man (microcosm) and Nature (macrocosm). He took an approach different from those before him, using this analogy not in the manner of soul-purification but in the manner that humans must have certain balances of minerals in their bodies, and that certain illnesses of the body had chemical remedies that could cure them. Most of his influence came after his death. Paracelsus is a highly controversial figure in the history of medicine, with most experts hailing him as a Father of Modern Medicine for shaking off religious orthodoxy and inspiring many researchers; others say he was a mystic more than a scientist and downplay his importance.

Padua and Bologna

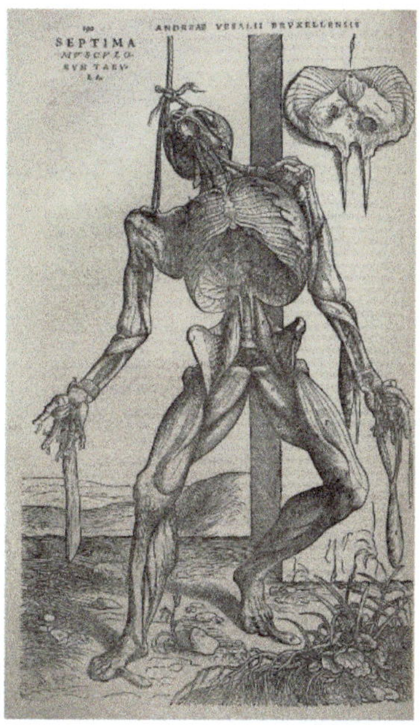

Vesalius's 1543 *De humani corporis fabrica* contained intricately detailed drawings of human dissections, often in allegorical poses.

University training of physicians began in the 13th century.

The University of Padua was founded about 1220 by walkouts from the University of Bologna, and began teaching medicine in 1222. It played a leading role in the identification and treatment of diseases and ailments, specializing in autopsies and the inner workings of the body. Starting in 1595, Padua's famous anatomical theatre drew artists and scientists studying the human body during public dissections. The intensive study of Galen led to critiques of Galen modeled on his own writing, as in the first book of Vesalius's *De humani corporis fabrica*. Andreas Vesalius held the chair of Surgery and Anatomy (*explicator chirurgiae*) and in 1543 published his anatomical discoveries in De Humani Corporis Fabrica. He portrayed the human body as an interdependent system of organ groupings. The book triggered great public interest in dissections and caused many other European cities to establish anatomical theatres.

At the University of Bologna the training of physicians began in 1219. The Italian city attracted students from across Europe. Taddeo Alderotti built a tradition of medical education that established the characteristic features of Italian learned medicine and was copied by medical schools elsewhere. Turisanus (d. 1320) was his student. The curriculum was revised and strengthened in 1560–1590. A representative professor was Julius Caesar Aranzi (Arantius) (1530–89). He became Professor of Anatomy and Surgery at the University of Bologna in 1556, where he established anatomy as a major branch of medicine for the first time. Aranzi combined anatomy with a description of pathological processes, based largely on his own research, Galen, and the work of his contemporary Italians. Aranzi discovered the 'Nodules of Aranzio' in the semilunar valves of the heart and wrote the first description of the superior levator palpebral and the coracobrachialis muscles. His books (in Latin) covered surgical techniques for many conditions, including hydrocephalus, nasal polyp, goitre and tumours to phimosis, ascites, haemorrhoids, anal abscess and fistulae.

Women

Catholic women played large roles in health and healing in medieval and early modern Europe. A life as a nun was a prestigious role; wealthy families provided dowries for their daughters, and these funded the convents, while the nuns provided free nursing care for the poor.

The Catholic elites provided hospital services because of their theology of salvation that good works were the route to heaven. The Protestant reformers rejected the notion that rich men could gain God's grace through good works—and thereby escape purgatory—by providing cash endowments to charitable institutions. They also rejected the Catholic idea that the poor patients earned grace and salvation through their suffering. Protestants generally closed all the convents and most of the hospitals, sending women home to become housewives, often against their will. On the other hand, local officials recognized the public value of hospitals, and some were continued in Protestant lands, but without monks or nuns and in the control of local governments.

In London, the crown allowed two hospitals to continue their charitable work, under nonreligious control of city officials. The convents were all shut down but Harkness finds that women—some of them former nuns—were part of a new system that delivered essential medical services to people outside their family. They were employed by parishes and hospitals, as well as by private families, and provided nursing care as well as some medical, pharmaceutical, and surgical services.

Meanwhile, in Catholic lands such as France, rich families continued to fund convents and monasteries, and enrolled their daughters as nuns who provided free health services to the poor. Nursing was a religious role for the nurse, and there was little call for science.

Age of Enlightenment

During the Age of Enlightenment, the 18th-century, science was held in high esteem and physicians upgraded their social status by becoming more scientific. The health field was crowded with self-trained barber-surgeons, apothecaries, midwives, drug peddlers, and charlatans.

Across Europe medical schools relied primarily on lectures and readings. The final year student would have limited clinical experience by trailing the professor through the wards. Laboratory work was uncommon, and dissections were rarely done because of legal restrictions on cadavers. Most schools were small, and only Edinburgh, Scotland, with 11,000 alumni, produced large numbers of graduates.

Britain

In Britain, there were but three small hospitals after 1550. Pelling and Webster estimate that in London in the 1580 to 1600 period, out of a population of nearly 200,000 people, there were about 500 medical practitioners. Nurses and midwives are not included. There were about 50 physicians, 100 licensed surgeons, 100 apothecaries, and 250 additional unlicensed practitioners. In the last category about 25% were women. All across Britain—and indeed all of the world—the vast majority of the people in city, town or countryside depended for medical care on local amateurs with no professional training but with a reputation as wise healers who could diagnose problems and advise sick people what to do—and perhaps set broken bones, pull a tooth, give some traditional herbs or brews or perform a little magic to cure what ailed them.

The London Dispensary opened in 1696, the first clinic in the British Empire to dispense medicines to poor sick people. The innovation was slow to catch on, but new dispensaries were open in the 1770s. In the colonies, small hospitals opened in Philadelphia in 1752, New York in 1771, and Boston (Massachusetts General Hospital) in 1811.

Guy's Hospital in 1820

Guy's Hospital, the first great British hospital opened in 1721 in London, with funding from businessman Thomas Guy. In 1821 a bequest of £200,000 by William Hunt in 1829 funded expansion for an additional hundred beds. Samuel Sharp (1709–78), a surgeon at Guy's Hospital, from 1733 to 1757, was internationally famous; his *A Treatise on the Operations of Surgery* (1st ed., 1739), was the first British study focused exclusively on operative technique.

English physician Thomas Percival (1740–1804) wrote a comprehensive system of medical conduct, *Medical Ethics, or a Code of Institutes and Precepts, Adapted to the Professional Conduct of Physicians and Surgeons* (1803) that set the standard for many textbooks.

Spain and Spanish Empire

In the Spanish empire, the viceregal capital of Mexico City was a site of medical training for physicians and the creation of hospitals. Epidemic disease had decimated indigenous populations starting with the early sixteenth-century Spanish conquest of the Aztec empire, when a black auxiliary in the armed forces of conqueror Hernán Cortés, with an active case of smallpox, set off a virgin land epidemic among indigenous peoples, Spanish allies and enemies alike. Aztec emperor Cuitlahuac died of smallpox. Disease was a significant factor in the Spanish conquest elsewhere as well.

Mexico City epidemic of 1737, with elites calling on the Virgin of Guadalupe

Medical education was instituted at the Royal and Pontifical University of Mexico, and their clientele was urban elites. Male and female *curanderos* or lay practitioners, attended to the ills of the popular classes. The Spanish crown began regulating the medical profession just a few years after the conquest, setting up the Royal Tribunal of the Protomedicato, a board for licensing medical personnel in 1527. It became more systematic after 1646, licensing physicians, druggists, surgeons, and bleeders, and requiring a license before public practice. Crown regulation of medical practice became more general in the Spanish empire.

Elites and the popular classes alike called on divine intervention in personal and society-wide health crises, such as the epidemic of 1737. The intervention of the Virgin of Guadalupe was depicted in a scene of dead and dying Indians, with elites on their knees praying for her aid. In the late eighteenth century, the crown began implementing secularizing policies on the Iberian peninsula and its overseas empire that sought to control disease more systematically and scientifically.

19th Century: Rise of Modern Medicine

The practice of medicine changed in the face of rapid advances in science, as well as new approaches by physicians. Hospital doctors began much more systematic analysis of patients' symptoms in diagnosis. Among the more powerful new techniques were anaesthesia, and the development of both antiseptic and aseptic operating theatres. Actual cures were developed for certain endemic infectious diseases. However the decline in many of the most lethal diseases was more due to improvements in public health and nutrition than to medicine.

Medicine was revolutionized in the 19th century and beyond by advances in chemistry and laboratory techniques and equipment, old ideas of infectious disease epidemiology were replaced with bacteriology and virology.

Germ Theory and Bacteriology

In the 1830s in Italy, Agostino Bassi traced the silkworm disease muscardine to microorganisms. Meanwhile, in Germany, Theodor Schwann led researches on alcoholic fermentation by yeast and proposed that they were alive—that is, microorganisms—a claim derided by leading chemists, such as Justus von Liebig, seeking solely physicochemical explanation, and alleging that Schwann's was regressing to vitalism. In 1847 in Vienna, Ignaz Semmelweis (1818–1865), by requiring physicians to clean their hands before attending childbirth, dramatically cut new mothers' death rate due to childbed fever, yet his principles were marginalized and attacked by professional peers.

Louis Pasteur experimenting on bacteria, c. 1870

Starting in 1857 by confirming Schwann's fermentation experiments, Louis Pasteur in France placed his eminent reputation behind the belief that yeast are microorganisms, and closed his paper by indicating that such process might also explain contagious diseases. In 1860, Pasteur's report on bacterial fermention to butyric acid motivated fellow Frenchman Casimir Davaine to establish a similar species, which he called *bacteridia*, as the pathogen of the disease anthrax, costly to the cattle industry. Yet *bacteridia* were found inconsistently and dismissed as a disease byproduct, not cause. British surgeon Joseph Lister, however, already took cue and introduced antisepsis to wound treatment in 1865.

German physician Robert Koch, noting fellow German Ferdinand Cohn's report of a spore stage of a certain bacterial species, traced the life cycle of Davaine's *bacteridia*, identified spores, inoculated laboratory animals with them, and reproduced anthrax—a breakthrough for experimental pathology and germ theory of disease. Pasteur's group added ecological investigations confirming spores' role in the natural setting, while Koch published a landmark treatise in 1878 on the bacterial pathology of wounds. In 1881, Koch reported discovery of the "tubercle bacillus", cementing germ theory and Koch's acclaim.

Upon the outbreak of a cholera epidemic in Alexandria, Egypt, two medical missions went to investigate and attend the sick, one was sent out by Pasteur and the other actually led by Koch.

Koch's group returned victorious in 1883, having discovered the cholera pathogen. In Germany, however, Koch's bacteriologists had to vie against Max von Pettenkofer, Germany's leading proponent of miasmatic theory. Pettenkofer conceded bacteria's casual involvement, but maintained that other, environmental factors were required to turn it pathogenic, and opposed water treatment as a misdirected effort amid more important ways to improve public health. The massive cholera epidemic in Hamburg in 1892 devastasted Pettenkoffer's position, and yielded German public health to "Koch's bacteriology".

On losing the 1883 rivalry in Alexandria, Pasteur switched research direction, and introduced his third vaccine—rabies vaccine—the first vaccine for humans since Jenner's for smallpox. From across the globe, donations poured in, funding the founding of Pasteur Institute, the globe's first biomedical institute, which opened in 1888. Along with Koch's bacteriologists, Pasteur's group—which preferred the term *microbiology*—led medicine into the new era of "scientific medicine" upon bacteriology and germ theory. Accepted from Jakob Henle, Koch's steps to confirm a species' pathogenicity became famed as "Koch's postulates". Although his proposed tuberculosis treatment, tuberculin, seemingly failed, it soon was used to test for infection with the involved species. In 1905, Koch was awarded the Nobel Prize in Physiology or Medicine, and remains renowned as the founder of medical microbiology.

Women

Women as Nurses

Florence Nightingale triggered the professionalization of nursing. Photograph c. 1860

Women had always served in ancillary roles, and as midwives and healers. The professionalization of medicine forced them increasingly to the sidelines. As hospitals multiplied they relied in Europe on orders of Roman Catholic nun-nurses, and German Protestant and Anglican deaconesses in the early 19th century. They were trained in traditional methods of physical care that involved little knowledge of medicine. The breakthrough to professionalization based on knowledge of advanced medicine was led by Florence Nightingale in England. She resolved to provide more advanced training than she saw on the Continent. At Kaiserswerth, where the first German nursing schools were founded in 1836 by Theodor Fliedner, she said, "The nursing was nil and the hygiene horrible.") Britain's male doctors preferred the old system, but Nightingale won out and her Nightingale Training School opened in 1860 and became a model. The Nightingale solution depended on the patronage of upper class women, and they proved eager to serve. Royalty became involved. In 1902 the wife of the British king took control of the nursing unit of the British army, became its president, and renamed it after herself as the Queen Alexandra's Royal Army Nursing Corps; when she died the next queen became president. Today its Colonel In Chief is the daughter-in-law of Queen Elizabeth. In the United States, upper middle class women who already supported hospitals promoted nursing. The new profession proved highly attractive to women of all backgrounds, and schools of nursing opened in the late 19th century. They soon a function of large hospitals, where they provided a steady stream of low-paid idealistic workers. The International Red Cross began operations in numerous countries in the late 19th century, promoting nursing as an ideal profession for middle class women.

The Nightingale model was widely copied. Linda Richards (1841 – 1930) studied in London and became the first professionally trained American nurse. She established nursing training programs in the United States and Japan, and created the first system for keeping individual medical records for hospitalized patients. The Russian Orthodox Church sponsored seven orders of nursing sisters in the late 19th century. They ran hospitals, clinics, almshouses, pharmacies, and shelters as well as training schools for nurses. In the Soviet era (1917–1991), with the aristocratic sponsors gone, nursing became a low-prestige occupation based in poorly maintained hospitals.

Women as Physicians

It was very difficult for women to become doctors in any field before the 1970s. Elizabeth Blackwell (1821–1910) became the first woman to formally study and practice medicine in the United States. She was a leader in women's medical education. While Blackwell viewed medicine as a means for social and moral reform, her student Mary Putnam Jacobi (1842–1906) focused on curing disease. At a deeper level of disagreement, Blackwell felt that women would succeed in medicine because of their humane female values, but Jacobi believed that women should participate as the equals of men in all medical specialties using identical methods, values and insights. In the Soviet Union although the majority of medical doctors were women, they were paid less than the mostly male factory workers.

Paris

Paris (France) and Vienna were the two leading medical centers on the Continent in the era 1750–1914.

In the 1770s-1850s Paris became a world center of medical research and teaching. The "Paris School" emphasized that teaching and research should be based in large hospitals and promoted the professionalization of the medical profession and the emphasis on sanitation and public health. A major reformer was Jean-Antoine Chaptal (1756–1832), a physician who was Minister of Internal Affairs. He created the Paris Hospital, health councils, and other bodies.

Louis Pasteur (1822–1895) was one of the most important founders of medical microbiology. He is remembered for his remarkable breakthroughs in the causes and preventions of diseases. His discoveries reduced mortality from puerperal fever, and he created the first vaccines for rabies and anthrax. His experiments supported the germ theory of disease. He was best known to the general public for inventing a method to treat milk and wine in order to prevent it from causing sickness, a process that came to be called pasteurization. He is regarded as one of the three main founders of microbiology, together with Ferdinand Cohn and Robert Koch. He worked chiefly in Paris and in 1887 founded the Pasteur Institute there to perpetuate his commitment to basic research and its practical applications. As soon as his institute was created, Pasteur brought together scientists with various specialties. The first five departments were directed by Emile Duclaux (general microbiology research) and Charles Chamberland (microbe research applied to hygiene), as well as a biologist, Ilya Ilyich Mechnikov (morphological microbe research) and two physicians, Jacques-Joseph Grancher (rabies) and Emile Roux (technical microbe research). One year after the inauguration of the Institut Pasteur, Roux set up the first course of microbiology ever taught in the world, then entitled *Cours de Microbie Technique* (Course of microbe research techniques). It became the model for numeous research centers around the world named "Pasteur Institutes."

Vienna

The First Viennese School of Medicine, 1750–1800, was led by the Dutchman Gerard van Swieten (1700–1772), who aimed to put medicine on new scientific foundations - promoting unprejudiced clinical observation, botanical and chemical research, and introducing simple but powerful remedies. When the Vienna General Hospital opened in 1784, it at once became the world's largest hospital and physicians acquired a facility that gradually developed into the most important research centre. Progress ended with the Napoleonic wars and the government shutdown in 1819 of all liberal journals and schools; this caused a general return to traditionalism and eclecticism in medicine.

Vienna was the capital of a diverse empire and attracted not just Germans but Czechs, Hungarians, Jews, Poles and others to its world-class medical facilities. After 1820 the Second Viennese School of Medicine emerged with the contributions of physicians such as Carl Freiherr von Rokitansky, Josef Škoda, Ferdinand Ritter von Hebra, and Ignaz Philipp Semmelweis. Basic medical science expanded and specialization advanced. Furthermore, the first dermatology, eye, as well as ear, nose, and throat clinics in the world were founded in Vienna. The textbook of ophthalmologist Georg Joseph Beer (1763–1821) *Lehre von den Augenkrankheiten* combined practical research and philosophical speculations, and became the standard reference work for decades.

Berlin

After 1871 Berlin, the capital of the new German Empire, became a leading center for medical research. Robert Koch (1843–1910) was a representative leader. He became famous for isolating *Bacillus anthracis* (1877), the *Tuberculosis bacillus* (1882) and *Vibrio cholerae* (1883) and for his development of Koch's postulates. He was awarded the Nobel Prize in Physiology or Medicine in 1905 for his tuberculosis findings. Koch is one of the founders of microbiology, inspiring such major figures as Paul Ehrlich and Gerhard Domagk.

U.S. Civil War

In the American Civil War (1861–65), as was typical of the 19th century, more soldiers died of disease than in battle, and even larger numbers were temporarily incapacitated by wounds, disease and accidents. Conditions were poor in the Confederacy, where doctors and medical supplies were in short supply. The war had a dramatic long-term impact on medicine in the U.S., from surgical technique to hospitals to nursing and to research facilities.

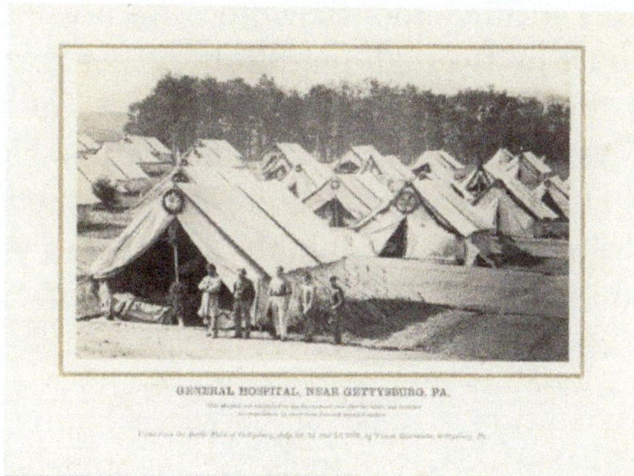

American Civil War hospital at Gettysburg, 1863

The hygiene of the training and field camps was poor, especially at the beginning of the war when men who had seldom been far from home were brought together for training with thousands of strangers. First came epidemics of the childhood diseases of chicken pox, mumps, whooping cough, and, especially, measles. Operations in the South meant a dangerous and new disease environment, bringing diarrhea, dysentery, typhoid fever, and malaria. There were no antibiotics, so the surgeons prescribed coffee, whiskey, and quinine. Harsh weather, bad water, inadequate shelter in winter quarters, poor policing of camps, and dirty camp hospitals took their toll.

This was a common scenario in wars from time immemorial, and conditions faced by the Confederate army were even worse. The Union responded by building army hospitals in every state. What was different in the Union was the emergence of skilled, well-funded medical organizers who took proactive action, especially in the much enlarged United States Army Medical Department, and the United States Sanitary Commission, a new private agency. Numerous other new agencies also targeted the medical and morale needs of soldiers, including the United States Christian Commission as well as smaller private agencies.

The U.S. Army learned many lessons and in August 1886, it established the Hospital Corps.

Statistical Methods

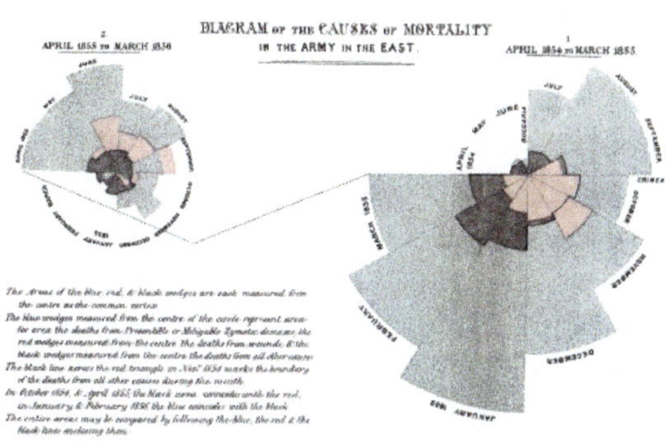

"Diagram of the causes of mortality in the army in the East" by Florence Nightingale.

A major breakthrough in epidemiology came with the introduction of statistical maps and graphs. They allowed careful analysis of seasonality issues in disease incidents, and the maps allowed public health officials to identifical critical loci for the dissemination of disease. John Snow in London developed the methods. In 1849, he observed that the symptoms of cholera, which had already claimed around 500 lives within a month, were vomiting and diahrrea. He concluded that the source of contamination must be through ingestion, rather than inhalation as was previously thought. It was this insight that resulted in the removal of The Pump On Broad Street, after which deaths from cholera plummeted afterwards. English nurse Florence Nightingale pioneered analysis of large amounts of statistical data, using graphs and tables, regarding the condition of thousands of patients in the Crimean War to evaluate the efficacy of hospital services. Her methods proved convincing and led to reforms in military and civilian hospitals, usually with the full support of the government.

By the late 19th and early 20th century English statisticians led by Francis Galton, Karl Pearson and Ronald Fisher developed the mathematical tools such as correlations and hypothesis tests that made possible much more sophisticated analysis of statistical data.

During the U.S. Civil War the Sanitary Commission collected enormous amounts of statistical data, and opened up the problems of storing information for fast access and mechanically searching for data patterns. The pioneer was John Shaw Billings (1838–1913). A senior surgeon in the war, Billings built the Library of the Surgeon General's Office (now the National Library of Medicine, the centerpiece of modern medical information systems. Billings figured out how to mechanically analyze medical and demographic data by turning facts into numbers and punching the numbers onto cardboard cards that could be sorted and counted by machine. The applications were developed by his assistant Herman Hollerith; Hollerith invented the punch card and counter-sorter system that dominated statistical data manipulation until the 1970s. Hollerith's company became International Business Machines (IBM) in 1911.

Worldwide Dissemination

United States

Johns Hopkins Hospital, founded in 1889, originated several modern medical practices, including residency and rounds.

Japan

European ideas of modern medicine were spread widely through the world by medical missionaries, and the dissemination of textbooks. Japanese elites enthusiastically embraced Western medicine after the Meiji Restoration of the 1860s. However they had been prepared by their knowledge of the Dutch and German medicine, for they had some contact with Europe through the Dutch. Highly influential was the 1765 edition of Hendrik van Deventer's pioneer work *Nieuw Ligt* ("A New Light") on Japanese obstetrics, especially on Katakura Kakuryo's publication in 1799 of *Sanka Hatsumo* ("Enlightenment of Obstetrics"). A cadre of Japanese physicians began to interact with Dutch doctors, who introduced smallpox vaccinations. By 1820 Japanese ranpô medical practitioners not only translated Dutch medical texts, they integrated their readings with clinical diagnoses. These men became leaders of the modernization of medicine in their country. They broke from Japanese traditions of closed medical fraternities and adopted the European approach of an open community of collaboration based on expertise in the latest scientific methods.

Kitasato Shibasaburō (1853–1931) studied bacteriology in Germany under Robert Koch. In

1891 he founded the Institute of Infectious Diseases in Tokyo, which introduced the study of bacteriology to Japan. He and French researcher Alexandre Yersin went to Hong Kong in 1894, where; Kitasato confirmed Yersin's discovery that the bacterium *Yersinia pestis* is the agent of the plague. In 1897 he isolates and described the organism that caused dysentery. He became the first dean of medicine at Keio University, and the first president of the Japan Medical Association.

Japanese physicians immediately recognized the values of X-Rays. They were able to purchase the equipment locally from the Shimadzu Company, which developed, manufactured, marketed, and distributed X-Ray machines after 1900. Japan not only adopted German methods of public health in the home islands, but implemented them in its colonies, especially Korea and Taiwan, and after 1931 in Manchuria. A heavy investment in sanitation resulted in a dramatic increase of life expectancy.

Psychiatry

The Quaker-run York Retreat, founded in 1796, gained international prominence as a centre for moral treatment and a model of asylum reform following the publication of Samuel Tuke's *Description of the Retreat* (1813).

Until the nineteenth century, the care of the insane was largely a communal and family responsibility rather than a medical one. The vast majority of the mentally ill were treated in domestic contexts with only the most unmanageable or burdensome likely to be institutionally confined. This situation was transformed radically from the late eighteenth century as, amid changing cultural conceptions of madness, a new-found optimism in the curability of insanity within the asylum setting emerged. Increasingly, lunacy was perceived less as a physiological condition than as a mental and moral one to which the correct response was persuasion, aimed at inculcating internal restraint, rather than external coercion. This new therapeutic sensibility, referred to as moral treatment, was epitomised in French physician Philippe Pinel's quasi-mythological unchaining of the lunatics of the Bicêtre Hospital in Paris and realised in an institutional setting with the foundation in 1796 of the Quaker-run York Retreat in England.

From the early nineteenth century, as lay-led lunacy reform movements gained in influence, ever more state governments in the West extended their authority and responsibility over the mentally ill. Small-scale asylums, conceived as instruments to reshape both the mind and behaviour of the disturbed, proliferated across these regions. By the 1830s, moral treatment, together with the asylum itself, became increasingly medicalised and asylum doctors began to establish a distinct medical identity with the establishment in the 1840s of associations for their members in France,

Germany, the United Kingdom and America, together with the founding of medico-psychological journals. Medical optimism in the capacity of the asylum to cure insanity soured by the close of the nineteenth century as the growth of the asylum population far outstripped that of the general population.[a] Processes of long-term institutional segregation, allowing for the psychiatric conceptualisation of the natural course of mental illness, supported the perspective that the insane were a distinct population, subject to mental pathologies stemming from specific medical causes. As degeneration theory grew in influence from the mid-nineteenth century, heredity was seen as the central causal element in chronic mental illness, and, with national asylum systems overcrowded and insanity apparently undergoing an inexorable rise, the focus of psychiatric therapeutics shifted from a concern with treating the individual to maintaining the racial and biological health of national populations.

Patient, Surrey County Lunatic Asylum, *c.* 1850–58.
The asylum population in England and Wales rose from 1,027 in 1827 to 74,004 in 1900.

Emil Kraepelin (1856–1926) introduced new medical categories of mental illness, which eventually came into psychiatric usage despite their basis in behavior rather than pathology or etiology. Shell shock among frontline soldiers exposed to heavy artillery bombardment was first diagnosed by British Army doctors in 1915. By 1916, similar symptoms were also noted in soldiers not exposed to explosive shocks, leading to questions as to whether the disorder was physical or psychiatric. In the 1920s surrealist opposition to psychiatry was expressed in a number of surrealist publications. In the 1930s several controversial medical practices were introduced including inducing seizures (by electroshock, insulin or other drugs) or cutting parts of the brain apart (leucotomy or lobotomy). Both came into widespread use by psychiatry, but there were grave concerns and much opposition on grounds of basic morality, harmful effects, or misuse.

In the 1950s new psychiatric drugs, notably the antipsychotic chlorpromazine, were designed in laboratories and slowly came into preferred use. Although often accepted as an advance in some ways, there was some opposition, due to serious adverse effects such as tardive dyskinesia. Patients often opposed psychiatry and refused or stopped taking the drugs when not subject to psychiatric control. There was also increasing opposition to the use of psychiatric hospitals, and attempts to move people back into the community on a collaborative user-led group approach ("therapeutic communities") not controlled by psychiatry. Campaigns against masturbation were done in the Victorian era and elsewhere. Lobotomy was used until the 1970s to treat schizophrenia. This was denounced by the anti-psychiatric movement in the 1960s and later.

20th Century and Beyond

Twentieth Century Warfare and Medicine

The ABO blood group system was discovered in 1901, and the Rhesus blood group system in 1937, facilitating blood transfusion.

During the 20th century, large-scale wars were attended with medics and mobile hospital units which developed advanced techniques for healing massive injuries and controlling infections rampant in battlefield conditions. During the Mexican Revolution (1910-1920), General Pancho Villa organized hospital trains for wounded soldiers. Boxcars marked *Servicio Sanitario* ("sanitary service") were re-purposed as surgical operating theaters and areas for recuperation, and staffed by up to 40 Mexican and U.S. physicians. Severely wounded soldiers were shuttled back to base hospitals. Canadian physician Norman Bethune, M.D. developed a mobile blood-transfusion service for frontline operations in the Spanish Civil War (1936-1939), but ironically, he himself died of blood poisoning. Thousands of scarred troops provided the need for improved prosthetic limbs and expanded techniques in plastic surgery or reconstructive surgery. Those practices were combined to broaden cosmetic surgery and other forms of elective surgery.

During the First World War, Alexis Carrel and Henry Dakin developed the Carrel-Dakin method of treating wounds with an irrigation, Dakin's solution, a germicide which helped prevent gangrene.

The Great War spurred the usage of Roentgen's X-ray, and the electrocardiograph, for the monitoring of internal bodily functions. This was followed in the inter-war period by the development of the first anti-bacterial agents such as the sulpha antibiotics.

Public Health

Public health measures became particular important during the 1918 flu pandemic, which killed at least 50 million people around the world. It became an important case study in epidemiology. Bristow shows there was a gendered response of health caregivers to the pandemic in the United States. Male doctors were unable to cure the patients, and they felt like failures. Women nurses also saw their patients die, but they took pride in their success in fulfilling their professional role of caring for, ministering, comforting, and easing the last hours of their patients, and helping the families of the patients cope as well.

From 1917 to 1923, the American Red Cross moved into Europe with a battery of long-term child health projects. It built and operated hospitals and clinics, and organized antituberculosis and antityphus campaigns. A high priority involved child health programs such as clinics, better baby shows, playgrounds, fresh air camps, and courses for women on infant hygiene. Hundreds of U.S. doctors, nurses, and welfare professionals administered these programs, which aimed to reform the health of European youth and to reshape European public health and welfare along American lines.

Second World War

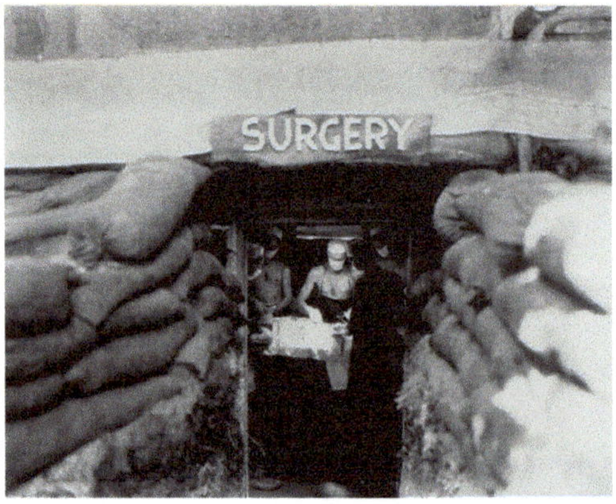

American combat surgery during the Pacific War, 1943.
Major wars showed the need for effective hygiene and medical treatment.

The advances in medicine made a dramatic difference for Allied troops, while the Germans and especially the Japanese and Chinese suffered from a severe lack of newer medicines, techniques and facilities. Harrison finds that the chances of recovery for a badly wounded British infantryman were as much as 25 times better than in the First World War. The reason was that:

> "By 1944 most casualties were receiving treatment within hours of wounding, due to the increased mobility of field hospitals and the extensive use of aeroplanes as ambulances. The care of the sick and wounded had also been revolutionized by new medical technologies, such as active immunization against tetanus, sulphonamide drugs, and penicillin."

Nazi and Japanese Medicine

Unethical human subject research, and killing of patients with disabilities, peaked during the Nazi era, with Nazi human experimentation and Aktion T4 during the Holocaust as the most significant examples. Many of the details of these and related events were the focus of the Doctors' Trial. Subsequently, principles of medical ethics, such as the Nuremberg Code, were introduced to prevent a recurrence of such atrocities. After 1937, the Japanese Army established programs of biological warfare in China. In Unit 731, Japanese doctors and research scientists conducted large numbers of vivisections and experiments on human beings, mostly Chinese victims.

Malaria

Starting in World War II, DDT was used as insecticide to combat insect vectors carrying malaria, which was endemic in most tropical regions of the world. The first goal was to protect soldiers, but it was widely adopted as a public health device. In Liberia, for example, the United States had large military operations during the war and the U.S. Public Health Service began the use of DDT for indoor residual spraying (IRS) and as a larvicide, with the goal of controlling malaria in Monrovia, the Liberian capital. In the early 1950s, the project was expanded to nearby villages. In 1953, the World Health Organization (WHO) launched an antimalaria program in parts of Liberia as a pilot project to determine the feasibility of malaria eradication in tropical Africa. However these projects encountered a spate of difficulties that foreshadowed the general retreat from malaria eradication efforts across tropical Africa by the mid-1960s.

Post-World War II

Smallpox vaccination in Niger, 1969.
A decade later, this was the first infectious disease to be eradicated.

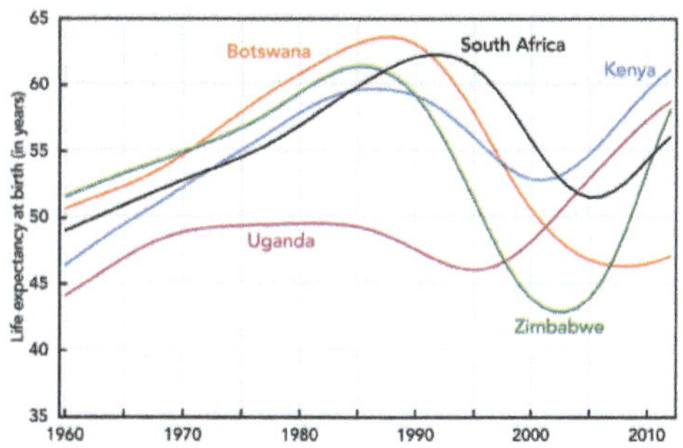

Most countries have seen a tremendous increase in life expectancy since 1945.
However, in southern Africa, the HIV epidemic beginning around 1990 has eroded national health.

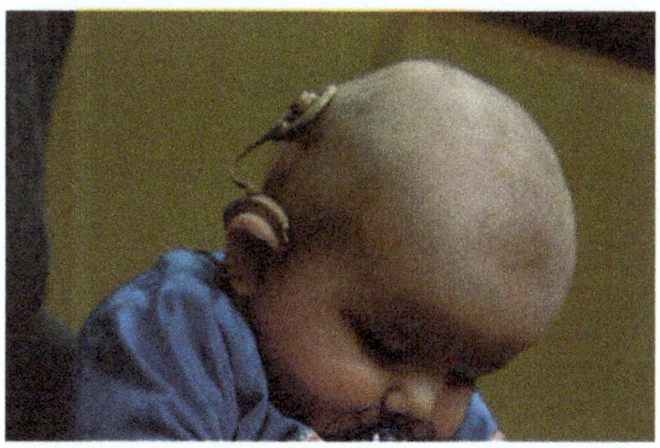

A cochlear implant is a common kind of neural prosthesis, a device replacing part of the human nervous system.

The World Health Organization was founded in 1948 as a United Nations agency to improve global health. In most of the world, life expectancy has improved since then, and was about 67 years as of 2010, and well above 80 years in some countries. Eradication of infectious diseases is an international effort, and several new vaccines have been developed during the post-war years, against infections such as measles, mumps, several strains of influenza and human papilloma virus. The long-known vaccine against Smallpox finally eradicated the disease in the 1970s, and Rinderpest was wiped out in 2011. Eradication of polio is underway. Tissue culture is important for development of vaccines. Though the early success of antiviral vaccines and antibacterial drugs, antiviral drugs were not introduced until the 1970s. Through the WHO, the international community has developed a response protocol against epidemics, displayed during the SARS epidemic in 2003, the Influenza A virus subtype H5N1 from 2004, the Ebola virus epidemic in West Africa and onwards.

As infectious diseases have become less lethal, and the most common causes of death in developed countries are now tumors and cardiovascular diseases, these conditions have received increased attention in medical research. Tobacco smoking as a cause of lung cancer was first researched in the 1920s, but was not widely supported by publications until the 1950s. Cancer treatment has been developed with radiotherapy, chemotherapy and surgical oncology.

Oral rehydration therapy has been extensively used since the 1970s to treat cholera and other diarrhea-inducing infections.

The sexual revolution included taboo-breaking research in human sexuality such as the 1948 and 1953 Kinsey reports, invention of hormonal contraception, and the normalization of abortion and homosexuality in many countries. Family planning has promoted a demographic transition in most of the world. With threatening sexually transmitted infections, not least HIV, use of barrier contraception has become imperative. The struggle against HIV has improved antiretroviral treat-ments, and in the late 2000s (decade), male circumcision was cited to diminish infection risk (circumcision and HIV).

X-ray imaging was the first kind of medical imaging, and later ultrasonic imaging, CT scanning, MR scanning and other imaging methods became available.

Genetics have advanced with the discovery of the DNA molecule, genetic mapping and gene therapy. Stem cell research took off in the 2000s (decade), with stem cell therapy as a promising method.

Evidence-based medicine is a modern concept, not introduced to literature until the 1990s.

Prosthetics have improved. In 1958, Arne Larsson in Sweden became the first patient to depend on an artificial cardiac pacemaker. He died in 2001 at age 86, having outlived its inventor, the surgeon, and 26 pacemakers. Lightweight materials as well as neural prosthetics emerged in the end of the 20th century.

Modern Surgery

Cardiac surgery was revolutionized in the 1948 as open-heart surgery was introduced for the first time since 1925.

In 1954 Joseph Murray, J. Hartwell Harrison and others accomplished the first kidney transplantation. Transplantations of other organs, such as heart, liver and pancreas, were also introduced during the later 20th century. The first partial face transplant was performed in 2005, and the first full one in 2010. By the end of the 20th century, microtechnology had been used to create tiny robotic devices to assist microsurgery using micro-video and fiber-optic cameras to view internal tissues during surgery with minimally invasive practices.

Laparoscopic surgery was broadly introduced in the 1990s. Natural orifice surgery has followed. Remote surgery is another recent development, with the Lindbergh operation in 2001 as a groundbreaking example.

References

- Bynum, W. F. (2006). "The Rise of Science in Medicine, 1850–1913". The Western Medical Tradition: 1800– 2000. Hardy, Anne; Jacyna, Stephen; Lawrence, Christopher; Tansey, E.M. Cambridge University Press. pp. 198–199. ISBN 978-0-521-47565-5.
- Porter, Roy (2008). The Cambridge History of Science: Volume 4: 18th Century Based Science. Cambridge: Cambridge University Press. pp. 661–3. ISBN 978-0-511-46833-9.
- Kenneth G. Zysk, Asceticism and Healing in Ancient India: Medicine in the Buddhist Monastery, Oxford University Press, rev. ed. (1998) ISBN 0-19-505956-5
- Loudon, Irvine (2002-03-07). Western Medicine: An Illustrated History. Oxford University Press. ISBN 978-0-19-924813-1. Retrieved 16 January 2013.
- Longrigg, James (1993). Greek Rational Medicine: Philosophy and Medicine from Alcmaeon to the Alexandrians. Psychology Press. ISBN 978-0-415-02594-2.
- Gunn, S. William A.; Masellis, Michele (23 October 2007). Concepts and Practice of Humanitarian Medicine. Springer. pp. 87–. ISBN 978-0-387-72264-1.
- Porter, Roy (1992). "Madness and its Institutions". In Wear, Andrew. Medicine in Society: Historical Essays. Cambridge: Cambridge University Press. pp. 277–302. ISBN 978-0-521-33639-0.
- Marx, Otto M. (1994). "The Beginning of Psychiatric Historiography in Nineteenth-Century Germany". In Micale, Mark S.; Porter Roy (eds). Discovering the History of Psychiatry. New York & Oxford: Oxford University Press. pp. 39–51. ISBN 0-19-507739-3.
- Hayward, Rhodri (2011). "Medicine and the Mind". In Jackson, Mark. The Oxford Handbook of the History of Medicine. Oxford University Press. pp. 524–542. ISBN 978-0-19-954649-7.
- Pick, Daniel (1993). Faces of Degeneration: A European Disorder, C.1848-1918. Cambridge University Press. ISBN 978-0-521-45753-8.
- Scull, Andrew (2005). Most Solitary of Afflictions: Madness And Society in Britain, 1700–1900. Yale University Press. p. 281. ISBN 978-0-300-10754-8.
- Jones, Kathleen (1993). Asylums and after: a revised history of the mental health services: from the early 18th century to the 1990s. Athlone Press. p. 116. ISBN 978-0-485-12091-2.
- Murray, Joseph E., "The First Successful Organ Transplants in Man", Nobel Lect., 1990, Nobelprize.org., Web, July 7, 2010.

Permissions

All chapters in this book are published with permission under the Creative Commons Attribution Share Alike License or equivalent. Every chapter published in this book has been scrutinized by our experts. Their significance has been extensively debated. The topics covered herein carry significant information for a comprehensive understanding. They may even be implemented as practical applications or may be referred to as a beginning point for further studies.

We would like to thank the editorial team for lending their expertise to make the book truly unique. They have played a crucial role in the development of this book. Without their invaluable contributions this book wouldn't have been possible. They have made vital efforts to compile up to date information on the varied aspects of this subject to make this book a valuable addition to the collection of many professionals and students.

This book was conceptualized with the vision of imparting up-to-date and integrated information in this field. To ensure the same, a matchless editorial board was set up. Every individual on the board went through rigorous rounds of assessment to prove their worth. After which they invested a large part of their time researching and compiling the most relevant data for our readers.

The editorial board has been involved in producing this book since its inception. They have spent rigorous hours researching and exploring the diverse topics which have resulted in the successful publishing of this book. They have passed on their knowledge of decades through this book. To expedite this challenging task, the publisher supported the team at every step. A small team of assistant editors was also appointed to further simplify the editing procedure and attain best results for the readers.

Apart from the editorial board, the designing team has also invested a significant amount of their time in understanding the subject and creating the most relevant covers. They scrutinized every image to scout for the most suitable representation of the subject and create an appropriate cover for the book.

The publishing team has been an ardent support to the editorial, designing and production team. Their endless efforts to recruit the best for this project, has resulted in the accomplishment of this book. They are a veteran in the field of academics and their pool of knowledge is as vast as their experience in printing. Their expertise and guidance has proved useful at every step. Their uncompromising quality standards have made this book an exceptional effort. Their encouragement from time to time has been an inspiration for everyone.

The publisher and the editorial board hope that this book will prove to be a valuable piece of knowledge for students, practitioners and scholars across the globe.

Index